Examining the Concepts, Issues, and Implications of Internet Trolling

Jonathan Bishop
Centre for Research into Online Communities and E-Learning Systems, UK

Managing Director:	Lindsay Johnston
Editorial Director:	Joel Gamon
Book Production Manager:	Jennifer Yoder
Publishing Systems Analyst:	Adrienne Freeland
Assistant Acquisitions Editor:	Kayla Wolfe
Typesetter:	Alyson Zerbe
Cover Design:	Jason Mull

Published in the United States of America by
Information Science Reference (an imprint of IGI Global)
701 E. Chocolate Avenue
Hershey PA 17033
Tel: 717-533-8845
Fax: 717-533-8661
E-mail: cust@igi-global.com
Web site: http://www.igi-global.com

Library of Congress Cataloging-in-Publication Data

Examining the concepts, issues, and implications of Internet trolling / Jonathan Bishop, editor.
    p. cm.
  Includes bibliographical references and index.
  Summary: "This book provides current research on the technical approaches as well as more social and behavioral involvements for gaining a better understanding of Internet trolling"--Provided by publisher.
    ISBN 978-1-4666-2803-8 (hbk.) -- ISBN 978-1-4666-2804-5 (ebook) -- ISBN 978-1-4666-2805-2 (print & perpetual access) 1. Internet--Moral and ethical aspects. 2. Internet--Social aspects. 3. Internet users. I. Bishop, Jonathan, 1979-
TK5105.878.E93 2013
302.23'1--dc23
                                2012033293

British Cataloguing in Publication Data
A Cataloguing in Publication record for this book is available from the British Library.

The views expressed in this book are those of the authors, but not necessarily of the publisher.

Table of Contents

Section 4
Possible Solutions for Dealing with Internet Trolling

Detailed Table of Contents

Section 1
Social, Legal, and Ethical Issues in Internet Trolling

In this information age, serious concerns with unethical behaviour in information technology (e.g., software piracy, deception, plagiarism, etc.) have cast doubts on the claims of the unmitigated success of rapid adoption of information technology. Surprisingly, there have been very few studies in information systems (IS) that have tried to understand the general rise in such unethical behaviours with respect to information technology. Especially, the question that remains to be understood is: Are these problems of unethical behaviour representative of the human nature at large, or are they related to characteristics of technology in any way? This article tries to partly answer this question. It looks at dyadic communicative behaviour using technology-mediated communication and proposes a conceptual model of unethical communicative behaviour.

The development of the World Wide Web has created new opportunities for interpersonal interaction. The Internet allows one-to-one (e-mail), one-to-many (Websites, e-mail lists) or many-to-many (online discussion forums) interaction, which represent a unique feature in comparison with traditional communication channels (Armstrong & Hagel, 1996). On the other hand, the Internet has specific characteristics, such as interactivity, transparency, and memory. These characteristics permit the development of online or virtual communities/groups of people with similar interests who communicate on the Web in a regular manner (Armstrong & Hagel, 1996; Goldsborough, 1999a, 1999b; Gordon, 2000). This article attempts to investigate, analyze and present the main patterns of the codes/rules of ethics used in the public discussion forums, otherwise known as Newsgroups, and their influence on the profile and functioning of the community.

Chapter 3

Georgios Michaelides, Budapest University of Technology and Economics, Hungary

Gábor Hosszú, Budapest University of Technology and Economics, Hungary

The importance of the virtual communities' privacy and security problems comes into prominence by the rapid development of online social networks. This article presents the multiple threats currently plaguing the virtual world, Internet privacy risks, and recommendations and countermeasures to avoid such problems. New generations of users feel comfortable publishing their personal information and narrating their lives. They are often unaware how vulnerable the data in their public profiles are, which a large audience daily accesses. A so-called digital friendship is built among them. Such commercial and social pressures have led to a number of privacy and security risks for social network members. The article presents the most important vulnerabilities and suggests protection methods and solutions that can be utilized according to the threat. Lastly, the authors introduce the concept of a privacy-friendly virtual community site, named CWIW, where privacy methods have been implemented for better user protection.

Chapter 4

Alok Mishra, Atilim University, Turkey

Deepti Mishra, Atilim University, Turkey

Cyber stalking is a relatively new kind of cyber terrorism crime. Although it often receives a lower priority then cyber terrorism it is an important global issue. Due to new technologies, it is striking in different forms. Due to the Internets provision of anonymity and security it is proliferating quickly. Technology and tools available to curb it have many limitations and are not easy to implement. Legal acts to protect people from cyber stalking are geographically limited to the concerned state or country. This chapter reviews cyber stalking, its approaches, impacts, provision of legal acts, and measures to be taken to prevent it. There is an immediate need for research in the various dimensions of cyber stalking to assess this social problem.

Section 2
Psychological and Wellbeing Issues in Internet Trolling

Chapter 5

Jonathan Bishop, Glamorgan Blended Learning Ltd. & GTi Suite & Valleys Innovation Centre & Navigation Park & Abercynon, UK

The rise of online communities in Internet environments has set in motion an unprecedented shift in power from vendors of goods and services to the customers who buy them, with those vendors who understand this transfer of power and choose to capitalize on it by organizing online communities and being richly rewarded with both peerless customer loyalty and impressive economic returns. A type of online community, the virtual world, could radically alter the way people work, learn, grow consume, and entertain. Understanding the exchange of social and economic capital in online communities could involve looking at what causes actors to spend their resources on improving someone else's reputation. Actors' reputations may affect others' willingness to trade with them or give them gifts. Investigating online communities reveals a large number of different characters and associated avatars. When an actor looks at another's avatar they will evaluate them and make decisions that are crucial to creating interaction between customers and vendors in virtual worlds based on the exchange of goods and services. This chapter utilizes the ecological cognition framework to understand transactions, characters and avatars in virtual worlds and investigates the exchange of capital in a bulletin board and virtual. The chapter finds strong evidence for the existence of characters and stereotypes based on the ecological cognition framework and empirical evidence that actors using avatars with antisocial connotations are more likely to have a lower return on investment and be rated less positively than those with more sophisticated appearing avatars.

The purpose of this chapter is to explore cyber-bullying from three different, but interrelated, perspectives: students, educators and parents. The authors also explore the opposite spectrum of online behaviour - that of "cyber-kindness" - and whether positive, supportive or caring online exchanges are occurring among youth, and how educators, parents and policy-makers can work collaboratively to foster a kinder online world rather than simply acting to curtail cyber-bullying. These proactive efforts tackle the deeper causes of why cyber-bullying occurs, provide students with tools for positive communication, open the door for discussion about longer term solutions, and get at the heart of the larger purposes of education – to foster a respectful and responsible citizenry and to further a more caring and compassionate society. In the course of this discussion, they highlight the findings from two studies they conducted in British Columbia, Canada, one on cyber-bullying and a later study, which addressed both cyber-bullying and cyber-kindness.

Scientific research on empathy started in the early 20th century. Only in 1992 did the development of cognitive neuroscience help di Pellegrino, Fadiga, Fogassi, Gallese and Rizzolatti to identify the mirror neurons related to representations of an Object from a Subject, verifying Lipps' (1903) and McDugall's (1908) suggestions on empathy. Primary empathy is related to the automatic matching of the feelings of the other person (Fischer, 1980). An example is the relationship newborns have between each other on their first days on the earth. Another verification of mirror neurons was made by Rizzolati and Arbib (1998), as well as identification of the areas where the mirror neurons are located, interacting with areas in both hemispheres (Broca area 44 and PE/PC).

This chapter describes how social politeness is relevant to computer system design. As the Internet becomes more social, computers now mediate social interactions, act as social agents, and serve as information assistants. To succeed in these roles computers must learn a new skill—politeness. Yet selfish software is currently a widespread problem and politeness remains a software design "blind spot." Using an informational definition of politeness, as the giving of social choice, suggests four aspects: 1. respect, 2. openness, 3. helpfulness, and 4. remembering. Examples are given to suggest how polite computing could make human-computer interactions more pleasant and increase software usage. In contrast, if software rudeness makes the Internet an unpleasant place to be, usage may minimize. For the Internet to recognize its social potential, software must be not only useful and usable, but also polite.

Section 3
Trust and Participation Issues in Web 2.0 Systems at Risk of Internet Trolling

Chapter 9

The rise of social networking services have furthered the proliferation of online communities, transferring the power of controlling access to content from often one person who operates a system (sysop), which they would normally rely on, to them personally. With increased participation in social networking and services come new problems and issues, such as trolling, where unconstructive messages are posted to incite a reaction, and lurking, where persons refuse to participate. Methods of dealing with these abuses included defriending, which can include blocking strangers. The Gamified Flow of Persuasion model is proposed, building on work in ecological cognition and the participation continuum, the chapter shows how all of these models can collectively be used with gamification principles to increase participation in online communities through effective management of lurking, trolling, and defriending.

Chapter 10

This chapter addresses how members of a blog-based community share problems and support each other in the problem solving process, both sharing knowledge and offering support. Problems are divided into three categories, unique, shared, and community, each having its own particular norms for presentation, knowledge sharing, and resolution. Additionally, processes may differ based on discourse that centers on one individual blog versus discourse that spans multiple blogs. Findings show that intersubjectivity, norms, roles, and individual ownership of virtual space all are important elements contributing to the problem sharing and solving process.

Chapter 11

Social media technologies such as blogs, social networking sites, microblogs, instant messaging, wikis, widgets, social bookmarking, image/video sharing, virtual worlds, and internet forums, have been identified to have played a role in crises. This chapter examines how social media technologies interact with formal and informal crises communication and information management. We first review the background and history of social media (Web 2.0) in crisis contexts. We then focus on the use of social media in the recent Gaza humanitarian crisis (12.2008-1.2009) in an effort to detect signs of a paradigm shift in crisis information management. Finally, we point to directions in the future development of collaborative intelligence systems for crisis management.

Chapter 12

This chapter discusses the concept of trust and how trust is used and modeled in online systems currently available on the Web or on the Internet. It starts by describing the concept of information overload and introducing trust as a possible and powerful way to deal with it. It then provides a classification of the systems that currently use trust and, for each category, presents the most representative examples. In these systems, trust is considered as the judgment expressed by one user about another user, often directly and explicitly, sometimes indirectly through an evaluation of the artifacts produced by that user or his/her activity on the system. We hence use the term "trust" to indicate different types of social relationships between two users, such as friendship, appreciation, and interest. These trust relationships are used by the systems in order to infer some measure of importance about the different users and influence their visibility on the system. We conclude with an overview of the open and interesting challenges for online systems that use and model trust information.

Chapter 13

Commerce performed electronically using the Internet (e-commerce) faces a unique and difficult problem, the anonymity of the Internet. Because the parties are not in physical proximity to one another, there are limited avenues for trust to arise between them, and this leads to the fear of cheating and promise-breaking. To resolve this problem, I explore solutions that are based on Thomas Hobbes's solutions to the problem of the free rider and apply them to e-commerce.

Section 4
Possible Solutions for Dealing with Internet Trolling

Organizations need effective and affordable software training. In face-to-face settings, behavior modeling is an effective, but expensive, training method. Can behavior modeling be employed effectively, and more affordably, for software training in the online environment? An experiment was conducted to compare the effectiveness of online behavior modeling with that of face-to-face behavior modeling for software training. Results indicate that online behavior modeling and face-to-face behavior modeling provide essentially the same outcomes in terms of knowledge near transfer, immediate knowledge for transfer, delayed knowledge for transfer, perceived ease of use, perceived usefulness, and satisfaction.

Observed differences were not significant, nor were their patterns consistent, despite sufficient power in the experimental design to detect meaningful differences, if any were present. These results suggest that organizations should consider online behavior modeling as a primary method of software training.

Lakshmi Goel, University of Houston, USA
Elham Mousavidin, University of Houston, USA

Despite considerable academic and practitioner interest in knowledge management, success of knowledge management systems is elusive. This chapter provides a framework which suggests that KM success can be achieved by designing sustainable communities of practice. Communities of practice have proven to have significant economic and practical implications on organizational practices. A growing body of literature in KM recognizes the importance of communities that foster collaborative learning in organizations and almost all KMS have a 'network' component that facilitates connecting people in communities of practice. Evidence has shown that communities have been a key element in KMS of many companies including Xerox PARC, British Petroleum Co., Shell Oil Company, Halliburton, IBM, Proctor and Gamble, and Hewlett Packard.

Ross A. Malaga, Montclair State University, USA

Online auctions are an increasingly popular avenue for completing electronic transactions. Many online auction sites use some type of reputation (feedback) system—where parties to a transaction can rate each other. However, retaliatory feedback threatens to undermine these systems. Retaliatory feedback occurs when one party in a transaction believes that the other party will leave them a negative feedback if they do the same. This chapter examines data gathered from E-Bay in order to show that retaliatory feedback exists and to categorize the problem. A simple solution to the retaliatory feedback problem—feedback escrow—is described.

Carlos Alberto Ochoa Ortiz-Zezzatti, University of Ciudad Juárez, Mexico
Julio Cesar Ponce Gallegos, Autonomous University of Aguascalientes, Mexico
José Alberto Hernández Aguilar, Autonomous University of Morelos, Mexico
Felipe Padilla Diaz, Autonomous University of Aguascalientes, Mexico

The contribution of this chapter is to present a novel approach to explain the performance of a novel Cyberbullying model applied on a Social Network using Multiagents to improve the understanding of this social behavior. This approach will be useful to answer diverse queries after gathering general information about abusive behavior. These mistreated people will be characterized by following each one of their tracks on the Web and simulated with agents to obtain information to make decisions to improve their life's and reduce their vulnerability in different locations on a social network and to prevent its retort in others.

Chapter 18

Zheng Yan, Nokia Research Center, Finland
Silke Holtmanns, Nokia Research Center, Finland

This chapter introduces trust modeling and trust management as a means of managing trust in digital systems. Transforming from a social concept of trust to a digital concept, trust modeling and management help in designing and implementing a trustworthy digital system, especially in emerging distributed systems. Furthermore, the authors hope that understanding the current challenges, solutions and their limitations of trust modeling and management will not only inform researchers of a better design for establishing a trustworthy system, but also assist in the understanding of the intricate concept of trust in a digital environment.

Preface

Internet trolling is one of the fastest spreading pieces of computer jargon of the 21st century. Barely a day goes by where trolling and "trolls" are not in the news. Internet trolling has come to refer to any form of abuse carried out online for the pleasure of the person causing the abuse or the audience to which they are trying to appeal. The simplest definition of Internet trolling is the posting of provocative or offensive messages on the Internet for humourous effect. The word 'troll' when used to refer to persons who try to provoke others originated in the US military in the 1960s prior to the realisation of the Internet for mass communication, with the term, *trolling for MiGs*. This was reputed to be used by US Navy pilots in Vietnam in their dog-fighting, popularised by the film starring Tom Cruise called Top Gun. Such a practice, of trying to provoke the opposing fighter pilots was not an authorised operation, but was defended by pilots in order to identify their "strengths and weaknesses." Following this military use of the term trolling, the advent of the Internet soon led to similar phrases like *trolling for newbies*, which became a common expression in the early 1990s in the Usenet group, alt.folklore.urban. One of the first recorded attempts to define trolling was in the mid-1990s with the launch of the book, *netlingo*, which is now also online where one can see their definition of trolling. Others have described a Troll as someone who mostly initiates threads with seemingly legitimate questions or conversation starters, but with the ultimate goal of drawing others into useless discussions.

We now know something about what Internet trolling is and where it came from, but what do we do about it? This book starts the process of answering this question. The first section on social, legal and ethical issues in relation to Internet trolling gets the ball rolling.

SOCIAL, LEGAL, AND ETHICAL ISSUES IN INTERNET TROLLING

Sutirtha Chatterjee's chapter on ethical behaviour in online environments is particularly poignant in the discussion of Internet trolling. Sutirtha describes a number of unethical behaviours that inhibit participation in online communities. This includes flaming, swearing, insults, and deception. Such behaviours are now grouped under the header of Internet trolling. Even though flaming is the posting of offensive messages online, they are not always as negative as one might first think. For instance, there are users known as Troller Rollers, who are trollers who troll other trollers. Sutirtha shows how flame trolling, which is the posting of messages which are not only offensive but provocative, is caused by a process called deindividuation. In this process, a person, such as a participant in an online community, becomes estranged from the mainstream members and this results in misbehaviours from them. Many of the mainstream trollers that are talked about in the media have been men in their 20s, who more generally have been hit the hardest by the lack of jobs due to the global recession and pension crisis. Serial flame troller, Sean Duffy, for instance was known to have difficulties in social interaction with others and being unable to develop social relationships among his peers and others. Two of Duffy's most prominent victims were the families of Natasha MacBryde, who died at 15, and Sophie Taylor, who died at 16. In the case of MacBryde, Duffy created a YouTube video called "Tasha the Tank Engine" which had MacBryde's face superimposed on a locomotive. Disturbed by Duffy's actions, MacBryde's father said, "This person was hiding behind a computer. For me you can't see him, you can't do anything. It is very hard for a father. You all try and protect your kids." The next chapter in this section by Calin Gurau discusses how the rules of online communities like message-boards often explicitly prohibit actions like "*flaming*" and "*trolling*," particularly where they start "*flame wars*" which is a barrage of abuse between members of online communities in a particular thread. Sometimes these are started by members of a website called *Elders* who are out-bound from the community, who "troll for newbies," by posting messages that will obviously incite a flame war from unsuspecting new members.

Calin's chapter also discusses how communities based around anonymity, such as the community called 4chan which has members who abuse others and inaccurately call themselves "trolls" is a perfect example of this. Calin says how users are often asked to refrain from providing information that might lead to them being identified. Indeed, in online communities like Wikipedia there is a detailed policy on Outing, where one can't reveal the identity of another user. This often results in another sin at Wikipedia, Conflict of Interest, and the two are rarely reconciled, resulting in flaming, like that seen on the Talk Page of the article about Barry Wellman displayed in Figure 1 where a user accuses the editors of the article of using it to promote the subject they are alleged to know.

Figure 1. An example of flame trolling on Wikipedia resulting from perceived conflict of interest

Promotional, windy, and of dubious value

I second (or third, or fourth) the remark that this article is absurdly promotional and needs extensive revision by someone who is refer to WP:COI, and expert editors who review the article should be skeptical of the neutrality of the article's routine editors. This on neutrality. Readers cannot and should not rely on it to provide NPOV information on the subject. Antiselfpromotion (talk) 18:0(

I've edited this article, and I'm not connected to Barry Wellman. He's an important academic, a fellow of the Royal Society of C to be an attempt to "out" another editor. Let me suggest that you not do that again. --Anthon.Eff (talk) 18:22, 8 May 2010 (UTC

I had believed that his and his friends' extensive editing of this article was public knowledge but later noticed that you and past. Thus, I have already contacted Oversight and requested that my comment in question be deleted. I continue to obje questioning the subject's notability, but the article is windy and of dubious value as currently written; I am however certain opposition from those with significant conflicts who have a personal interest in inflating the subject's reputation. Wikipedi like many of those who are responsible for this page. Like the promotional article for Danah Boyd, which has also been e Wikipedia. Antiselfpromotion (talk) 18:54, 8 May 2010 (UTC)

You are right that self-promotion is a huge problem here--anyone edits, and administrators are over-stretched and ca that Wellman is an egregious case. I'm an academic economist, and Wellman is a sociologist, but even I have read s people who know him. He's that well-known. The more complete his article, the better. Danah Boyd, on the other hanc you are going after other articles related to Wellman, and I would advise you to broaden your interests a bit.--Anthon.E

I have nothing against Prof Wellman personally, and my other edits were aiming to correct the same problem as th avoid the appearance of bias, but I earnestly wish that Prof Wellman himself would adhere to that same standard. promotion compared to other examples of it, but the resultant article here is more of a CV than an encyclopedia en reputation he has earned; I am instead claiming that the article about him needs to be guided by a purpose other t people who know him. If his reputation is sound, the article should be able to survive without such a firm guiding h economist, compare the entry for Roy Harrod with this one. I am personally inclined to think the length there is mor of the subject, but length is not even the main issue. Compare the modesty and fairness of that entry with this one. K. Merton is longer, but it is fairer and more appropriately modest in context for someone whose academic signific Wellman. The 'concepts' discussed in the introduction to Merton's article are known to the public and have influenc citations in Wellman's article do not even begin to demonstrate that he is known, or that the 'concepts' claimed to t degree. I do not doubt he is a moderately well-known academic, particularly in his field. As I wrote previously, I am particular choice of content. The conflicts themselves, moreover, are embarrassing and reflect poorly on the proce May 2010 (UTC)

Socialism wanted everyone to be equal, and made everyone equally poor. To say that Wellman should have an the way to rectify the imbalance would be to make the Harrod article fuller, not the Wellman article thinner. --Antl

Calin's chapter is followed very appropriately by Georgios Michaelides and Gabor Hosszu's chapter on privacy and security in virtual worlds and the threats to look out for. They raise important issues in the study of Internet trolling, including cyberstalking and cyberbullying, and the effect availability of personal information in online communities and on social networking services can be used to aid Internet abuse, including by flame trollers. The victims of these trollers, who are often variants of "Snerts" (i.e. Sexually nerdish egotistically rude trolls) if they have a grudge against a general group that person is in, or "E-Vengers" who have a grudged against them personally. Herogios and Gabor explain the effects Snerts and E-Vengers have on users of online communities, including loss of self-esteem, increased suicidal ideation, and feelings of being scared, angry, frustrated, and depressed. A notorious flame troller who carried out such abuses over a four-year-long period against a number of victims was Maria Marchese of Bow, London. Her campaign as a Hater troller resulting in the break-up of an engagement of a respected psychiatrist, Dr. Jan Falkowski following allegations of rape he was exonerated from with forensic evidence. She also drove an accountant, Miss Deborah Pemberton to the brink of suicide telling her to "dig her own grave." Judge John Price at Southwark Crown Court sentenced Marchese for three counts of harassment under the Protection from Harassment Act 1997, including a restraining order. He told Marchese, "It is difficult to imagine a more serious case," as her "uniquely disturbing" obsession as a cyberstalker caused "enormous pain and suffering." Whilst they are at risk of 'Reporter Trollers' who report any content they disagree with, Georgios and Gabor recommend in their chapter using "Report Abuse" buttons in the same way one would normally use "Contact Us."

Georgios and Gabor's chapter leads suitably on to cyberstalking more specifically, with Alok Mishra and Deepti Mishra taking the problem head on by identifying it not only as a social problem, but a risk to web security more generally that needs to be taken seriously. Alok and Deepti concur with the findings of Georgios and Gabor about the long-term damaging effects cyberstalking has on the victims, who are mostly young women, they find. This is certainly evident in the news coverage on Internet trolling in general. Colm Coss was one of the first to be sentenced for Internet trolling for the posting of offensive content on a memorial website of celebrity and cancer victim Jade Goody, aged 28 when she died. Coss defaced the memorial page of former Big Brother star Jade Goody, becoming one of the first recorded cases of "RIP Trolling." Coss's actions shocked the nation, where he was reported to have ransacked memorial pages, including with boasts that he'd had sex with dead bodies. Manchester Magistrates Court was presented with clear and convincing evidence in the form of photographs that Coss shows neighbours as proof of him being a "*troll*." Magistrate Pauline Salisbury said to Coss, "You preyed on bereaved families who were suffering trauma and anxiety," and "We know you gained pleasure and you aren't sorry for what you did."

PSYCHOLOGICAL AND WELLBEING ISSUES IN INTERNET TROLLING

It is clear from the first section that as well as a number of legal, ethical, and security issues, the outcome of the flame trolling has significant effects on the psychology and wellbeing of the victims which is explored in the second section, along with chapters that seek to understand the trollers themselves. The first chapter in this section presents an ethnographic study that identifies the different troller character types that become apparent in online communities where trolling occurs within. It followed a single case study by early career researcher Whitney Phillips of the 4chan website whose members identified personally as trolls to justify their abusive behaviour for their own entertainment. An interview with Phillips, who wrote that case study for an online journal, citing mainly hyperlinks, led to the convenient adoption of the term "troll" in the British media to refer to anyone involved in any form of abuse or harassment online, which then spread across the world. This chapter on the other hand shows that there is in fact a rich array of trollers, some of them who act provocatively to entertain others such as the Trolls and also those who act offensively for their own sick entertainment, like the ones interviewed on 4chan, who are primarily made up of the obnoxious type of troller known as a Snert, as they try to provoke humour like a Troll but in an offensive way for their own sick purposes. A depiction of the users on 4chan who call themselves 'Anonymous' is in Figure 2.

The next chapter in this section is by Wanda Cassidy, Karen Brown, and Margaret Jackson who seek out ways to address problems created by Snerts and other flame trollers in educational settings. They suggest it is not only important to try to curtail the flame trolling that forms part of cyber-bullying, but also to foster kudos trolling, through what they term, *cyber-kindness*. They show that through what has become known as Web 2.0, such as blogs and other social media websites, young people who are often both the victims and perpetrators of flame trolling can make positive contributions to Cyberspace to promote self-esteem and confidence in them and their peers. A notable example of such trolling is that Liam Stacey, who was 21 at the time of his flame trolling offence. Stacey was convicted for a racially-aggravated public order offence following a tirade of racist abuse on Twitter after he was rebuked by others after he mocked the cardiac arrest of Bolton player, Fabrice Muamba. Stacey was sentenced to 56 days in jail by district judge John Charles, who summed up saying, "It was racist abuse via a social networking site instigated as a result of a vile and abhorrent comment about a young footballer who was fighting for his life."

This chapter is followed appropriately by one on online empathy by Niki Lambropoulos. Niki finds that trollers lack the same propensity towards forming factions as other participants and that those participants who empathises with others, such as having similar outlooks are most likely to participate. This might explain why online communities have traditionally been founded around mutual social ties, such as a topical interest or common geographical location. In this regard her finding that those most likely to troll are more remote from other users shows that flame trolling is more likely to occur where participants lack empathy with other members, which in the case of E-Vengers can be down to the community not tolerating their personality or other aspects of their identity in the first place. A notable example of such a flame trolling is that of Sean Duffy who first started trolling in his 20s, and is considered one of the

Figure 2. Anonymous trollers on 4chan justify their abuse as Snerts by calling themselves trolls (Courtesy: Wikipedia)

most prolific flame trollers to have surfaced on the Internet. He targets his victims, the family members of teenagers who have died, by posting offensive pictures and videos on their memorial pages, making him a *R.I.P. Troller*. Duffy has been sentenced to 13-week prison terms a number of times under the Malicious Communications Act 1988, but this has not stopped him from flame trolling his many victims.

The final chapter in this section, by Brian Whitworth and Tong Liu, takes the issue of empathy a step further. They argue that not only should users try to be more polite and considerate of others, but the software which they utilise should be designed for promoting sociability as much as traditional factors like designing for usability. "If software can be social, it should be designed accordingly" is how they put it. They argue that by designing out factors that promote anti-social behaviour typical in flame trolling, benefits can be achieved for all participants. Polite computing as they call it will, they argue, increase the number of legitimate interactions, including by kudos trollers, reduce anti-social attacks by flame trollers, and increase synergistic trade and software use. They present research which shows that politeness can prevent some of the worst types of flame trolling, like resentment where some are given greater opportunities to participate than others. Lack of politeness can breed mistrust and can sometimes result in claims of harassment, even if they are unfounded. One example is that of Angela Martindale, who was 39 at the time, was found not guilty of harassment at Prestatyn Magistrates Court following taking retaliatory action against former cagefigher Adam Finnigan. Finnigan told the court that he and his partner, Suzanne Rogovskis, had been bombarded with abusive text messages and voicemails between August and October 2010, following Martindale and Finnigan ending their relationship. Rogovskis told the court she recognised Martindale in one call shouting, "I'm going to rip your head off." Clearing her of harassment, Magistrate Peter Oakley said there was no clear evidence that she had made the calls, and said the police had not investigated the matter fully.

TRUST AND PARTICIPATION ISSUES IN WEB 2.0 SYSTEMS AT RISK OF INTERNET TROLLING

This third section extends on the last section about psychology and wellbeing by looking at the related issues of trust and participation. It begins with a chapter by the editor looking at the role of defriending and gamification for increasing participation in online communities, where trolling and lurking are both commonplace. Trust is not just an issue in e-commerce based online communities, but in all of them. It is this lack of trust that can result in many people staying, or becoming, Lurkers –people who don't feel they can fully participate beyond browsing. Fear of being flame trolled is one of the barriers to posting in an online community, and this needs to be overcome. Chapter 9 extends upon the character theory in Chapter 5 by introducing a twelfth type of character - the Elder. This addition is important as means one has to recognise that Lurkers are a type of participant - who take part in online communities for the enjoyment of surveillance. Elders are naturally their opposite, as they take part for escapism using their expertise to 'troll for newbies,' knowing they are on an outbound basis. The gamified flow of persuasion model in this chapter seeks to show the different transitional stages that a user has to go through to attain their optimal level of participation. In Web 2.0 systems based around technologies such as the Circle of Friends method used on Facebook and MySpace, online community participation can be dashed by defriending, which is where a friend stops someone access their page by deleting or blocking them from their buddy list. In the paper, the author presents an analysis of the typical things said on weblogs when someone has defriended another, to make it easier for both participants and systems operators to look out for the tell-tale signs of what can lead to defriending.

The next chapter, by Vanessa P. Dennen, continues the theme of understanding the narrative structure weblogs and their impact online participation. Vanessa looks at the unique, shared, and community problems that need to be addressed in order to negotiate a shared meaning and purpose on blog-based platforms. She finds that blogs are often based on a reflective process, like as discussed in the last chapter, but also they can go beyond this and seek advice. Such blogs are ripe environments for flame trollers who may seek to dash someone's confidence when they are looking for sympathy and empathy for the situation they are in. The problems are not as apparent as in message boards because in many cases the owner of the weblog can delete comments in ways in which they could not if it was on such an open commenting platform. Vanessa also shows how it is possible that the visitors to blogs can reduce lurking and encourage kudos trolling by linking the different blogs they visit together, where others are going through a similar narrative.

In the next chapter, Miranda Dandoulaki and Matina Halkia look at the role Web 2.0 system play in the reporting and misreporting of crisis situations. They show how the global village that exists through the availability of multiple mass media via the Internet means what happens in one part of the globe can have an effect in another part. A disturbing example of Internet trolling that has an effect on the Gaza and wider world is that of depictions of desecration of religious texts like the Qur'an. Whether they are videos of their burning or the posting of this text being vandalised, what might seem a joke or protest has resulted in unnecessary incitements of hate. Probably the most notable example of religious hate posted in the blogosphere was that by Paul Z Myers, pictured in Figure 3, who was a well-known Junior Lecturer that posted pictures on his website of banana skins over the Qur'an. Both this and the Gaza case presented by Miranda Matine demonstrate the influence of social media technologies on global events and in the last of the latter, even when the disaster or crisis area has huge inadequacies in technological infrastructure.

Trying to justify his actions, Myers said, "By the way, I didn't want to single out just the cracker, so I nailed it to a few ripped-out pages from the Qur'an and The God Delusion. They are just paper. Nothing must be held sacred. Question everything. God is not great, Jesus is not your lord, you are not dis-

Figure 3. Paul Myers commits regular and serious acts of flame trolling (Courtesy: Wikipedia)

ciples of any charismatic prophet." Myers' university refused to take action, saying that as it was on his personal website they have no jurisdiction. Myers' abuse towards women and people of a faith other than Atheism led to a whole raft of *counter-trolling*, where people posted equally abusive or insightful comments about him. One person who was banned from Myers' free thoughts blog for free thinking made an entire video, posted to YouTube, showing a lot of compelling evidence and concluding that he had an "utter lack of academic integrity." The whole episode shows how the effects of trolling go beyond the environment in which they originated.

In the following chapter, Paolo Massa describes how using reputation systems on blogging platforms can resolve such abuse by involving the website users. They talk about the Slashdot website and how a moderation point can be spent within 3 days for increasing the score of a comment by 1 point, choosing from a list of positive adjectives (insightful, interesting, informative, funny, underrated), for decreasing the score of a comment by 1 point. They can also choose from a list of negative adjectives (off-topic, flame-bait, troll, redundant, overrated) with moderation points being added or subtracted to the reputation of the user who submitted their opinion on a comment.

In the final chapter of this section, Eric M. Rovie continues the theme of trust, with particular reference to anonymity. Anonymity often allows users to create pseudonyms for the online community they are in to hide their true identity. A pseudonym acts to remove barriers to flame trolling as these trollers think they are safe from prosecution and detection. Eric points out however, that with the right technology or a court order the details are obtainable. This was found to be the case in Great Britain where a victim of obscene flame trolling, Nicola Brooks, brought a case in the UK High Court after she suffered "vicious and depraved" abuse following posting a comment on Facebook supporting a former reality TV contestant Frankie Cocozza when he left the X-Factor show the year before. The High Court, in the first judgement of its kind in the UK, ordered Facebook to reveal the identity of those anonymous users who flame trolled Brooks. Eric explains how trust is the key to any online transaction, especially on an e-commerce site where people have to trust another to fulfil their side of the bargain. Eric shows how people who do not wish to take part in give-and-take, which he calls Hobbesian Fooles, can remove a lot of the trust in an online community. Built around the Chatroom Bob character, these users get what they want from others, regardless of the consequences. In an e-commerce website, Eric suggests that guarantee schemes can act as a disincentive, as it will put opportunists off as they could lose any money they make and their trading status in general. Above all he suggests that removing anonymity like that seen in 4chan will help those who would otherwise not use the community feel confident in taking part by knowing the identities of the other participants. Eric's chapter leads nicely on to the fourth and final chapter on possible solutions to reducing flame trolling and encouraging kudos trolling.

POSSIBLE SOLUTIONS FOR DEALING WITH INTERNET TROLLING

Finally, the fourth part of the book looks at what options are available for tackling the problem of flame trolling and encouraging kudos trolling. The first chapter of this section, by Chen, Ryan, & Olfman, takes psychological theories around social cognitive theory. They suggest that behaviour changes online, including learning to tolerate others, and taking part in kudos training involves four processes – attentional, retention, production, and motivational. They make a clear distinction between face-to-face behaviour modelling and online behaviour. To put their theories into action they present an experiment that assesses the effectiveness of an online behaviour management system on educational outcomes. They show that comparable levels of behaviour change between participants in the online group as the offline group, providing a strong basis for the development of behaviour management systems to combat flame trolling.

Lakshmi Goel and Elham Mousavidin discuss the importance of knowledge management for designing sustainable communities online in the second chapter of this section. They argue strongly that moderation is essential to combating flame trolling. To achieve this, they argue that every online community needs a dedicated community manager with the authority and resources to manage and act for the benefit of the community. These managers, also known as systems operators, or sysops, should seek out independent experts to provide their experiences to the community in order to reduce flame trolling and general conflict. These are two among a number of recommendations that Lakshmi and Elham make, and their chapter is an important read for supporting the running of online communities so as to design out the likelihood of flame trolling.

The next chapter in this section is by Ross A Malaga, which looks at the role of *Retaliatory Feedback* in affecting behaviour in online communities. Retaliatory feedback can result in discouraging honest feedback and the unfortunate encouragement of flame trolling in response to honest feedback. Ross proposes an innovative solution to the problem, which is a *feedback escrow*. One of the most noteworthy cases dealing with the effects of retaliatory feedback and lack of reliable feedback was that of Shumon Ullah, who was jailed for 32 months at Burnley Crown Court for fraudulently receiving £38,500 in three weeks from eBay. Ullah built up a base of kudos from other uses by selling many affordable goods to become a "reliable trader." He then used his neighbour's details to "break out" and fraudulently offer electrical equipment worth nearly £230,000, but failed to deliver them. Customers started flaming him by giving him negative feedback, which resulted in *fight flame with flame* abuse from him, including claiming he was bankrupt. He was then reported by customers to the police who carried out what the presiding judge, David Heaton, QC, called a "a very comprehensive, time-consuming and no doubt expensive operation." Sentencing him, Judge Heaton said it was necessary to "punish" Ullah in order to "deter" others. The case shows the importance of reputation systems, but like Alberto Ochoa, Julie Ponce, Alberto Hernandez, and Felipe Padilla say in the next chapter in this section, artificial intelligence can be used to pick up on the tell-tale signs sooner.

Alberto, Alberto, and Felipe present a novel multi-agent based system that can help detect and respond to flame trolling. They show that abusive behaviour need not always be about the posting of abusive message but social exclusion also. Called a *Social Blockade*, people, especially young people, have their contributions ignored, which can then have severe consequences. They could either become a Lurker and not participate, or become an E-Venger and wreak havoc in the name of justice to avenge the way they were treated. A prominent case of this kind was that of Gavin Brent from Holywell, Flintshire. When he was aged 24, Brent was fined £150 with £364 costs by Mold Magistrates Court under the Telecommunications Act 1984. The case came about after Brent suffered what he called "mis-treatment" by the police who refused to listen to his concerns, the court heard. The court was told that this instigated the online abuse by Brent towards police detective, D.C. Lloyd, resulting in Brent posting "menacing" messages against Lloyd, which Brent thought was a legitimate expression of "freedom of speech" in the circumstances. Brent's messages against Lloyd included, "P.S. - D.C. Lloyd, God help your newborn baby" and were later removed. Alberto, Alberto, and Felipe discuss how multi-agent systems can intervene to resolve such postings, as by operating autonomously according to a fixed set of routines, they can help intervene to achieve desired behaviours.

In the final chapter of this section and the book, Zheng Yang and Silke Holtmanns look at trust modelling and management to enable the movement from social trust to digital trust. They introduce a comprehensive solution to reducing the barriers caused by flame trolling or fear of flame trolling called, *trust modelling and management*. They suggest that someone is most likely to gain a sense of trust from a system if they feel a sense of confidence, belief, gratification, and disposition. They argue that an online community that is most able to convey that it is able to show benevolence, motivation, honesty and faith is most likely to generate trust in participants.

CONCLUSION

It is clear for all to see that Internet Trolling is not going to go away, as while it is possible for the human race to exist, we will be seeking out ways to be humourous and mischievous to entertain both ourselves and others. Transgressive forms of humour, like that done by Snerts who abuse others for their own sick entertainment, are a reality for most types of online community. This book provides many ways for understanding and responding to these problems, varying from technical approaches to more social and behavioural interventions. It is clear that the impact of the offensive kind of Internet trolling – flame trolling – can have huge effects on the wellbeing and psychology of Internet users, which the mangers of online communities, known as sysops need to take more seriously. However achieved, online empathy will be essential to building a shared meaning about users of an online community, whether a traditional message board or more recent innovations like weblogs. As Internet trolling is likely to stay, practitioners should read this book to understand the problem more and the potential solutions available.

This book, with its descriptions and solutions for dealing with flame trolling in controlled environments, like e-commerce websites and learning and social networking platforms, will raise a number of other issues. For instance, based on the findings of the chapters in this book, organisations will have to ask about the effectiveness of their information security policies, and how to implement statute and case law for dealing with flame trolling. In organisations based around a strong Web presence they are going to have to review the extent and limitations of their webmaster's powers over trolling (i.e., sysop prerogative) and potential new technologies for dealing with the problem. Organisations need to consider the experiences of those users who have been subjected to flame trolling in assessing consumer satisfaction as their brand will be as dependent on customer experiences on the Web as much as in the real-world settings.

The issues raised in this book will be challenging for political parties and politicians. For example, the use and misuse of "trolling" for electoral gain, and mishaps in electoral campaigns due to social networking trolling *bloopers* will become an important consideration that can't be avoided. Technologies like those discussed in the final chapters will have to be taken seriously. Information systems, particularly on the Internet, will have to be designed so as to encourage kudos trolling and at the same time reduce flame trolling. The political, ethical, security, privacy, and legal issues discussed in this book will need to be put into practice. New guidelines will have to be developed in order for systems operators to exercise proper control over their members who want to take part in flame trolling, and the role of state intervention and regulation on flame trolling needs to be considered.

Jonathan Bishop
Centre for Research into Online Communities and E-Learning Systems, UK

Section 1
Social, Legal, and Ethical Issues in Internet Trolling

Chapter 1
Ethical Behaviour in Technology–Mediated Communication

Sutirtha Chatterjee
Washington State University, USA

INTRODUCTION AND HISTORICAL PERSPECTIVE

In this information age, serious concerns with unethical behaviour in information technology (e.g., software piracy, deception, plagiarism, etc.) have cast doubts on the claims of the unmitigated success of rapid adoption of information technology. Surprisingly, there have been very few studies in information systems (IS) that have tried to understand the general rise in such unethical behaviours with respect to information technology. Especially, the question that remains to be understood is: Are these problems of unethical behaviour representative of the human nature at large, or are they related to characteristics of technology in any way? This article tries to partly answer this question. It looks at dyadic communi-

nicative behaviour using technology-mediated communication and proposes *a conceptual model* of unethical communicative behaviour. To summarize, the question(s) that this article tries to address are:

In a dyadic technology-based communication between two individuals, what characteristics of technology-based media influence unethical behaviour for an individual? Does individual difference have a role to play in affecting such unethical behaviour? If so, how does it do so?

In answering these questions, the article poses arguments based on literature on media richness, social presence, and deindividuation, and also philosophy-based ethical outlooks of an individual.

DOI: 10.4018/978-1-4666-2803-8.ch001

BACKGROUND AND LITERATURE REVIEW

Unethical Communicative Behaviour

Chatterjee (2005) defined unethical usage of information technology as the violation of privacy, property, accuracy, and access of an individual or an organization by another individual or organization. Since violations of property and access might not be directly relatable to a communicative scenario, this article defines *unethical communicative behaviour between two individuals as the violation of the privacy and/or accuracy of an individual by another individual*. It should be noted that commonly identified forms of unethical communicative behaviour mentioned in the literature (e.g., flaming, swearing, insults, deception, etc.) fall within the scope of violation of either *privacy* (e.g., insults) or *accuracy* (e.g., deception).

Technology-Based Media Characteristics

The key features of technology-based communicative media have been addressed in the media richness literature in IS. Richness of media is the ability to unequivocally transfer the message from the sender to the recipient (Daft & Lengel, 1986). The ability to do this depends on numerous characteristics that the media possesses. Kumar and Benbasat (2002) provide a nice review summary of the key *media characteristics* identified in the media richness literature over the years. These are presented in the following:

- **Modality:** The degree to which a media can support a variety of symbols to present rich information.
- **Synchronicity:** The ability of the media to support communication in real time.
- **Contingency:** The extent to which the communication responses are pertinent to previous responses.

- **Participation:** The extent to which the media supports the active engagement of senders and receivers in the communication.
- **Identification:** The extent to which the senders and receivers are identified (as opposed to being anonymous) by the media.
- **Propinquity:** The extent to which the media supports communication between geographically dispersed senders and receivers
- **Anthromorphism:** The degree to which the interface simulates or incorporates characteristics pertinent to human beings.
- **Rehearsability:** The extent to which the media supports fine tuning of the message before sending.
- **Reprocessability:** The extent to which the media supports messages to be reexamined within the same context.

This summarization forms the fundamental set of antecedents in this article. Latter sections of the article argue how these key media characteristics ultimately influence unethical communicative behaviour.

Media Richness and Social Presence

Technology-based communication is a mediated experience (Biocca, Burgoon, Harms, & Stoner, 2001) with the aim to emulate face-to-face (FTF) communication. The primary aim of technology-mediated communication is to make the mediation disappear (as in FTF) in order to result in a perception of "being there" and "being together" (Biocca et al., 2001, p. 1). Social presence—defined as the extent of perception (of users of a media) that the media conveys the communicators' physical presence in terms of humanness, sociability, personalness, and warmth (Baker, 2002)—also revolves around the ideas of "being there" and being "together."

Existing literature on media richness (e.g., Rice, 1992; Kinney & Dennis, 1994) has always linked media richness to social presence. It has bccn argucd that FTF differs significantly than other environments because it exudes a greater perception of presence than other media. Media that are not sufficiently rich have limited capability to transfer information from the sender to the receiver and have a lower social presence than media that are high in richness. Media richness and social presence are essentially two sides of the same coin and can be defined individually in terms of the other. For example, a rich media is one that exudes a greater social presence, and a higher social presence implies a richer communicative media. Evidence of this fact can be found in the literature (Carlson & Davis, 1998), and the fact that social presence and media richness have been grouped together under the "Trait Theories of Media Selection"(Kumar & Benbasat 2002).

Following Kinney and Dennis (1994) and Dennis and Kinney (1998), this article argues that *media characteristics are the key influencers of media richness (and thus, of social presence)*. This thought is also echoed by Kumar and Benbasat (2002), where they say that it can be reasonable to argue that a media being perceived as being high on the communication characteristics would result in a richer and more socially present media.

PROPOSITION DEVELOPMENT

This section develops the model and propositions. For the benefit of the reader, we present the entire model *a priori* in Figure 1.

Media Characteristics and Social Presence

This section develops propositions on how each of the abovementioned media characteristics influences social presence. FTF has the most positive influence on social presence (Chidambaram & Jones, 1993). Hence, each of the media characteristics is compared against the "baseline" of FTF. Any media characteristic that is different from those of FTF has a negative influence on social presence; on the other hand, any characteristic that helps closely approximate FTF has a positive influence social presence.

• **Modality:** A media high in modality can present information in alternate ways so as to suit the receiver of the information. As in FTF, the sender of the message can try alternate means to get the message

Figure 1. Model of unethical communicative behavior

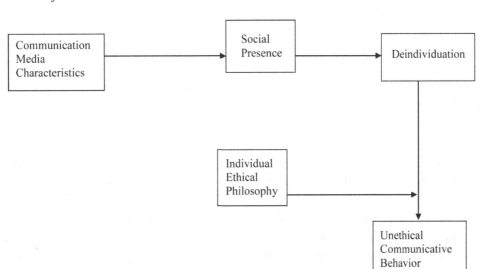

through to the receiver. For example, if the language is not well understood by the receiver, the sender can resort to using visual signs or gestures in order to make the receiver understand. Since, the characteristic of (high) modality enables a sender to present such alternate ways to send the message, it becomes a close approximation of this characteristic of FTF. We already know that FTF leads to a high degree of social presence. Therefore:

Proposition 1a: Media high in modality leads to a high degree of social presence.

- **Synchronicity:** A media high in synchronicity implies that there is no time delay between the sending and receiving of messages. This again is a scenario very similar to a FTF. The immediacy of the communication makes the sender and receiver of the messages feel that they are in a "real" conversation, with the sense of "being there and "being together." Media high in synchronicity gives a high illusion of physical co-presence (even though the individuals are not physically co-present) and stimulates the feeling that the other entity in the conversation is also real. This results in a need to reply immediately and creates a sense of social urgency, just like FTF, thus leading to a high degree of social presence. Therefore:

Proposition 1b: Media high in synchronicity leads to a high degree of social presence.

- **Contingency:** A media high in contingency (e.g., technology-based chat) retains the relevance of conversation. In such a case, there is little scope to "avoid" any message that is received and reply back in an entirely different manner. While chatting, if A asks B about an issue, then B is "forced" to

reply back and does not realistically have a chance to conveniently ignore the question that A posed. However, in a medium like e-mail, B can possibly conveniently ignore A's question (or not reply to A's question at all). This is because typically an e-mail is more assimilative (many issues can be discussed and one of them can be ignored). In a chat, any discussion is atomic (one issue at a time), so responses must be pertinent. Even in FTF, the discussion is more atomic and thus there is no scope of "avoidance." Hence media high in contingency approximates this feature of FTF closely. Therefore:

Proposition 1c: Media high in contingency leads to a high degree of social presence.

- **Participation:** Participation and the sense of "being part of the conversation" is an important contributor to social presence. If one is not actively engaged (psychologically at least) in a communication process, then the feelings of "being there" or "being together" are absent. FTF requires the active participation of both parties, else the communication would stop immediately. Similarly, media high in the participation characteristic (e.g.. videoconferencing, or even a voice chat) help to approximate this effect. Therefore:

Proposition 1d: Media high in participation leads to a high degree of social presence.

- **Identification:** Identifiability (as opposed to anonymity) has been shown to be a key positive influencer of social presence. FTF provides absolute identification. Some media as compared to others may be more appropriate for identification (e.g., videoconferencing as compared to e-mail). For example, videoconferencing or video chats improve the possibility of identification, making it very similar to FTF. Therefore:

Proposition 1e: Media high in identification leads to a high degree of social presence.

- **Propinquity:** A high degree of propinquity implies that individuals who are geographically far away can communicate. However, propinquity is more concerned with actually delivering the message between geographically dispersed communicators. The actual understanding of the message by the receiver and the corresponding influence on social presence is not directly related to propinquity. When an individual is already interacting with an individual, the geographical distance between them does not matter. Therefore:

Proposition 1f: Media high in propinquity would have no effect on the degree social presence.

- **Anthromorphism:** If a media interface can simulate anthromorphism or human-like characteristics, then an individual would feel as if s/he is interacting with an animate object and not an inanimate object like the computer. Literature has argued that individuals can perceive computers and technology to be similar to them and respond using social norms (Lee & Nass, 2002). This compulsion to behave according to the social norms implies that the individuals are in a similar scenario to FTF. Therefore:

Proposition 1g: Media high in anthromorphism leads to a high degree of social presence.

- **Rehearsability:** Rehearsability is not an attribute of FTF. In FTF, messages cannot

be fine tuned before they are sent. Hence, a media high in rehearsability cannot approximate FTF. Too much of rehearsability destroys the "naturalness" of a communication. Being contradictory to FTF, a high amount of rehearsability characteristic of a media leads to a low level of social presence. Therefore:

Proposition 1h: Media high in rehearsability leads to a low degree of social presence.

- **Reprocessability:** The characteristic of reprocessability helps to reanalyze messages. This is an important feature of FTF. During FTF, both the sender and the receiver can go back and reassess and reanalyze any previous discussion. A media that can support such reprocessability simulates an FTF-like communication. Therefore:

Proposition 1i: Media high in reprocessability leads to a high degree of social presence.

Having presented our propositions on how media characteristics can influence social presence, we next turn our attention to the downstream effect of social presence. In particular, we concentrate on a phenomenon called deindividuation.

Social Presence and Deindividuation

Deindividuation is defined as a feeling of being estranged from others, leading to behaviour violating appropriate norms (Zimbardo, 1969). During deindividuation, one loses awareness for others, feels more anonymous, and has less inhibition for socially unacceptable acts (Sproull & Kiesler, 1991). Classic deindividuation theory (Diener, 1980; Zimbardo, 1969) posits that a decrease of social presence leads to greater deindividuation characterized by negative outcomes such as reduction in awareness, greater anonymity, and lesser group cohesion. Within a dyadic communication, it can therefore be proposed:

Proposition 2: In a dyadic technology-mediated communicative discourse, a higher level of social presence would lead to a lower level of deindividuation. Conversely, a lower level of social presence would lead to a higher level of deindividuation.

Having seen how social presence causes deindividuation, we next turn our attention to the effects of deindividuation on human behaviour.

Deindividuation and Communicative Behaviour

Deindividuation theory argues that when a person is deindividuated, then the individual senses more opportunities to carry out behaviour that are unacceptable to society. Deindividuation results in an inability to identify other stakeholders and leads to ethically questionable acts (Lin, Hsu, Kuo, & Sun, 1999). Hsu and Kuo (2003) mention that deindividuation results in a high probability of unethical conduct. From all these arguments, we can say that the probability of unethical communicative behaviour for a deindividuated person is high. Therefore, it can be proposed:

Proposition 3: In a dyadic technology-mediated communicative discourse, a high level of deindividuation will lead to a high level of probability to commit unethical communicative behaviour.

However, even though deindividuation affects unethical behaviour, this article argues that unethical behaviour is not only affected by deindividuation, but also by individual differences. We next address this individual difference in the form of the different ethical philosophies of the individual.

Individual Ethical Philosophy

Individual ethical philosophy can be defined as the individual belief about the acceptability of right or wrong. In order to understand our arguments based on individual ethical philosophy, a brief introduction to the philosophy of *ethics* is warranted.

Ethics has two broad views: the consequentialist and the categorical schools. The consequentialist school (e.g., Mill, 1861) views that the rightness of an action is determined by how much consequential benefit comes out of the action. The categorical school of thought (e.g., Kant, 1804) views that rightness or wrongness of a behaviour is guided by certain objective rules in place. For example, it is objectively necessary to speak the truth, and hence it would be "unethical" to lie, even to help somebody.

Unethical communicative behaviour can be thought to be a violation, either in a categorical sense (violating norms of privacy and accuracy) or in a consequential sense (violating benefits of privacy and accuracy). An individual might subscribe to the consequentialist or categorical view above. However, within the consequentialist view, the benefits could be either self-interest based (benefits for self only) or altruistic (benefits for others). Depending on the view that an individual subscribes to, an individual might find an act to be ethical for one of the three: (a) it satisfied established norms of conduct, (b) it satisfied benefits for self, or (c) it satisfied benefits for others.

For example, a categorical individual would never violate another individual with respect to privacy and accuracy. This follows similar to the categorical imperatives: privacy and accuracy should not be violated. A self-interest-based individual, on the other hand, might be much more favorable towards such a violation if it promoted benefits for self. The benefits could be hedonistic (e.g., one could get satisfied by simply calling another person names) or non-hedonistic (e.g., one could deceive another to achieve a selfish monetary end). An altruistic individual would not be favorable towards the abovementioned violations, as they have the increasing possibility of causing harm to the victim.

Though higher deindividuation leads to the possibility of unethical behaviour (as previously argued), the final commitment of an unethical behaviour would differ according to the ethical philosophy of individuals due to the abovementioned reasons. Based on our arguments, we therefore propose:

Proposition 4: Individual ethical philosophy will moderate the positive effects of deindividuation on unethical behaviour such that self-interested individuals will have higher levels of unethical behaviour as compared to either altruistic individuals or categorical individuals.

CONCLUSION AND FUTURE IMPLICATIONS

This article attempted to propose a conceptual model of how technology-based media characteristics can influence unethical behaviour and how individual differences in terms of ethical philosophy can also influence such behaviour. The foremost contribution of this article is in the fact that such a model has been previously non-existent in the literature. Future empirical studies on this model should give us a deeper understanding of technology-mediated unethical communicative behaviour. If propositions regarding the technology-induced effects were supported, it would mean that technology is an important driver of unethical behaviour. If not, it would mean that unethical behaviour is more due to individual characteristics (especially ethical philosophy) than

technological characteristics. Either way, it would provide us with a deeper insight into the ethical communicative aspect of IS ethics.

Furthermore, the results of the empirical study can have important practical considerations. If the unethical behaviour is mainly based on individual characteristics, the answer to a lowering of such unethical behaviour might lie in improved moral education. If the unethical behaviour is dependent on technology, then the answer might lie in improving technology so that the communicating participants feel a high degree of social presence. This would result in lower deindividuation and thus a lower possibility of unethical behaviour. Thus, the validation of the model proposed in this article might have important implications for the framing of public policy (improved moral education) or the designing of technological improvements.

REFERENCES

Baker, G. (2002). The effects of synchronous collaborative technologies on decision making: A study of virtual teams. *Information Resources Management Journal, 15*(4), 79–93. doi:10.4018/irmj.2002100106

Biocca, F., Burgoon, J., Harms, C., & Stoner, M. (2001). Criteria and scope conditions for a theory and measure of social presence. *Proceedings of Presence 2001,* Philadelphia, PA.

Carlson, P. J., & Davis, G. B. (1998). An investigation of media selection among directors and managers: From "self" to "other" orientation. *Management Information Systems Quarterly, 22*(3). doi:10.2307/249669

Chatterjee, S. (2005). A model of unethical usage of information technology. *Proceedings of the 11th Americas Conference on Information Systems,* Omaha, NE.

Chidambaram, L., & Jones, B. (1993). Impact of communication medium and computer support on group perceptions and performance: A comparison of face-to-face and dispersed meetings. *Management Information Systems Quarterly, 17*(4). doi:10.2307/249588

Daft, R. L., & Lengel, R. H. (1986). Organizational information requirements, media richness and structural determinants. *Management Science, 32*(5), 554–571. doi:10.1287/mnsc.32.5.554

Dennis, A. R., & Kinney, S. T. (1998). Testing media richness theory in the new media: The effects of cues, feedback, and task equivocality. *Information Systems Research, 9*(3). doi:10.1287/isre.9.3.256

Diener, E. (1980). Deindividuation: The absence of self-awareness and self-regulation in group members. In Paulus, P. B. (Ed.), *Psychology of group influence*. Hillsdale, NJ: Lawrence Erlbaum.

Hsu, M., & Kuo, F. (2003). The effect of organization-based self-esteem and deindividuation in protecting personal information privacy. *Journal of Business Ethics, 42*(4). doi:10.1023/A:1022500626298

Kant, I. (1981). *Grounding for the metaphysics for morals*. Indianapolis, IN: Hackett. (Original work published 1804)

Kinney, S. T., & Dennis, A. R. (1994). Re-evaluating media richness: Cues, feedback and task. *Proceedings of the 27th Annual Hawaii International Conference on System Sciences.*

Kumar, N., & Benbasat, I. (2002). Para-social presence and communication capabilities of a Website: A theoretical perspective. *E-Service Journal, 1*(3).

Lee, E., & Nass, C. (2002). Experimental tests of normative group influence and representation effects in computer-mediated communication: When interacting via computers differs from interacting with computers. *Human Communication Research, 28*, 349–381. doi:10.1093/hcr/28.3.349

Lin, T. C., Hsu, M. H., Kuo, F. Y., & Sun, P. C. (1999). An intention model based study of software piracy. *Proceedings of the 32nd Annual Hawaii International Conference on System Sciences.*

Mill, J. S. (1979). *Utilitarianism*. Indianapolis, IN: Hackett. (Original work published 1861)

Rice, R. E. (1992). Task analyzability, use of new media, and effectiveness: A multi-site exploration of media richness. *Organization Science, 3*(4). doi:10.1287/orsc.3.4.475

Sproull, L., & Kiesler, S. (1991). *Connections: New ways of working in the networked organization*. Cambridge, MA: MIT Press.

Zimbardo, P. G. (1969). The human choice: Individuation, reason, and order vs. deindividuation, impulse and chaos. In Arnold, W. J., & Levine, D. (Eds.), *Nebraska symposium on motivation (Vol. 17*, pp. 237–307). Lincoln, NE: University of Nebraska Press.

KEY TERMS AND DEFINITIONS

Categorical School of Ethics: The rightness or wrongness of a behaviour is guided by certain rules in place.

Consequentialist School of Ethics: The rightness (or wrongness) of an action is determined by how much consequential benefit (or loss) comes out of the action.

Deindividuation: A feeling of being estranged from others, leading to behaviour violating appropriate norms.

Individual Ethical Philosophy: The individual belief about the acceptability of right or wrong.

Media Richness: The ability of media to unequivocally transfer the message from the sender to the recipient.

Social Presence: Extent of perception (of users of a medium) that the medium conveys the communicators' physical presence in terms of humanness, sociability, personalness, and warmth.

Unethical Communicative Behaviour (Dyadic): The violation of the privacy and accuracy of an individual by another individual.

This work was previously published in Encyclopedia of Information Ethics and Security, edited by Marian Quigley, pp. 201-207, copyright 2007 by Information Science Reference (an imprint of IGI Global).

Chapter 2
Codes of Ethics in Discussion Forums

Cãlin Gurãu
GSCM – Montpellier Business School, France

INTRODUCTION

The development of the World Wide Web has created new opportunities for interpersonal interaction. The Internet allows one-to-one (e-mail), one-to-many (Websites, e-mail lists) or many-to-many (*online discussion forums*) interaction, which represent a unique feature in comparison with traditional communication channels (Armstrong & Hagel, 1996). On the other hand, the Internet has specific characteristics, such as:

- **Interactivity:** The Internet offers multiple possibilities of interactive communication, acting not only as an interface, but also as a communication agent (allowing a direct interaction between individuals and software applications).
- **Transparency:** The information published online can be accessed and viewed by any Internet user, unless this information is specifically protected.

- **Memory:** The Web is a channel not only for transmitting information, but also for storing information in other words, the information published on the Web remains in the memory of the network until it is erased.

These characteristics permit the development of online or virtual communities/groups of people with similar interests who communicate on the Web in a regular manner (Armstrong & Hagel, 1996; Goldsborough, 1999a, 1999b; Gordon, 2000). Many studies deal with the ethics of research in Cyberspace and Virtual Communities (Bakardjieva, Feenberg, & Goldie, 2004), but very few publications relate with the *Codes of Ethics* used in Public Discussion Forums (Belilos, 1998; Johnson, 1997). Other specialists have analyzed specific categories or uses of online discussion forums, such as online learning (Blignaut & Trollip, 2003; DeSanctis, Fayard, Roach, & Jiang, 2003) or the creation of professional communities of

DOI: 10.4018/978-1-4666-2803-8.ch002

practice (Bickart & Schindler, 2001; Kling, Mc-Kim & King, 2003; Millen, Fontaine, & Muller, 2002), and in this context, have also discussed briefly the importance of netiquette and forum monitoring (Fauske & Wade, 2003, 2004). The difference between these online communities and public discussion forums is the degree of control exercised on the functioning and purpose of the forum by a specific individual or organization. This article attempts to investigate, analyze and present the main patterns of the codes/rules of ethics used in the public discussion forums, otherwise known as Newsgroups, and their influence on the profile and functioning of the community.

THE ORGANIZATION OF DISCUSSION FORUMS

The discussion forum is a Web-based application that functions as a worldwide bulletin board (Fox & Roberts, 1999). Each discussion forum has a specific topic, or a series of related topics, and there are literally thousands of newsgroups available on the Internet, covering virtually any issue (Preece, 2001; Rheingold, 2000).

Typically, online discussion forums use a three-tiered structure (Bielen, 1997):

1. **Forums:** Focus on individual topic areas, such as classifieds or current news.
2. **Threads:** Created by end users to narrow a discussion to a particular topic, such as a car someone is looking to buy or a comment on a previously posted message. A thread opens a new topic of conversation. Once the topic is created, anyone can continue the ongoing conversation.
3. **Messages:** Individual topic postings. A message is often a reply to someone else's message, or users can post a message to initiate a conversation (thread).

An interested person can access the messages transmitted by other members of the discussion forum, post messages on the same discussion forum or start a new thread of discussion. Usually, in order to post a message or start a new thread, participants are asked to register first; however, many discussion forums are totally transparent, since anyone (members or visitors) can access the archived messages and read them.

Most discussion forums are monitored by people (monitors and administrators) and/or software applications (e.g., programs that automatically censor specific words from the posted messages). The monitors are usually volunteers that have a good knowledge of and interest in the topics discussed (Preece, 2000).

CODES OF ETHICS IN DISCUSSION FORUMS

Ethical rules of discussion forums are usually displayed in two formats:

1. **Explicit Codes of Ethics:** These are presented under titles such as Terms of Use, Guidelines, Forum Rules, Terms and Conditions, Web Site User Agreement or Forum Policy. Very often, the ethical rules are just a topic among other legal disclaimers and definitions; in other cases they are a stand-alone text that does not include any other legal information. The link to these guidelines is easily identifiable, as the members of the discussion forum are expected to read them before engaging in online interaction.
2. **Implicit Ethical Rules:** In this case, no clear indication is provided regarding the ethical rules that have to be respected by forum members; however, indirect references are made in the frequently asked questions

(FAQ) section regarding the possibility of censoring members' messages by replacing specific words with a string of "*." In other sites, the ethical rules are limited to a general principle or "Golden Rule," such as "We should do unto others as we would have them do unto us," from which the members can derive the desirable rules of ethical behavior.

When a site hosts multiple discussion forums (e.g., Google Groups), the ethical guidelines officially published by the site usually have a standardized style that might not properly apply to every group active on the site. Also, the site may indicate that it does not monitor the content of specific postings. Sometimes, members of a particular group attempt to create and enforce a specific ethical code for their group, in order to fill in this organizational gap. The attempt is seldom successful, since the member initiating this action does not have any official, recognized authority. The reactions of fellow members to such an initiative are very diverse, ranging from constructive dialogue to ironic criticism.

THE CONTENT OF ETHICAL CODES FOR DISCUSSION FORUMS

The ethical rules used in discussion forums usually fall into one of five general categories:

1. Rules that concern the good technical functioning of the forum. Since the discussion forum is supposed to be an open (and in most cases, free) board for expressing and exchanging ideas and opinions, some members might have the tendency to abuse it. To counter this tendency, participants are forbidden from:
   a. Posting multiple messages.
   b. Inserting in their message long quotes from previous posts of other members, or entire articles downloaded from the Web—as a rule, the member is expected to edit the text, providing only the relevant quotes or an abstract; and if absolutely necessary, providing a hyperlink to the article of interest.
   c. Inserting pictures or sound files in their messages.
   d. Posting files that are corrupted or contain viruses.

   All these actions occupy substantial computer memory, and slow down or damage the general functioning of the discussion forum.

2. Rules concerning the content of the posted message. Members should not include in their post:
   a. Content that is not relevant for the subject of the discussion forum (crossposting).
   b. Defamatory, obscene or unlawful material.
   c. Information that infringes patent, trademark, copyright or trade secret rights.
   d. Advertisements or commercial offers, other than those accepted by the group (since some discussion forums have a commercial purpose).
   e. Questionnaires, surveys, contests, pyramid schemes or chain letters, other than those accepted by the group (since some discussion forums have a data collection purpose).

3. Rules concerning the purpose of use. The members must not use the site to:
   a. Defame, abuse, harass, stalk or threaten other members of the same group.
   b. Encourage hatred or discrimination of racial nature.
   c. Practice flaming or trolling.
   d. Engage in flame wars with fellow members.
   e. Excessively criticize the opinion of other participants (although some sites do not include this advice for restraint).
   f. Advertise products.

g. Conduct surveys.

4. Rules pertaining to personal identification issues. The members should not:
    a. Refrain from providing all the information required for their identification within the discussion forum.
    b. Impersonate another person or entity.
    c. Falsify or delete any author attribution.
    d. Falsely state or otherwise misrepresent the affiliation with a person or entity.
    e. Delete or revise any material posted by any other person or identity.

5. Rules concerning the writing style of the message. Forum members are expected to post messages that:
    a. Are in the official language of the Web site.
    b. Have a title that accurately describes the topic of the message.
    c. Are not excessively using capital letters (no shouting).
    d. Are free from spelling and grammatical mistakes.
    e. Are not highly emotional, so that they might disrupt the exchange of opinions.
    f. Do not attack people, but opinions.

THE ENFORCEMENT OF ETHICAL RULES

Usually, the ethical codes governing the participation in discussion forums are enforced by the forum's monitors or administrators. This can be done proactively or reactively.

The proactive enforcement of ethical rules usually takes place in smaller forum communities, where the monitors have the possibility to read every posted message and approve it before it becomes publicly available.

The reactive mode is implemented when a participant identifies or has a problem related with the unethical behavior of another member. In this situation the participant can alert the monitor, who can take direct measures against the participant infringing the ethical rules.

In some cases, when a site hosts a huge number of forums, it is clearly specified that site monitors/administrators do not monitor the content of specific postings, although a number of ethical rules might be provided. However, this case can be considered as a particular application of the reactive monitoring system, since the site monitors/administrators will probably react to complaints concerning a blatant breach of ethical rules.

Monitors can take a number of measures progressively to enforce the ethical rules of the forum, such as:

1. A warning for the first-time breach of ethical rules.
2. Suspension of posting privileges for repeated infringement of rules.
3. Withdrawal of posting rights and deactivation of a member's account for repeated and flagrant violations of ethical rules.

In parallel with these actions, the monitors using the proactive mode of surveillance can require a forum participant to edit a message or withdraw it. Alternatively, in some cases, the monitors themselves have the specific right to edit or completely erase posts that might breach the ethical rules of the Forum, and to close the threads that are out of focus or that have degenerated in a flame war. These measures are also taken when monitors manage the forum using a reactive mode, but in this case, the action will be initiated by a complaint sent by a forum participant regarding a breach of ethical rules.

Any complaint or opposition to these measures should be discussed outside the forum with the moderator, and, if the problem cannot be solved satisfactorily, the member can eventually send an e-mail message to the forum administrator.

The advantage of the proactive mode of surveillance is that the messages published online are cleaner and better edited in the first instance.

However, this type of monitoring is difficult to implement when the forum is very popular and dynamic, publishing hundreds of messages daily.

A POSSIBLE FRAMEWORK FOR ANALYZING AND DESIGNING ETHICAL RULES FOR PUBLIC DISCUSSION FORUMS

The implementation and enforcement of ethical codes in discussion forums represent a complex and multidimensional process (see Figure 1). The main purpose of these rules is to regulate the exchange of opinions within the forum by establishing reasonable limits to participants' behavior. The effectiveness of these rules will be determined by two related variables: the clarity of the rules and their enforcement.

The online exploratory survey of 200 discussion forums has provided multiple values for these two variables, indicating the use of a continuous scale as the best tool for their evaluation. The degree of clarity of rules will vary between implicit and explicit ethical codes, and the enforcement of rules between a proactive and reactive monitoring style. It is therefore possible to design a two-dimensional graph on which every Discussion Forum can be represented in a position defined by the specific values given to the two variables (see Figure 2).

To evaluate the effectiveness of each of the four Ethical Systems, 10 messages have been randomly selected and accessed in each of the

Figure 1. The implementation and enforcement of ethical rules in the interactive environment of discussion forums

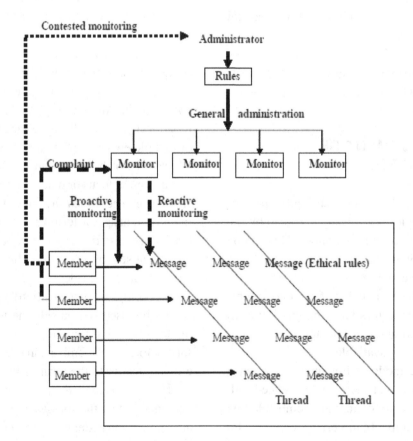

Figure 2. A bi-dimensional framework for the representation of ethical codes implementation in discussion forums

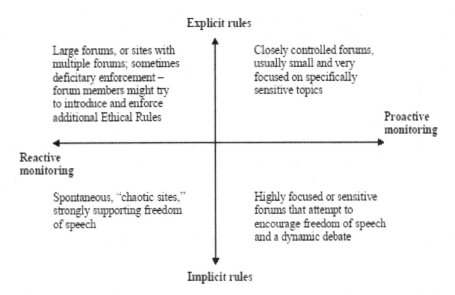

surveyed discussion forums, and their level of ethical compliance has been evaluated on a scale from 1 (unethical message) to 10 (fully ethical message). The mean of these 10 measurements of ethical compliance was calculated, and then was used to calculate the general mean for the sites included in each of the four possible categories.

Figure 3 represents the level of ethical compliance for each of the four categories of sites (the size of the star is directly related with the general mean of ethical compliance measurements - writ-

Figure 3. The results of the empirical study regarding ethical codes implementation in discussion forums

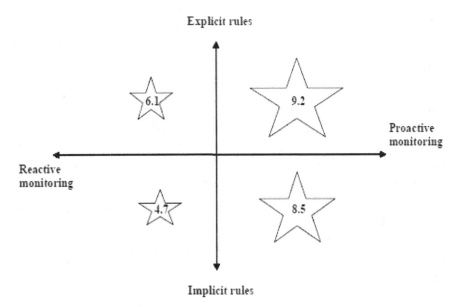

ten inside each star). As it can be seen, a proactive style of monitoring has a strong influence on the ethical dimension of the posted messages, even when the rules are only implicit. On the other hand, the combination of implicit rules with reactive monitoring creates a highly anarchic discussion forum environment, which in some cases might be the desired option.

The proposed framework can be applied by creators/administrators of discussion forums in order to identify the best format for the design and implementation of ethical rules that are consistent with the specific profile and purpose of the site.

Another important consideration is the content of the ethical code in terms of coverage and generality of the ethical issues that are presented, as well as the clarity of the penalties and the specific power/authority of monitors.

All these elements can be combined and dynamically tuned to create the desired environment in the discussion forum. The interactive and dynamic nature of a discussion forum introduces evolutionary features in relation to the style of the posted messages, which consequently determines the specific audience profile of the forum. The "personality" and audience of a discussion forum can therefore be changed over time by manipulating the variables presented above.

CONCLUSION

The implementation of ethical codes represents an essential element for the good functioning of virtual communities (Herring, Job-Sluder, Scheckler, & Barab, 2002; Preece & Maloney-Krichmar, 2002). The discussion forums are very diverse from the point of view of their topic and organization, and these characteristics introduce variability at the level of ethical rules.

The content, structure, mode of presentation and style of enforcement of ethical rules in a discussion forum can represent an important tool for defining the type of the community and the style of interpersonal interaction. The results of the study presented outline the relation among the mode of presentation and the style of enforcement of the ethical code, and the style/values/ profile of the online community.

The design and enforcement of ethical codes do not and cannot represent an exact science. However, the dynamics of interaction in a discussion forum permit an evolutionary adaptation of the ethical code to the desired profile of the discussion forum. In this context, more research is necessary to identify, define and measure the influence of ethical codes on the specific organization of a discussion forum. Future studies may concentrate on particular case studies (forums) in order to analyze the parallel evolution of ethical rules and online community, emphasizing the relation between a particular style of ethical rules and the behavior of participants in the community.

REFERENCES

Armstrong, A., & Hagel, J. III. (1996). The real value of online communities. *Harvard Business Review, 74*(3), 134–141.

Bakardjieva, M., Feenberg, A., & Goldie, J. (2004). Usercentered Internet research: The ethical challenge. In Buchanan, E. (Ed.), *Readings in virtual research ethics: Issues and controversies* (pp. 338–350). Hershey, PA: Idea Group Publishing. doi:10.4018/978-1-59140-152-0.ch018

Belilos, C. (1998). *Networking on the net.* Professionalism, ethics and courtesy on the net. Retrieved November, 2004, from www.easytraining.com/networking.htm

Bickart, B., & Schindler, R. M. (2001). Internet forums as influential sources of consumer information. *Journal of Interactive Marketing, 15*(3), 31–41. doi:10.1002/dir.1014

Bielen, M. (1997). Online discussion forums are the latest communication tool. *Chemical Market Reporter, 252*(7), 9.

Blignaut, A. S., & Trollip, S. R. (2003). Measuring faculty participation in asynchronous discussion forums. *Journal of Education for Business, 78*(6), 347–354. doi:10.1080/08832320309598625

DeSanctis, G., Fayard, A.-L., Roach, M., & Jiang, L. (2003). Learning in online forums. *European Management Journal, 21*(5), 565–578. doi:10.1016/S0263-2373(03)00106-3

Fauske, J., & Wade, S. E. (2003/2004). Research to practice online: Conditions that foster democracy, community, and critical thinking in computer-mediated discussions. *Journal of Research on Technology in Education, 36*(2), 137–154.

Fox, N., & Roberts, C. (1999). GPs in cyberspace: The sociology of a 'virtual community.'. *The Sociological Review, 47*(4), 643–672. doi:10.1111/1467-954X.00190

Goldsborough, R. (1999a). Web-based discussion groups. *Link-Up, 16*(1), 23.

Goldsborough, R. (1999b). Take the 'flame' out of online chats. *Computer Dealer News, 15*(8), 17. Gordon, R.S. (2000). Online discussion forums. *Link-Up, 17*(1), 12.

Herring, S., Job-Sluder, K., Scheckler, R., & Barab, S. (2002). Searching for safety online: Managing "trolling" in a feminist forum. *The Information Society, 18*(5), 371–385. doi:10.1080/01972240290108186

Johnson, D. G. (1997). Ethics online. *Communications of the ACM, 40*(1), 60–65. doi:10.1145/242857.242875

Kling, R., McKim, G., & King, A. (2003). A bit more to it: Scholarly communication forums as socio-technical interaction networks. *Journal of the American Society for Information Science and Technology, 54*(1), 47–67. doi:10.1002/asi.10154

Millen, D. R., Fontaine, M. A., & Muller, M. J. (2002). Understanding the benefit and costs of communities of practice. *Communications of the ACM, 45*(4), 69–74. doi:10.1145/505248.505276

Preece, J. (2000). *Online communities: Designing usability, supporting sociability.* Chichester, UK: John Wiley & Sons.

Preece, J. (2001). Sociability and usability in online communities: Determining and measuring success. *Behaviour & Information Technology, 20*(5), 347–356. doi:10.1080/01449290110084683

Preece, J., & Maloney-Krichmar, D. (2002). Online communities: Focusing on sociability and usability. In Jacko, J., & Sears, A. (Eds.), *The human-computer interaction handbook* (pp. 596–620). Mahwah: Lawrence Earlbaum Associates.

Rheingold, H. (2000). *The virtual community: Homesteading on the electronic frontier.* Cambridge: MIT Press.

KEY TERMS AND DEFINITIONS

Crossposting: Posting the same message on multiple threads of discussion, without taking into account the relevance of the message for every discussion thread.

Flame War: The repetitive exchange of offensive messages between members of a Discussion Forum, which can eventually escalate and degenerate in exchange of injuries.

Flaming: Posting a personally offensive message, as a response to an opinion expressed on the Discussion Forum.

Monitor: A person who is monitoring the good functioning of a Public Discussion Forum. It is usually a volunteer who is specialised and interested in the specific topic of the forum.

Public Discussion Forum: Internet-based application which permits an open exchange of opinions and ideas among various Internet users, on a specific topic of interest, and that can be easily accessed by interested individuals.

Shouting: Posting a message written entirely or partially in capital letters.

Trolling: Posting a controversial message on a Discussion Forum, with the purpose to attract or instigate a flaming response, mainly targeting the inexperienced members of the Forum.

This work was previously published in Encyclopedia of Virtual Communities and Technologies, edited by Subhashish Dasgupta, pp. 22-28, copyright 2006 by Idea Group Reference (an imprint of IGI Global).

Chapter 3
Privacy and Security for Virtual Communities and Social Networks

Georgios Michaelides
Budapest University of Technology and Economics, Hungary

Gábor Hosszú
Budapest University of Technology and Economics, Hungary

ABSTRACT

The importance of the virtual communities' privacy and security problems comes into prominence by the rapid development of online social networks. This article presents the multiple threats currently plaguing the virtual world, Internet privacy risks, and recommendations and countermeasures to avoid such problems. New generations of users feel comfortable publishing their personal information and narrating their lives. They are often unaware how vulnerable the data in their public profiles are, which a large audience daily accesses. A so-called digital friendship is built among them. Such commercial and social pressures have led to a number of privacy and security risks for social network members. The article presents the most important vulnerabilities and suggests protection methods and solutions that can be utilized according to the threat. Lastly, the authors introduce the concept of a privacy-friendly virtual community site, named CWIW, where privacy methods have been implemented for better user protection.

INTRODUCTION

Social networks are pre-technological creations that sociologists have been analyzing for years. (Lohr, 2006). The Internet brought a new era for social networking that most of us could never imagine before. Today we have organizational and software procedures that control the exchange of interpersonal information in social networking sites, text messaging, instant messenger programs, bulletin boards, online role-playing games, Computer-Supported Collaborative Work (CSCW), and online education. All of these applications fit into the larger category of social media, or media that support social collaboration. The term social media is an umbrella concept that describes social

DOI: 10.4018/978-1-4666-2803-8.ch003

software and social networking. "Social software refers to various, loosely connected types of applications that allow individuals to communicate with one another and to track discussions across the Web as they happen" (Tepper, 2003). Defining the term *Virtual Community* is a hard thing to be done, because we cannot just specify it, as it can exists in different forms. Explaining what a *Virtual Community* is, we can describe that it is a group of individuals that uses the Internet Technology and the Social Network Services for sharing common interests, ideas and even feelings. These services can be accessed through newsgroups, Web forums, Web chats, IRC channels, weblogs and private messages.

Security and privacy are one of the most common and discussed issues which lacks in social network services. The idea of privacy in social network could be described as a paradox; because information about you is shared with other members of the network, thus there is little or no privacy. Privacy is not just about hiding things. It is about self–possession, autonomy, and integrity. The comprehension of privacy is something that the users do not understand and in many cases it is not well implement between the user privacy and the privacy network that is held on the service, as there are so many sites holding millions of members.

Users are often not aware of how vulnerable is when their public profile data are accessed by a large size of audience, so that digital "friends" are created (Higgins, 2008). Such commercial and social pressures have led to a number of privacy and security risks for social network members (Gross & Acquitsi, 2005).

Easy access for uploading any kind of personal information, data including age, location, contact information (home address and telephone numbers), even identity card number, pictures and special notes for interests and hobbies, giving a clear idea how a person looks like and how he behaves. It is a deep societal problem emerging people giving up their privacy without realizing it. Also, many users can fall into trap surveys of popular Social Network Sites (SNSs) by answering naively questions as: "Have you ever cheated on your boyfriend/girlfriend?", "Do you download illegal content from Internet?" or "Have you stole something from a stranger?" (De Rosa, Cantrell, Havens, Hawk, & Jenkins, 2007). Surveys with these trick questions can easily harm the individuals who have participated.

In the following sections, we have reviewed the current stages of security problems of virtual communities. Authors have shared their experiences and highlighted the significant vulnerabilities and introduced countermeasures and solutions to confront them. To prove and test that these security methods work and that the users can be really comfort and secure, we have created the concept of a basic privacy-friendly Social Network Site. Finally, we should think and predict the future horizon of virtual communities. Problems are fast growing and arising social connections must be taken into consideration.

RESEARCH FOR SOCIAL NETWORK SITES

Digital Lives

Even though the new generations of younger people are born and grow up with their personal computers and Internet technology, it is rather difficult to distinguish and compare the digital behavior and attitude of people born before and after the Internet era. The new technology has invaded and conquered a major and requisite part of our lives. Figure 1 shows the comparison of ages according to Internet usage in a scale of time.

Sharing Information Online

The online information sharing is a common feature that allows the currents users to adopt different roles in social networking community.

Figure 1. Length of time using the Internet by age (De Rosa et al., 2007)

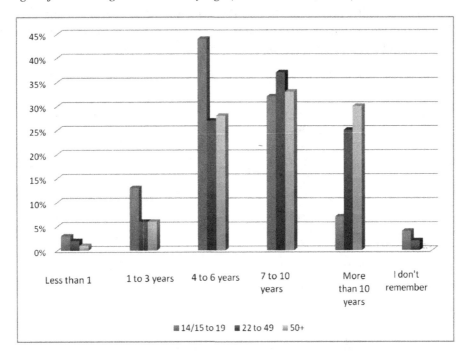

Based on researches, users exchange personal and general information when they want to fill in online registration forms at random websites. More than half of the general public mentions that they *always, often* or *sometimes* complete fully the online registration form and not just the required indicated information. Specifically 11% *always* complete the entire registration form.

Based on the OCLC (Online Computer Library Center) report, users have been categorized according to their answers on what information is giving online. The *contact information* group holds information of name, address, email and phone. The *individual information* group refers to the marital status, personality and physical attributes birthday and sexual liking. The last is the *interest information* group, which provides information about different subjects of interest like books, political beliefs, and religious affiliations. See Figure 2.

Country Comparison

Users from different places in the planet are classified according the information shared online. The first prize goes to United States and Canada as they like sharing most of their *personal and physical attributes,* the *books read* and their *first names.* Japanese and French are tighter and more careful in giving information online, as they take the last place in charts. Germany and United Kingdom walk on the same level of sharing. The most common information shared online among the several social network sites is the *birthday, marital status, subjects of interest* and *photos/ videos.* See Figure 3 and Figure 4.

True Online Personalities

All the known social network sites require filling a registration form online. In another research, users have been asked if they really share their identities during registration. Almost the two-thirds of the online community responded that

Figure 2. (a) Information shared on commercial and social sites by total general public for contact information (De Rosa et al., 2007) (b) Information shared on commercial and social sites by total general public for individual information (De Rosa et al., 2007) (c) Information shared on commercial and social sites by total general public for interest information (De Rosa et al., 2007)

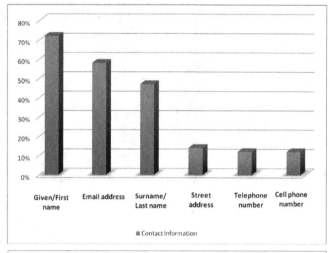

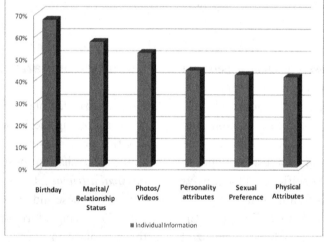

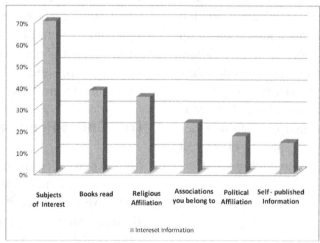

Figure 3. Information shared on social network sites by country (De Rosa et al., 2007)

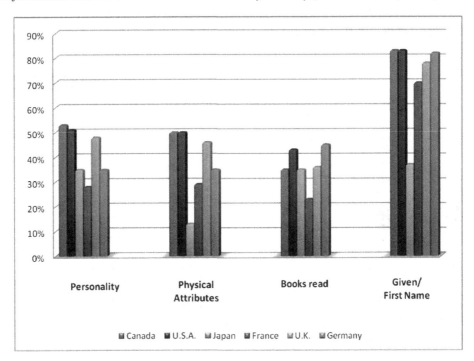

they provide their real name, email address and age. Also over the half of the public gives also their real telephone number, a large amount of audience. See Figure 5.

The answer on why so much people disclose that sensible information online, lays on the trust and good communication the members gain between them and the respective social networks site. In bigger and more famous online communities' networks, keeping in track millions of users like Facebook and MySpace, users' easier feel confident and protected for the integrity of their personal information. Privacy policies must be included and revised in every website that holds these types of personal data. Nevertheless, how often do people read these policies? Are they aware the danger and threats providing real information of them online?

THREATS

As user benefits increase, so will opportunities for attackers. Spammers and scammers will look to exploit this treasure trove of information and will more easily construct convincing social engineering attacks with all this data. Users will be taken off guard by the level of detail and personalization in attack messages. Social botnets will also have the potential to disrupt seriously the ecosystem, poisoning the network with solicitations and false testimonials. Site administrators will have their work cut out for them to keep the content quality high, while blocking the bad people and still allowing everyone else to use the site as it is intended. Securing future social networks will depend more heavily on server-side defences. Back-end systems will need to scan large amounts of incoming and outgoing data, searching for evidence of mischief or malicious code. Site and content reputation services may help balance usability and security. The trust relationship between sites and users is key

Figure 4. (a)Information shared on social network sites in detail by country on contact information (De Rosa et al., 2007) (b) Information shared on social network sites in detail by country on individual information(De Rosa et al., 2007)

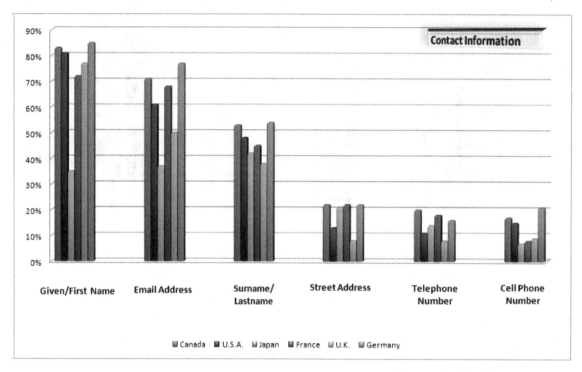

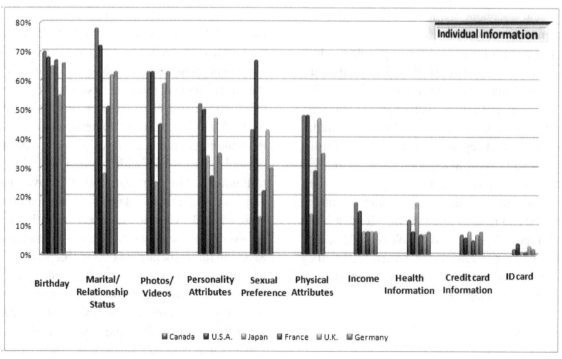

Figure 5. Information provided while registering by country (De Rosa et al., 2007)

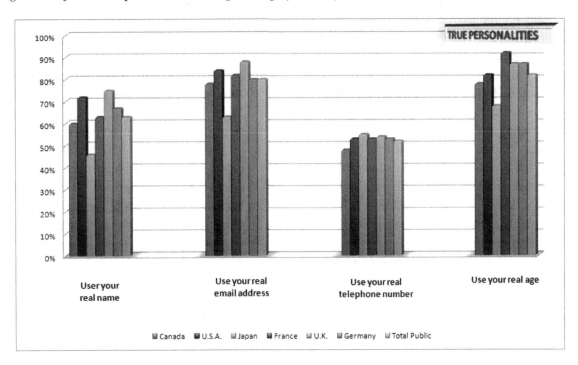

to the success of tomorrow's networks. Violation of that trust could lead to the failure of an entire community (Schmugar, 2008).

Privacy Related Threats

The following discussed threats are covered mostly for the Social Network Sites (SNSs) and Virtual Communities rather not generalizing for all the Web applications.

- **Digital Dossier Aggregation:** The public profiles can be over SNSs can be easily downloaded by third parties creating a digital dossier aggregation. The information contained in individual profiles can be accumulated easily in order to track and highlight changes. A most common vulnerability is that more private attributes that are directly accessible by profile browsing can be accessed via search (e.g., a person's name in Facebook). *Risks:* This

information can be manipulated, deleted or even threatening individuals by publishing personal data for bad purposes.

- **Secondary Data Collection:** A user despite the personal data stored on a SNS database, there are some more tracing clues that can leave behind (e.g., IP address, logging time, private messages and e-mails). It is not certain whether this information or a part of it is public or not. *Risks:* An example is from the IP address you can easily find the location of the user using tracing tools which calculate the specific parameters. E-mails can be used for spamming when this person can be popular or a celebrity.

- **Face Recognition:** One really popular property on most SNSs. The members are allowed to upload their profile picture or creating personal galleries. Facebook hosts billions of photos which many are labeled and noted using face recognition. *Risks:*

The photograph is, in effect, a binary identifier for the user, enabling linking across profiles, for example, a fully identified Bebo profile and a pseudo-anonymous dating profile.

- **Content-Based Image Retrieval (CBIR):** It is related to face recognition but it is linked to a place or paintings. It was firstly discovered for digital forensics. This kind of threat can specify the location of a person and his surroundings. *Risks*: CBIR can recognize an individual's place (e.g., home, work place) and can lead to stalking, blackmailing and other unwanted behaviors.

- **Difficulty of Complete Account Deletion:** Although it looks easy to change our personal data like age, name day or birthday, deleting your personal account seems harder. Facebook offers account deletion but prompts that some user data are still active but not visible to others. Also an e-mail it is sent on how to reactivate the account. Users have to remove manually all public notes or comments on their profiles, which it is feasible impossible due to a large number of steps involved. *Risks*: The user loses control over his or her identity. The fundamental right to control their personal information it is lost.

SNS Variant of Traditional Network and Information Security Threats

- **Social Network Spam:** SN spam can be unwanted messages, e-mails, friend invitations that include also comment posting, even bots trying to register automatically on a site with random personal information. Also it can include embedded links for *Warez* (generally refers to illegal releases by organized groups, as opposed to peer-to-peer file sharing between friends or large groups of people with similar inter-

est, where refers primarily to copyrighted works traded in violation of copyright law), pornographic or any other commercial sites. Mostly members' personal information and especially passwords are vulnerable on these kinds of spamming attacks. *Risks:* Traffic overload. Diversions to unwanted sites that can be are malicious for the user. Creation of false profiles on SNSs reduces the values of legitimate users.

- **Cross Site Scripting, Aggregators, Viruses and Worms:** Third party applications and HTML codes that users can post within their profile are vulnerable to SNSs. As we can recall SAMY virus hit MySpace infected million of profiles within 20 hours. *Risks:* User account compromise, identity theft, *Denial of Service* (DoS) attacks (an attempt to make a computer resource unavailable to its intended users by preventing an Internet site or a service to functional efficiently or at all), and loss of privacy.

Identity-Related Threats

- **Spear Phishing Using SNSs and SN-Specific Phishing:** Spear phishers send e-mail that appears genuine to all the employees or members within a certain company, government agency, organization, or group. Spear phishers often customize e-mails with information they have found on Web sites, blogs, or social networking sites like Facebook or MySpace. They also might create fake social networking login pages to lure people into sites where they are used to entering personal information. *Risks:* Compromised logins can turn into increase the speed of spread of a phishing attacks, in addition to identity theft (the co-option of another person's personal information for example, name, Social Security number, credit card number, passport, without that person's knowledge and the fraudulent use of such knowledge), financial and reputation damage.

- **Infiltration of Networks Leading to Information Leakage:** Some information is only available to a restricted group or network of friends, which should provide the first line of defense in protecting privacy on SNSs. However, since it is often easy to become someone's "friend" under false pretences, this mechanism is not effective. On many SNSs it is even possible to use scripts to invite friends. *Risks:* Able to view members' personal information. The predators can have physical contact with a member if they become trustful.

Social Threats

- **Stalking:** Cyber stalking is threatening behavior in which a perpetrator repeatedly contacts a victim by electronic means such as e-mail, Instant Messenger and messaging on SNSs. Statistics suggest that stalking using SNSs is increasing. *Risks:* These criminal activities can consequence into loss of privacy or even physical and psychological damage.
- **Bullying:** Cyber bullying involves the use of information and communication technologies to support deliberate, repeated, and hostile behavior by an individual or group, which is intended to harm others. *Risks:* Victims have lower self-esteem, increased suicidal ideation, and a variety of emotional responses, cyber-bullying back, being scared, frustrated, angry, and depressed. Sometimes the effects of cyber-bullying are so strong that victims committed suicide; for example, the suicide of Megan Taylor Meier.

RECOMMENDATIONS AND COUNTERMEASURES

- **Awareness and User Education:** Social Network Sites should publish and point out some user-friendly guidance about SNSs threats and how-to avoid them. Policies about should be implemented about bullying like in schools and universities. Inform the parents, an adult or even the local police if someone is threatened with harm. Never agree to meet a person or someone you have met online. Software companies should consider better security conscious practices and corporate policies.
- **Review and Reinterpret Regulatory Framework and Increase Transparency of Data-Handling:** Some scenarios of legal data protection are not so clarified. These issues affect the legal position on imaging tagging, privacy policies of third-party applications and widgets for user interaction, location data might be illegal in some cases. The purpose for which that data is used should be clarified. Particular third parties and groups should have access restrictions. Users should take in their consideration and be given accurate information of what is done with their data before and after account deletion. All these should be done in a user-friendly way instead of publishing in Terms and Conditions.
- **Implementation of Better Security Policies and Access-Control:** There are a number of additional authentication factors that SNSs could use to enhance their offer by reducing the level of fake and troublesome memberships. These range from basic e-mail verification through CAPTCHAs and recommendation-only networks to physical devices such as mobile phones and identity card readers. In addition, the scheduled basis of security

holes identification is another issue that most SNSs should be considered.

- **Maximize Possibilities for Reporting and Detecting Abuse:** SNSs should make it as easy as possible to report abuse and concerns. "Report Abuse" buttons should be as ubiquitous as "Contact Us" options on classic Web sites.

- **Setting Appropriate Default Values:** User profile change is rarely used, especially on secondary data information. Default values should be implemented as safe as possible for the sake of the individual. It can be good if some policies can be implemented for specific age groups like teenagers and above 21 years old. Also guided approaches and explanations of how the profile values work and what are the consequences by manipulating these values.

- **Built-In Automated Filters:** Smart filters are advised and a must for a proper functionality of a Virtual Community. Examples of these filters is when a single user is spamming the site by making automated submissions or e-mail spamming, filtering any illegal content such as cracks or pornographic material. Limited user posts for small intervals can reduce the traffic of the server.

- **Detain Spider-Bots, Bulk Downloads, and Search Results:** Bots can attack Virtual Communities and collect information based on members' data and other Web services the SNS is offering to its audience. For the interest of both users and service providers, these threats should be restricted through measurements like CAPTCHAs (a type of challenge-response test used in computing to ensure that a computer does not generate the response). The process usually involves one computer (a server) asking a user to complete a simple

test which the computer is able to generate and grade. Because other computers are unable to solve the CAPTCHA, any user entering a correct solution is presumed to be human) and network traffic. Users will be well protected from blackmailing and lack of control over their profiles.

In addition, SNSs should limit the visibility while someone is searching personal information of an individual. This can be done by implementing access-rights to groups of users and their relation with their friendly groups. Data should either be anonymised, not displayed, or the user should be clearly informed that it will appear in search results and given the choice to opt out.

- **Promotion of Anonymisation Techniques and Best Practices:** Users should be aware that an image data file is a rectangular grid of pixels. It has a definite height and a definite width counted in pixels. Each pixel is square and has a fixed size on a given display. Each pixel's color sample has three numerical RGB components (Red, Green, Blue) to represent the color. These three RGB components are three 8-bit numbers for each pixel. Face recognition algorithms are still under development, but still the audience of SNSs should take notice of restricting the possibilities of their identification and linking to other sites. Not publishing direct face pictures can do this. Advisable is to post pictures which are hard to be recognized such non-frontal images, under non-standard illumination, displaying varied facial expressions. Network operators, some of which already actively encourage users to upload face images, could give similar recommendations and check for compliance.

EXAMPLE OF A PRIVACY-FRIENDLY COMMUNITY SITE

We have developed a method to experiment a privacy-friendly Virtual Community site based on PHP programming language and MySQL rational database called CWIW to test the different security methods mentioned above (Michaelides, 2009). In order to minimize the security threats that could enquire on the community, we reduced by having only people who share specific interests, not for general purposes like Facebook. With this approach the joined audience is minimal and easier to handle.

Upon registration process, we have implemented a confirmation e-mail and a CAPTCHA image. The user is unable to sign up if he or she has not entered the correct string that is provided by the CAPTCHA mechanism. Later he has to activate his account by the received e-mail from the site, where it contains an activation link.

Another mechanism is the "Forgot Password" feature. Again, the user should enter his username and e-mail address and the CAPTCHA string to proceed. If any field in misspelled, the procedure fails. An e-mail with expiration time in 24 hours will be sent provided with a recovery password link and a temporary password.

We minimized the profile fields, uploading pictures limit for public and personal use that we believed are necessary for simple social interaction.

We have also included and attached "Policy Privacy Rules" so the users can read what the site is about and what principles should take in consideration if they want to join. The rules are easy to read and follow, eye friendly not to get tired and explained in simple words.

We have asked from a group of individuals to join the site and report how was their interaction with our site and if it was difficult to adjust according to our rules. Most of the users felt confident and comfortable about their privacy and they were still able to communicate freely with their friendly group.

FUTURE TRENDS

Social Network Sites have the potential to change fundamentally the character and take part in our social lives. Until now we have seen that social connections are getting wider and spread rapidly especially among to the young audience, who are the heaviest users of these sites. An ecosystem changes its volume, variety, velocity, venues and availability. The future holds more and better technological approaches so we should be prepared and ready to handle them in a proper way.

It could sound a bit excessive but risking lowering or eliminating all the barriers for social interaction can cause various problems that mentioned before. By creating a Virtual Community, it is advisable and recommended to take in consideration the security, privacy and risk factors. It is much easier through the communities to get the telephone number of another person, find where he lives but at the same time to flatten personal privacies and human rights.

Everything nowadays it is standardized, even the smallest things in our everyday life are following standard procedures to become in their final form.

SNSs development products should be standardized and planned carefully based on the security factors not only for the sake of money, fame and glory but also for the greater use of general inapprehensive public.

CONCLUSION

The article focused and described mainly the security threats on Social Network Sites. Its aim is to help the developers on creating better and safer virtual community sites related products. It should attract and awake the audience who uses social networking interactions online in a regular daily basis. Still the cyber world lacks the mentioned facts but we hope and expect that the new introduced future SNSs be subjected to a better frequent beneficial, secure and safe manner for the greatest good.

We have talked about the experiences and guidelines of authors that highlight the most important vulnerabilities and suggested protection methods, countermeasures and solutions that can be handled according to that threats.

Lastly, we have introduced the concept of a privacy friendly virtual community site, named CWIW, where some basic privacy methods have been implemented for the better user protection.

Future trends in privacy awareness of Social Network Sites are also highlighted, and what future will arise to the new era of virtual communities.

REFERENCES

Acquisti, A., Carrara, E., Stutzman, F., Callas, J., Schimmer, K., Nadjm, M., et al. (2007, October). *Security issues and recommendations for online social networks* (ENISA Position Paper No. 1). Crete, Greece: ENISA.

Acquisti, A., & Gross, R. (2005). *Information revelation and privacy in online social networks (The Facebook case)*. Pittsburgh, PA: Carnegie Mellon University.

Agraz, D. (2004). *Making friends through social networks: A new trend in virtual communication*. Palo Alto, CA: Stanford University.

Campbell, J., Sherman, R. C., Kraan, E., & Birchmeier, Z. (2005, June). *Internet privacy awareness and concerns among college students*. Paper presented at APS, Toronto, Ontario, Canada.

Cavoukian, A. (2005). *Identity theft revisited: Security is not enough*. Toronto, Ontario, Canada: Information and Privacy Commissioner of Ontario.

De Rosa, C., Cantrell, J., Havens, A., Hawk, J., & Jenkins, L. (2007). *Sharing, privacy and trust in our networked world: A report to the OCLC membership*. Dublin, OH: OCLC Online Computer Library Center.

Donath, S. J. (1998). *Identity and deception in the virtual community*. Cambridge, MA: MIT Media Lab.

EDUCAUSE Learning Initiative. (2006). *7 things you should know about... Facebook*. Retrieved February 25, 2009, from http://educause.edu/LibraryDetailPage/666?ID=ELI7017

Goettke, R., & Joseph, C. (2007). *Privacy and online social networking websites*. Retrieved from http://www.eecs.harvard.edu/cs199r/fp/RichJoe.pdf

Govani, T., & Pashley, H. (2005). *Student awareness of the privacy implications when using Facebook*. Retrieved from http://lorrie.cranor.org/courses/fa05/tubzhlp.pdf

Jackson, H. K. (2008). *The seven deadliest social network attacks*. Retrieved August 26, 2008, from http://www.darkreading.com/security/app-security/showArticle.jhtml?articleID=211201065

Joinson, A. (2003). *Understanding the psychology of Internet behavior: Virtual worlds, real lives*. New York: Palgrave MacMillan.

Jones, H., & Hiram, S. J. (2005). *Facebook: Threats to privacy*. Cambridge, MA: MIT.

McCandlish, S. (2005). *EFF's top 12 ways to protect your online privacy*. Electronic Frontier Foundation. Retrieved from http://www.eff.org/Privacy/ eff_privacy_top_12.html

Michaelides, G. (2009). *The project page of the CWIW*. Budapest University of Technology and Economics, Department of Electron Devices. Retrieved February 18, 2009, from http://turul.eet.bme.hu/~adam30

Schmugar, C. (2008). *The future of social networking sites.* Retrieved December 13, 2008 from http://www.mcafee.com

State Goverment of Victoria, Department of Justice. (2008). *Privacy and security tips for using social network sites.* Retrieved in 2009 from http://www.justice.vic.gov.au

Stutzman, F. (2005). *An evaluation of identity-sharing behavior in social network communities.* Retrieved from http://www.ibiblio.org/fred/pubs/stutzman_pub4.pdf

Chapter 4
Cyber Stalking:
A Challenge for Web Security

Alok Mishra
Atilim University, Turkey

Deepti Mishra
Atilim University, Turkey

ABSTRACT

Cyber stalking is a relatively new kind of cyber terrorism crime. Although it often receives a lower priority then cyber terrorism it is an important global issue. Due to new technologies, it is striking in different forms. Due to the Internets provision of anonymity and security it is proliferating quickly. Technology and tools available to curb it have many limitations and are not easy to implement. Legal acts to protect people from cyber stalking are geographically limited to the concerned state or country. This chapter reviews cyber stalking, its approaches, impacts, provision of legal acts, and measures to be taken to prevent it. There is an immediate need for research in the various dimensions of cyber stalking to assess this social problem.

INTRODUCTION

A survey of Fortune 1000 companies found an annual 64% growth rate in cyber attacks being carried out through the Internet (Bagchi & Udo, 2003). The New York State Police cyber terrorism unit takes into account cyber stalking as a part of their cyber crime investigation. The behaviour of stalking has been reported since the 19th-century (Lewis, Fremouw, Ben, & Farr, 2001). The Internet has provided users with new opportunities (Miller, 1999) yet, many users are unaware that the same qualities found off-line exist online (Lancaster, 1998). Cyber stalking is when a person is followed and pursued online. Their privacy is invaded, their every move watched. It is a form of harassment, and can disrupt the life of the victim and leave them feeling very afraid and threatened. Many authors, have defined cyber stalking, as the use of electronic communication including, pagers, cell phones, e-mails and the Internet, to bully, threaten, harass, and intimidate a victim (CyberAngels, 1999; Dean, 2000; Ellison & Akdeniz, 1998; Laughren, 2000; Ogilvie, 2000).

DOI: 10.4018/978-1-4666-2803-8.ch004

Thus it is a kind of cyber attack which may lead to cyber terrorism. With the growing economic dependency on information technology (IT), civilian infrastructures are increasingly the primary targets of cyber attacks. This growing reliance on IT has increased exposure to diverse sources of cyber war threats. Cyber stalking is an important global issue and an increasing social problem (CyberAngels, 1999; Ellison, 1999; Ellison & Akdeniz, 1998; Report on Cyberstalking, 1999) creating new offenders' and victims' (Wallace, 2000). For instance, in *Stalking and Harassment,* one of a series of Research Notes published on behalf of The Scottish Parliament in August 2000, stated: "Stalking, including cyberstalking, is a much bigger problem than previously assumed and should be treated as a major criminal justice problem and public health concern." (Bocij, 2004). Another detailed definition of cyber stalking that includes organisations by Bocij and McFarlane (2002) is:

A group of behaviours in which an individual, group of individuals or organisation, uses information and communications technology (ICT) to harass one or more individuals. Such behaviours may include, but are not limited to, the transmission of threats and false accusations, identity theft, data theft, damage to data or equipment, computer monitoring, the solicitation of minors for intimidation purposes and confrontation. Harassment is defined as a course of action that a reasonable person, in possession of the same information, would think causes another reasonable person to suffer emotional distress.

This definition shows cyber stalking may sometimes involve harassment carried out by an organisation also. Such behaviour is often termed corporate cyber stalking. This may lead to cyber warfare within the corporate world.

Typically, the cyber stalker's victim is new on the Web, and inexperienced with the rules of netiquette and Internet safety. Their targets are mostly females, children, emotionally weak, or unstable persons. It is believed that over 75% of the victims are female, but sometimes men are also stalked. These figures are assumed and the actual figures may never be known since most crimes of this nature go unreported ("Cyber Crime," 2004). To date, there is no empirical research to determine the incidence of cyber stalking (Ogilvie, 2000).

However depending on the use of the internet, there are three primary ways of cyber stalking (Ogilvie, 2000):

- **E-Mail Stalking:** This is direct communication through e-mail. Which is the most easily available form for harassment. It is almost similar to traditional stalking in some aspects. One may send e-mail of a threatening, hateful, or obscene nature, or even send spam or viruses to harass others. For example, in India in 2004 two MBA students sent e-mails to their female classmate to intimidate her. The free availability of anonymisers and anonymous remailers (which shield the sender's identity and allow the e-mail content to be concealed) provide a high degree of protection for stalkers seeking to cover their tracks more effectively.

- **Internet Stalking:** There is global communication through the Internet. Here the domain is more wide and public in comparison to e-mail stalking. Here stalkers can use a wide range of activities to harass their victims. For example, a woman was stalked for a period of six months. Her harasser posted notes in a chat room that threatened to intimidate and kill her, and posted doctored pornographic pictures of her on the net together with personal details (Dean, 2000).

- **Computer Stalking:** This is unauthorised control of another person's computer. In this type of stalking, the stalker exploits the working of the Internet and the

Windows operating system in order to to assume control over the computer of the targeted victim. Here the cyber stalker can communicate directly with their target as soon as the target computer connects in any way to the Internet. The stalker can assume control of the victim's computer and the only defensive option for the victim is to disconnect and relinquish their current Internet "address." In this way, an individuals Windows-based computer connected to the Internet can be identified, and connected to, by another computer connected to the Internet. This "connection" is not the link via a third-party characterising typical Internet interactions; rather, it is a computer-to-computer connection allowing the interloper to exercise control over the computer of the target. At present, a reasonably high degree of computer savvy is required to undertake this form of explotiation of the Internet and the Windows operating system. However, instructions on how to use technology in this way are available on the Internet. It is likely that in the future easier scripts will be made freely available for anyone inclined to download them.

Furthermore cyber stalkers can be categorized into three types:

- **The Common Obsessional Cyber Stalker:** These stalkers refuse to believe that their relationship is over.
- **The Delusional Cyber Stalker:** They may be suffering from some mental illness like schizophrenia, and so forth, and have a false belief that keeps them tied to their victims. They assume that the victim loves them even though they have never met. A delusional stalker is usually a loner and most often chooses victims such as a married woman, a celebrity, a doctor, a teacher, and so forth. Those in the noble and help-

ing professions like doctors, teachers, and so forth, are often at risk for attracting a delusional stalker. They are very difficult to shake off.
- **The Vengeful Stalker:** These cyber stalkers are angry at their victim due to some minor reason—either real or imagined. Typical example are disgruntled employee or ex-spouse, and so forth.

Cyber stalking can take many forms. However, Ellison (1999) suggests, cyber stalking can be classified by the type of electronic communication used to stalk the victim and the extent to which the communication is private or public. Ellison (1999) has classified cyber stalking as either "direct" or "indirect." For example, direct cyber stalking includes the use of pagers, cell phones and e-mail to send messages of hate, obscenities, and threats to intimidate a victim. Direct cyber stalking has been reported to be the most common form of cyber stalking with a close resemblance to off-line stalking (Wallace, 2000). Direct cyber stalking is claimed to be the most common way in which stalking begins. For instance, Working to Halt Online Abuse (2003) show the majority of online harassment or cyber stalking begins with e-mail.

Indirect cyber stalking includes the use of the Internet to display messages of hate and threats, or used to spread false rumours about a victim (Ellison & Akdeniz, 1998). Messages can be posted on Web pages, within chat groups, or bulletin boards. Working to Halt Online Abuse (2003) statistics show chat rooms, instant messages, message boards, and newsgroups to be the most common way that indirect cyber stalking begins. Ogilvie (2000) claims indirect cyber stalking has the greatest potential to transfer into real-world stalking. Messages placed within the public space of the Internet can encourage third parties to contribute in their assault (Report on Cyberstalking, 1999). Therefore, indirect cyber stalking can increase the risk for victims by limiting the geographical boundaries of potential threats. Consequently, indirect cyber stalking can have a greater potential than direct cyber stalking to transfer into the real world as it increases the potential for third parties to become involved (Maxwell, 2001). According to Halt Online Abuse (2003) in the year 2000, 19.5% of online harrassment or cyber stalking cases became off-line stalking. Cyber stalking can vary in range and severity and often reflects off-line stalking behaviour. It can be seen as an extension to off-line stalking however, cyber stalking is not limited by geographic boundaries.

OFFENDERS AND THEIR BEHAVIOUR

What motivates a cyber stalker? Most studies have focused on the off-line stalking offender. Studies, (Farnham, James, & Cantrell, 2000; Meloy, 1996; Meloy & Gothard, 1995; Mullen, Pathe, Purcell, & Stuart, 1999) of off-line stalking offenders have placed offenders into three main groups. Zona, Sharma, and Lone (1993) grouped off-line stalkers into either the "simple obsessional," the "love obsessional," or the "erotomanic" group. The majority of stalkers are simple obsessional they have had a prior relationship with the victim and are motivated to stalk with the aim to reestablish the relationship or gain revenge once the relationship has been dissolved. Mullen, et al. (1999) claims the majority of simple obsessional stalkers have some form of personality disorder and as a group have the greatest potential to become violent. The love obsessional stalkers are those who have never met their victim. The erotomanic group is the smallest among stalkers and is motivated by the belief that the victim is in love with them, as a result of active delusions (Zona et al., 1993). Studies show that irrespective of the groups, male offenders account for the majority of off-line stalking (Meloy & Gothard, 1995; Mullen et al., 1999). Furthermore, according to Working to Halt Online Abuse (2003) statistics show that in the

year 2000, 68% of the online harassers or cyber stalkers were male. But now that trend is reversing and male harassers have decreased (52% in 2003) while female harassers have increased (from 27% in 2000 to 38% in 2003). Another interesting factor which has been found in common with off-line stalking offenders is that social factors such as the diversity in socioeconomic backgrounds and either underemployment or unemployment are significant (Meloy, 1996). While Kamphuis and Emmelkamp (2000) investigated psychological factors and found social isolation, maladjustment and emotional immaturity, along with an inability to cope with failed relationships common with off-line stalking groups.

Furthermore, off-line stalkers were above intelligence and older in comparison to other criminal offenders (McCann, 2000). According to (Maxwell, 2001) studies of off-line stalking offenders can present insights to cyber stalkers with some limitations. As earlier observed, only 50% of stalkers are reported to authorities, furthermore, only 25% will result in the offenders being arrested and only 12% will be prosecuted (Kamphuis & Emmelkamp, 2000). Researchers have claimed that cyber stalkers have similar characteristics to the off-line stalkers and most of them are motivated to control the victim (Jenson, 1996; Ogilvie, 2000; Report on Cyberstalking, 1999).

VICTIMS AND THEIR CHARACTERISTICS

Studies have shown that the majority of victims are females of average socioeconomic status, and off-line stalking is primarily a crime against young people, with most victims between the ages of 18 and 29 (Brownstein, 2000). Stalking as a crime against young people may account for the high prevalence of cyber stalking within universities. For example, the University of Cincinnati study showed 25% of college women had been cyber stalked (Tjaden & Thoennes, 1997). Also according to Working to Halt Online Abuse (2003) the majority of victims of online harassment or cyber stalking are between 18 and 30 years of age. Studies that have investigated offenders of off-line stalking found some common symptoms regarding victims, for instance, most are regular people rather than the rich and famous (Brownstein, 2000; McCann, 2000; Sinwelski & Vinton, 2001). Goode (1995) also supports that up to 80% of off-line stalking victims are from average socioeconomic backgrounds. Another important observation by Hitchcock (2000) is that 90% of off-line stalking victims are female. While Halt Online Abuse (2003) reports it to 78% as gender of victim's cumulative figure between 2000 and 2003. Zona, et al. (1993) reported, 65% of off-line victims had a previous relationship with their stalker. However according to Working to Halt Online Abuse (2003) statistics it is 51% but not enough to support this reason as a significant risk factor for cyber stalking.

SOCIAL AND PSYCHOLOGICAL EFFECTS

The studies, which have looked at off-line stalking and its effects on victims by and large, are of the university populations (Maxwell, 2001). For instance, the Fremauw, Westrup, and Pennypacker (1997) study explored coping styles of university off-line stalking victims. They found that the most common way of coping with a stalker was to ignore the stalker and the second most common way was to confront the stalker. According to them, victims were least likely to report the off-line stalker to the authorities. Many victims felt ashamed or were of the belief that the stalking was their fault (Sheridan, Davies, & Boon, 2001). Working to Halt Online Abuse (2003) reports that the majority of online cyber stalking was handled by contacting the Internet service provider (ISP), which accounted for 49% of cases, followed by, 16% contacting the police. Furthermore, 12% coped by other means

including ignoring messages, taking civil action, or not returning to the forum in which the cyber stalking took place. The Report on Cyberstalking (1999) mentions that many victims of cyber stalking claimed they did not think that they would be listened to if they reported the cyber stalking to authorities. Mostly victims of cyber stalking were unaware that a crime had been committed. Currently there are only a few studies on the psychological impact on victims (Maxwell, 2001). Westrup, Fremouw, Thompson, and Lewis (1999) studied the psychological effects of 232 female off-line stalking victims and found the majority of victims had symptoms of depression, anxiety, and experienced panic attacks. In another study by Mullen and Pathe (1997) found that 20% of victims showed increased alcohol consumption and 74% of victims suffered sleep disturbances. However social and psychological effects are interrelated. In a separate study, David, Coker, and Sanderson (2002) found that the physical and mental health effects of being stalked were not gender-related. Both male and female victims experienced impaired health, depression, injury, and were more likely to engage in substance abuse than their nonstalked peers.

TECHNICAL APPROACHES FOR AVOIDANCE

Although tools and techniques are available to protect users, their implementation is not easy and there are number of limitations. For example, answering machines and caller identification are two technologies that help to protect against telephone harassment, although these are of limited effectiveness. In contrast, the potential exists online to completely block contact from unwanted mailers with tools for different online media (Spertus, 1996):

- Programs to read Usenet news support *kill files*, used to automatically bypass messages listed as being from a certain individual or meeting other criteria specified by the user. This allows an individual to choose not to see further messages in a given discussion "thread" or posted from a specified user account or machine. People can choose to share their kill files with others in order to warn them about offensive individuals.

- Real-time discussion forums, such as MUDs and Internet relay chat (IRC), allow a user to block receiving messages from a specified user. Similar technology could be used to allow blocking messages containing words that the user considers unwelcome. Individuals can also be banned from forums at the operators' discretion.

- Programs have existed for years to automatically discard (file, or forward) e-mail based on its contents or sender and are now coming into widespread use. The second generation of filtering tools is being developed. The LISTSERV list maintenance software (Lsoft 96) contains heuristics to detect impermissible advertisements, and an experimental system, Smokey, recognizes "flames" (insulting e-mail).

- Numerous tools exist to selectively prevent access to World Wide Web sites. While the simplest ones, such as SurfWatch, maintain a central database of pages that they deem unsuitable for children, others are more sophisticated. SafeSurf rates pages on several different criteria. Net Nanny provides a starting dictionary of offensive sites, which the user can edit. The user can also specify that pages containing certain words or phrases should not be downloaded.

One of the biggest limitations to the above techniques is the computer's difficulty in determining whether a message is offensive. Many of the above tools use string matching and will not recognize a phrase as offensive if it is misspelled or restated in other words. Few systems use more sophisticated techniques. Smokey recognizes that "you" followed by a noun phrase is usually insulting, but such heuristics have limited accuracy, especially if they are publicly known.

LEGAL ACTS, PROVISIONS, AND PROTECTION

Law enforcement agencies now know that cyber stalking is a very real issue that needs to be dealt with, from local police departments to state police, the FBI, and the U.S. postal inspection service, among others. Many are asking their officers to learn how to use the Internet and work with online victim groups such as WHOA (Women Halting Online Abuse), SafetyED, and CyberAngels. Others are attending seminars and workshops to learn how to track down cyber stalkers and how to handle victims (Hitchcock, 2000).

Legal acts aimed to protect other from off-line stalking are relatively new. Only in the past ten years have off-line antistalking laws been developed (Goode, 1995). The first "Antistalking" law was legislated in California, in 1990 and in 1998 the antistalking law, specified cyber stalking as a criminal act. However, less than a third of the states in the U.S. have antistalking laws that encompass cyber stalking (Miller, 1999). According to Hitchcock (2000) in the U.S. almost 20 states with cyber stalking or related laws, a federal cyber stalking law is waiting for senate approval. Several other states with laws pending, cyber stalking is finally getting noticed, not only by law enforcement, but by media too. To protect against off-line stalking or cyber stalking the UK has the "Protections Against Harassment Act 1997" and the "Malicious Communication Act 1998" (ISE, n.d.). In New Zealand the "Harassment Act 1997," the "Crimes Act 1961," the "Domestic Violence Act 1995," and the "Telecommunication Act

1987" can apply to online harassment or cyber stalking (Computers and Crime, 2000). While in Australia, Victoria and Queensland are the only states to include sending electronic messages to, or otherwise contacting, the victim, as elements of the offence for most states cover activities which "could" include stalking.

These activities are the following (Ogilive, 2000):

- Keeping a person under surveillance.
- Interfering with property in the possession of the other person, giving or sending offensive material.
- Telephoning or otherwise contacting a person.
- Acting in a manner that could reasonably be expected to arouse apprehension or fear in the other person.
- Engaging in conduct amounting to intimidation, harassment, or molestation of the other person.

Two possible exceptions here are New South Wales and Western Australia, which have far narrower definitions of what constitutes stalking. Hence, both states identify specific locations such as following or watching places of residence, business, or work, which may not include cyber space. While cyber stalking could be included within "any place that a person frequents for the purposes of any social or leisure activity," the prosecution possibilities seem limited. Other difficulties may occur in South Australia and the Australian Capital Territory, where there is a requirement that offenders intend to cause "serious" apprehension and fear. Thus, the magistrates may dismiss cases of cyber stalking, given the lack of physical proximity between many offenders and victims (Ogilive, 2000).

There is a significant growth in cyber stalking cases in India, primarily because people still use the Internet to hide their identities and indulge in online harassment. It is important to note that though cyber stalking has increased, the number of cases reported is on the decline. This could be because of the failure of the law in dealing with this crime. The Information Technology Act 2000 does not cover cyber stalking and the Indian Penal Code 1860 does not have a specific provision that could help victims of cyber stalking. The government has now thought it fit to enact a distinct provision relating to cyber squatting. The provision is mentioned in the proposed Communications Convergence Bill 2001 which has been laid before Parliament, and the Parliamentary Standing Committee on Information Technology has already given its detailed report and recommendations on the proposed law to the government. The relevant provision relating to cyber stalking in the convergence bill is as follows:

Punishment for sending obscene or offensive messages:

Any person who sends, by means of a communication service or a network infrastructure facility:

a. any content that is grossly offensive or of an indecent obscene or menacing character;

b. for the purpose of causing annoyance, inconvenience, danger, obstruction, insult, injury, criminal intimidation, enmity, hatred or ill-well, any content that he knows to be false or persistently makes use for that purpose of a communication service or a network infrastructure facility, shall be punishable with imprisonment which may be extended up to three years or with fine which may be extended to approximate USD 4,25000 or with both. This is one of the heaviest fines known in criminal jurisprudence in India.

It is hoped that when it does come into effect, victims of cyber stalking can breathe a sigh of relief ("No Law," 2004).

Currently, there is no global legal protection against cyber stalking (Ellison & Akdeniz, 1998). Within the cyber world the lack of global legal protection further adds to an increasing problem. This is even true in the case of cyber warfare and cyber terrorism. Unlike offline stalking there are no geographical limitations to cyber stalking. Although some countries and/or states have responded to the increase of cyber stalking by the modification of current antistalking laws, laws criminalizing cyber stalking by and large are limited to the country and/or state and are ineffective within the cyber world. Furthermore according to Ogilvie (2000) while the criminalisation of threatening e-mails would be a reasonably easy fix, it does not overcome the primary difficulties in legislating against cyber stalking, which are the inter-jurisdictional difficulties. While in many ways cyber stalking can be considered analogous to physical world stalking, at other times the Internet needs to be recognised as a completely new medium of communication. It is at this point that legislating against cyber stalking becomes difficult. For example, according to Ogilvie (2000) if a stalker in California uses an international service provider in Nevada to connect to an anonymiser in Latvia to target a victim in Australia, which jurisdiction has responsibility for regulating the cyber stalking? This is a major constraint to be taken into consideration while formulating laws to curb cyber stalking. Nevertheless, the implementation of legal acts to protect from off-line stalking or cyber stalking remains dependent on victims to report the offence and the concerned authorities ability to gain adequate evidence (Maxwell, 2001).

PREVENTION STRATEGIES

As we know, prevention is always better than the cure and just a little care makes accidents rare. The best way to avoid being stalked is to always maintain a high level of safety awareness. The suggestions regarding staying safe online by Hitchcock (2000) are as follows:

1. Use your primary e-mail account only for messages to and from people you know and trust.
2. Get a free e-mail account from someplace like Hotmail, Juno, or Excite, and so forth, and use that for all of your other online activities.
3. When you select an e-mail username or chat nickname, create something gender-neutral and like nothing you have elsewhere or have had before. Try not to use your name.
4. Do not fill out profiles for your e-mail account, chat rooms, IM (instant messaging), and so forth.
5. Do set your options in chat or IM to block all users except for those on your buddy list.
6. Do learn how to use filtering to keep unwanted e-mail messages from coming to your e-mailbox.
7. If you are being harassed online, try not to fight back. This is what the harasser wants—a reaction from you. If you do and the harassment escalates, do the following:
   a. Contact the harasser and politely ask them to leave you alone.
   b. Contact their ISP and forward the harassing messages.
   c. If harassment escalates, contact your local police.
   d. If they cannot help, try the State Police, District Attorney's office and/or State Attorney General.
   e. Contact a victims group, such as WHOA, SafetyED or CyberAngels.

CONCLUSION

It is estimated that there are about 200,000 real-life stalkers in the U.S. today. Roughly 1 in 1,250 persons is a stalker—and that is a large ratio. Out of the estimated 79 million population worldwide on the Internet at any given time, we could find 63,000 Internet stalkers travelling the information superhighway, stalking approximately 474,000 victims (Cyber Crime in India, 2004; Hitchcock, 2000). It is a great concern for all Internet users. Cyber stalking may lend support to cyber warfare and cyber terrorism. Present laws to tackle cyber stalking are geographically limited to the concerned state or country. Therefore, there is an urgent need to make global legislation for handling cyber warfare and cyber terrorism. Organizations like the UN and Interpol should initiate this. In addressing cyber stalking, new and innovative legislations, technologies, and investigative countermeasures will almost certainly be mandatory. We hope that information system security professionals will move in this direction. Researchers will also put their efforts for empirical studies in various aspects of cyber stalking to know more about it, which will help technologist, lawmakers and others to make a real assessment.

REFERENCES

Bagchi, K., & Udo, G. (2003). An analysis of the growth of computer and internet security breaches. *Communications of the Association for Information Systems, 12*(46), 129.

Bocij, P. (2004). *Corporate cyberstalking: An invitation to build theory*. http://www.firstmonday. dk/issues/issues7_11/bocij/

Bocij, P., & McFarlane, L. (2002, February). Online harassment: Towards a definition of cyberstalking. *Prison Service Journal.* HM Prison Service, London, 139, 31-38.

Brownstein, A. (2000). In the campus shadows, women are stalkers as well as the stalked. *The Chronicle of Higher Education, 47*(15), 4042.

Computers and Crime. (2000). IT law lecture notes (Rev. ed.). http://www.law.auckland.ac.nz/itlaw/ itlawhome.htm

Cyber Crime in India. (2004). *Cyber stalking—online harassment.* http://www.indianchild.com/ cyberstalking.htm

CyberAngels. (1999). http://cyberangels.org

Davis, K. E., Coker, L., & Sanderson, M. (2002, August). Physical and mental health effects of being stalked for men and women. *Violence and Victims, 17*(4), 429–443. doi:10.1891/ vivi.17.4.429.33682

Dean, K. (2000). The epidemic of cyberstalking. *Wired News* http://www.wired.com/news/ politics/0,1283,35728,00.html

Ellison, L. (1999). Cyberspace 1999: Criminal, criminal justice and the internet. *Fourteenth BILETA Conference*, York, UK. http://www.bileta. ac.uk/99papers/ellison.html

Ellison, L., & Akdeniz, Y. (1998). Cyber-stalking: The regulation of harassment on the internet (Special Edition: Crime, Criminal Justice and the Internet). *Criminal Law Review*, 2948. http://www. cyber-rights.org/documents/stalking

Farnham, F. R., James, D. V., & Cantrell, P. (2000). Association between violence, psychosis, and relationship to victim in stalkers. *Lancet, 355*(9199), 199. doi:10.1016/S0140-6736(99)04706-6

Fremauw, W. J., Westrup, D., & Pennypacker, J. (1997). Stalking on campus: The prevalence and strategies for coping with stalking. *Journal of Forensic Sciences, 42*(4), 666–669.

Goode, M. (1995). Stalking: Crime of the nineties? *Criminal Law Journal, 19*, 21–31.

Hitchcock, J. A. (2000). Cyberstalking. *Link-Up, 17*(4). http://www.infotoday.com/lu/ju100/ hitchcock.htm

ISE. (n.d.). *The internet no1 close protection resource.* http://www.intel-sec.demon.co.uk

Jenson, B. (1996). *Cyberstalking: Crime, enforcement and personal responsibility of the on-line world.* S.G.R. MacMillan. http://www.sgrm.com/art-8.htm

Kamphuis, J. H., & Emmelkamp, P. M. G. (2000). Stalking—A contemporary challenge for forensic and clinical psychiatry. *The British Journal of Psychiatry, 176,* 206–209. doi:10.1192/bjp.176.3.206

Lancaster, J. (1998, June). Cyber-stalkers: The scariest growth crime of the 90's is now rife on the net. *The Weekend Australian,* 20-21.

Laughren, J. (2000). *Cyberstalking awareness and education.* http://www.acs.ucalgary.ca/~dabrent/380/webproj/jessica.html

Lewis, S. F., Fremouw, W. J., Ben, K. D., & Farr, C. (2001). An investigation of the psychological characteristics of stalkers: Empathy, problem-solving, attachment and borderline personality features. *Journal of Forensic Sciences, 46*(1), 8084.

Maxwell, A. (2001). *Cyberstalking.* Masters' thesis, http://www.netsafe.org.nz/ie/downloads/cyberstalking.pdf

McCann, J. T. (2000). A descriptive study of child and adolescent obsessional followers. *Journal of Forensic Sciences, 45*(1), 195–199.

Meloy, J. R. (1996). Stalking (obsessional following): A review of some preliminary studies. *Aggression and Violent Behavior, 1*(2), 147–162. doi:10.1016/1359-1789(95)00013-5

Meloy, J. R., & Gothard, S. (1995). Demographic and clinical comparison of obsessional followers and offenders with mental disorders. *The American Journal of Psychiatry, 152*(2), 25826.

Miller, G. (1999). Gore to release cyberstalking report, call for tougher laws. *Latimes.com.* http://www.latimes.com/news/ploitics/elect2000/pres/gore

Mullen, P. E., & Pathe, M. (1997). The impact of stalkers on their victims. *The British Journal of Psychiatry, 170,* 12–17. doi:10.1192/bjp.170.1.12

Mullen, P. E., Pathe, M., Purcell, R., & Stuart, G. W. (1999). Study of stalkers. *The American Journal of Psychiatry, 156*(8), 1244–1249.

No Law to Tackle Cyberstalking. (2004). *The Economic Times.* http://economictimes.indiatimes.com/articleshow/43871804.cms

Ogilvie, E. (2000). *Cyberstalking, trends and issues in crime and criminal justice.* 166. http://www.aic.gov.au

Report on Cyberstalking. (1999, August). *Cyberstalking: A new challenge for law enforcement and industry.* A Report from the Attorney General to The Vice President. http://www.usdoj.gov/criminal/cybercrime/cyberstalking.htm

Sheridan, L., Davies, G. M., & Boon, J. C. W. (2001). Stalking: Perceptions and prevalence. *Journal of Interpersonal Violence, 16*(2), 151–167. doi:10.1177/088626001016002004

Sinwelski, S., & Vinton, L. (2001). Stalking: The constant threat of violence. *Affilia, 16,* 46–65. doi:10.1177/08861090122094136

Spertus, E. (1996). *Social and technical means for fighting on-line harassment.* http://ai.mit.edu./people/ellens/Gender/glc

Tjaden, P., & Thoennes, N. (1997). Stalking in America: Findings from the National Violence. Retrieved May 25, 2007 from http://www.ncjrs.gov/txtfiles/169592.txt

Wallace, B. (2000, July 10). Stalkers find a new tool—The Internet e-mail is increasingly used to threaten and harass, authorities say. *SF Gate News.* http://sfgate.com/cgi-bin/article_cgi?file=/chronicle/archive/2000/07/10/MN39633.DTL

Westrup, D., Fremouw, W. J., Thompson, R. N., & Lewis, S. F. (1999). The psychological impact of stalking on female undergraduates. *Journal of Forensic Sciences, 44*, 554–557.

Working to Halt Online Abuse. (WHOA). (2003). *Online harrassment statistics.* Retrieved May 25, 2007 from http://www.haltabuse.org/resources/stats/index.shtml

Zona, M. A., Sharma, K. K., & Lone, J. (1993). A comparative study of erotomanic and obsessional subjects in a forensic sample. *Journal of Forensic Sciences, 38*, 894–903.

KEY TERMS AND DEFINITIONS

Netiquette: The etiquette of computer networks, especially the Internet.

SPAM: Unsolicited e-mail, often advertising a product or service. Spam can occasional "flood" an individual or ISP to the point that it significantly slows down the data flow.

Stalking: To follow or observe (a person) persistently, especially out of obsession or derangement.

Virus: A malicious code added to an e-mail program or other downloadable file that is loaded onto a computer without the users knowledge and which runs often without their consent. Computer viruses can often copy themselves and spread themselves to a users e-mail address book or other computers on a network.

This work was previously published in Cyber Warfare and Cyber Terrorism, edited by Lech J. Janczewski and Andrew M. Colarik, pp. 216-225, copyright 2007 by Information Science Reference (an imprint of IGI Global).

Section 2
Psychological and Wellbeing Issues in Internet Trolling

Chapter 5
Increasing Capital Revenue in Social Networking Communities:
Building Social and Economic Relationships through Avatars and Characters

Jonathan Bishop

Centre for Research into Online Communities and E-Learning Systems, UK

ABSTRACT

The rise of online communities in Internet environments has set in motion an unprecedented shift in power from vendors of goods and services to the customers who buy them, with those vendors who understand this transfer of power and choose to capitalize on it by organizing online communities and being richly rewarded with both peerless customer loyalty and impressive economic returns. A type of online community, the virtual world, could radically alter the way people work, learn, grow consume, and entertain. Understanding the exchange of social and economic capital in online communities could involve looking at what causes actors to spend their resources on improving someone else's reputation. Actors' reputations may affect others' willingness to trade with them or give them gifts. Investigating online communities reveals a large number of different characters and associated avatars. When an actor looks at another's avatar they will evaluate them and make decisions that are crucial to creating interaction between customers and vendors in virtual worlds based on the exchange of goods and services. This chapter utilizes the ecological cognition framework to understand transactions, characters and avatars in virtual worlds and investigates the exchange of capital in a bulletin board and virtual. The chapter finds strong evidence for the existence of characters and stereotypes based on the ecological cognition framework and empirical evidence that actors using avatars with antisocial connotations are more likely to have a lower return on investment and be rated less positively than those with more sophisticated appearing avatars.

DOI: 10.4018/978-1-4666-2803-8.ch005

INTRODUCTION

The rise of online communities has set in motion an unprecedented power shift from goods and services vendors to customers according to Armstrong and Hagel (1997). Vendors who understand this power transfer and choose to capitalize on it are richly rewarded with both peerless customer loyalty and impressive economic returns they argue. In contemporary business discourse, online community is no longer seen as an impediment to online commerce, nor is it considered just a useful Website add-on or a synonym for interactive marketing strategies. Rather, online communities are frequently central to the commercial development of the Internet, and to the imagined future of narrowcasting and mass customization in the wider world of marketing and advertising (Werry, 2001). According to Bressler and Grantham (2000), online communities offer vendors an unparalleled opportunity to really get to know their customers and to offer customized goods and services in a cost executive way and it is this recognition of an individual's needs that creates lasting customer loyalty. However, if as argued by Bishop (2007a) that needs, which he defines as pre-existing goals, are not the only cognitive element that affects an actor's behavior, then vendors that want to use online communities to reach their customers will benefit from taking account of the knowledge, skills and social networks of their customers as well.

According to Bishop (2003) it is possible to effectively create an online community at a click of a button as tools such as Yahoo! Groups and MSN Communities allow the casual Internet user to create a space on the Net for people to talk about a specific topic of interest. Authors such as Bishop have defined online communities based on the forms they take. These forms range from special interest discussion Web sites to instant messaging groups. A social definition could include the requirement that an information system's users go through the membership lifecycle identified

by Kim (2000). Kim's lifecycle proposed that individual online community members would enter each community as visitors, or "Lurkers." After breaking through a barrier they would become "Novices," and settle in to community life. If they regularly post content, they become "Regulars." Next, they become "Leaders," and if they serve in the community for a considerable amount of time, they become "Elders." Primary online community genres based on this definition are easily identified by the technology platforms on which they are based. Using this definition, it is possible to see the personal homepage as an online community since users must go through the membership lifecycle in order to post messages to a 'guestbook' or join a 'Circle of Friends'. The Circle of Friends method of networking, developed as part of the VECC Project (see Bishop, 2002) has been embedded in social networking sites, some of which meet the above definition of an online community. One of the most popular genres of online community is the bulletin board, also known as a message board. According to Kim (2000), a message board is one of the most familiar genres of online gathering place, which is asynchronous, meaning people do not have to be in the same place at the same time to have a conversation. An alternative to the message board is the e-mail list, which is the easiest kind of online gathering place to create, maintain and in which to participate (ibid). Another genre of online community that facilitates discussion is the Chat Group, where people can chat synchronously, communicating in the same place at the same time (Figallo, 1998). Two relatively new types of online community are the Weblog and the Wiki. Weblogs, or blogs, are Web sites that comprise hyperlinks to articles, news releases, discussions and comments that vary in length and are presented in chronological order (Lindahl & Blount, 2003). The community element of this technology commences when the owner, referred to as a 'blogger', invites others to comment on what he/she has written. A Wiki, which is so named through taking the first letters from the axiom, 'what I know is', is a collaborative page-editing tool with which users may add or edit content directly through their Web browser (Feller, 2005). Despite their newness, Wikis could be augmented with older models of hypertext system. A genre of online community that has existed for a long time, but is also becoming increasingly popular is the Virtual World, which may be a multi-user dungeon (MUD), a massively multiplayer online role-playing game (MMORG) or some other 3-D virtual environment. See Table 1.

Encouraging Social and Economic Transactions in Online Communities

According to Shen et al. (2002), virtual worlds could radically alter the way people work, learn, grow, consume and entertain. Online communities such as virtual worlds are functional systems that exist in an environment. They contain actors, artifacts, structures and other external representations that provide stimuli to actors who respond (Bishop, 2007a; 2007b; 2007c). The transfer of a response into stimuli from one actor to another is social intercourse and the unit of this exchange is

the transaction (Berne, 1961; 1964). A transaction is also the unit for the exchange of artifacts between actors and is observed and measured in currency (Vogel, 1999). Transactions can be observed in online communities, most obviously in virtual worlds, where actors communicate with words and trade goods and services. Research into how consumers trade with each other has considered online reputation, focusing on how a trader's reputation influences trading partner's trust formation, reputation scores' impact on transactional prices, reputation-related feedback's effect on online service adoption and the performance of existing online reputation systems (Li et al., 2007). According to Bishop (2007a), encouraging participation is one of the greatest challenges for any online community provider. There is a large amount of literature demonstrating ways in which online communities can be effectively built (Figallo, 1998; Kim, 2000; Levine-Young & Levine, 2000; Preece, 2000). However, a virtual community can have the right tools, the right chat platform and the right ethos, but if community members are not participating the community will not flourish and encouraging members to change from lurkers into elders is proving to be a challenge for community

Table 1. Advantages and disadvantages of specific online community genres

Genre	Advantages/Disadvantages
Personal Homepage	Advantages: Regularly updated, allows people to re-connect by leaving messages and joining circle of friends. Disadvantage: Members often need to re-register for each site and cannot usually take their 'Circle of Friends' with them.
Message Boards	Advantages: Posts can be accessed at any time. Easy to ignore undesirable content. Disadvantages: Threads can be very long and reading them time consuming.
E-mail Lists and Newsletters	Advantages: Allows a user to receive a message as soon as it is sent. Disadvantages: Message archives not always accessible.
Chat Groups	Advantages: Synchronous. Users can communicate in real time. Disadvantages: Posts can be sent simultaneously and the user can become lost in the conversation.
Virtual Worlds and Simulations	Advantages: 3-D metaphors enable heightened community involvement. Disadvantages: Requires certain hardware and software that not all users have.
Weblogs and Directories	Advantages: Easily updated, regular content. Disadvantages: Members cannot start topics, only respond to them.
Wikis and Hypertext Fiction	Advantages: Can allow for collaborative work on literary projects. Disadvantages: Can bring out the worst in people, for example, their destructive natures.

providers. Traditional methods of behavior modification are unsuitable for virtual environments, as methodologies such as operant conditioning would suggest that the way to turn lurkers into elders is to reward them for taking participatory actions. The ecological cognition framework proposed Bishop (2007a; 2007c) proposes that in order for individuals to carry out a participatory action, such as posting a message, there needs to be a desire to do so, the desire needs to be consistent with the individual's goals, plans, values and beliefs, and they need to have to abilities and tools to do so. Some individuals such as lurkers, may have the desire and the capabilities, but hold beliefs that prevent them from making participatory actions in virtual communities. In order for them to do so, they need to have the desire to do so and their beliefs need to be changed. Traditional methods, such as operant conditioning may be able to change the belief of a lurker that they are not being helpful by posting a message, but it is unlikely that they will be effective at changing other beliefs, such as the belief they do not need to post. In order to change beliefs, it is necessary to make an actor's beliefs dissonant, something that could be uncomfortable for the individual. While changing an actor's beliefs is one way of encouraging them to participate in a virtual community, another potential way of increasing their involvement is to engage them in a state of flow which might mean that they are more likely to act out their desires to be social, but there is also the possibility that through losing a degree of self-consciousness they are also more likely to flame others (Orengo Castellá et al., 2000).

A CHARACTER THEORY FOR ONLINE COMMUNITIES

Kim's membership lifecycle provides a possible basis for analyzing the character roles that actors take on in online communities. Existing character theories could be utilized to explore specific types of online community (e.g., Propp, 1969) or explain to dominance of specific actors in online communities (e.g., Goffman, 1959). Propp suggested the following formula to explain characters in media texts:

$$\alpha a^5 D^1 E^1 M F^1 T a^5 B K N T o Q W$$

Propp's character theory suggests that in media texts eight characters can be identified; the villain who struggles against the hero; the donor who prepares the hero or gives the hero an artifact of some sort; the helper who helps the hero in their expedition; the princess who the hero fights to protect or seeks to marry; her father the dispatcher; and the false hero who takes credit for the hero's actions or tries to marry the princess. While Propp's theory might be acceptable for analyzing multi-user dungeons or fantasy adventure games, it may not be wholly appropriate for bulletin board-based online communities. Goffman's character theory according to Beaty et al. (1998) suggests that there are four main types of characters in a media text: the protagonists who are the leading characters; the deuteragonists who are the secondary characters; the bit players who are minor characters whose specific background the audience are not aware of; and the fool who is a character that uses humor to convey messages. Goffman's model could be useful in explaining the dominance of specific types of online community members, but does not explain the different characteristics of those that participate online, what it is that drives them, or what it is that leads them to contribute in the way they do.

Bishop's (2007a; 2007c) ecological cognition framework (ECF) provides a theoretical model for developing a character theory for online communities based on bulletin board and chat room models. One of the most talked about types of online community participant is the troll. According to Levine-Young and Levine (2000), a troll posts provocative messages intended to start a flame war. The ECF would suggest that chaos drives these trolls, as they attempt to provoke other members into responding. This would also suggest there is a troll opposite, driven by order, which seeks to maintain control or rebuke obnoxious comments. Campbell et al. (2002) found evidence for such a character, namely the big man, existing in online communities. Salisbury (1965) suggests big men in tribes such as the Siane form a *de facto* council that confirms social policy and practices. Campbell et al. (2002) point out that big men are pivotal in the community as, according to Breton (1999), they support group order and stability by personally absorbing many conflicts. Actors susceptible to social stimuli activate one of two forces, either social forces or anti-social forces. Actors who are plainly obnoxious and offend other actors through posting flames, according to Jansen (2002) are known as snerts. According to Campbell, these anti-social characters are apparent in most online communities and often do not support or recognize any of the big men unless there is an immediate personal benefit in doing so. Campbell et al. (2002) also point out that the posts of these snerts, which they call sorcerers and trolls, which they call tricksters, could possibly look similar. Differentiating between when someone is being a troll and when they are being a snert although clear using the ECF, may require interviewing the poster to fully determine. Someone whose intent is to provoke a reaction, such as through playing 'devil's advocate' could be seen theoretically to be a troll, even if what they post is a flame. An actor who posts a flame after they were provoked into responding after interacting with another, could be seen theoretically to be a snert, as their intention is to be offensive. Another actor presented with the same social stimuli may respond differently. Indeed, Rheingold (1999) identified that members of online communities like to flirt. According to Smith (2001), some online community members banned from the community will return with new identities to disrupt the community. These actors could be labeled as e-vengers, as like Orczy's (1904) character the scarlet pimpernel, they hide

their true identities. Driven by their emotions, they seek a form of personal justice. A character that has more constructive responses to their emotions exists in many empathetic online communities according to Preece (1998), and may say things such as "MHBFY," which according to Jansen (2002) means "My heart bleeds for you," so perhaps this character type could be known as a MHBFY Jenny. Using the ecological cognition framework there should be also those actors that are driven by gross stimuli, with either existential or thanatotic forces acting upon them. Jansen (2002) identified a term for a member of an online community that is driven by existential forces, known to many as the chat room Bob, who is the actor in an online community that seeks out members who will share nude pictures or engage in sexual relations with them. While first believed to be theoretical by Bishop (2006), there is evidence of members of online communities being driven by thanatotic forces, as reported by the BBC (Anon., 2003). Brandon Vedas, who was a 21-year-old computer expert, killed himself in January 2003. This tragic death suggests strongly that those in online communities should take the behavior of people in online communities that may want to cause harm to themselves seriously. The existence of this type of actor is evidence for the type of online community member who could be called a Ripper, in memory of the pseudonym used by Mr. Vedas.

There are two more characters in online communities, driven by action stimuli that results in them experiencing creative or destructive forces. Surveying the literature reveals a type of actor that uses the Internet that are prime targets for "sophisticated technical information, beta test software, authoring tools [that] like software with lots of options and enjoy climbing a learning curve if it leads to interesting new abilities" (Mena, 1999), who are referred to as wizards. There is also the opposite of the wizard who according to Bishop (2006) seeks to destroy content in online communities, which could be called the iconoclast, which according to Bernstein and Wagner (1976) can mean to destroy and also has modern usage in Internet culture according to Jansen (2002) as a person on the Internet that attacks the traditional way of doing things, supporting Mitchell's (2005) definition of an iconoclast being someone that constructs an image of others as worshippers of artifacts and sets out to punish them by destroying such artifacts.

These eleven character types, summarized in Table 2, should be evident in most online communities, be they virtual worlds, bulletin boards, or wiki-based communities.

Investigating the Proposed Character Theory

Some of the most widely used methods for researching online are interviewing, observation and document analysis (Mann & Stewart, 2000). Ethnography offers a rigorous approach to the analysis of information systems practices using observational techniques, with the notion of context being one of the social construction of meaning frameworks and as a research method, ethnography is well suited to providing information systems researchers with rich insights into the human, social and organizational aspects of information systems development and application because as ethnography deals with actual practices in real-world situations, it allows for relevant issues to be explored and frameworks to be developed which can be used by both practitioners and researchers and also means that researchers can deal with real situations instead of having to contrive artificial situations for the purpose of quasi-experimental investigations (Harvey & Myers, 1995). While Yang (2003) argues that it is not feasible to spend a year or two investigating one online community as part of an ethnography, this is exactly the type of approach that was taken to evaluate the proposed character theory, partially due to the

Table 2. A character theory for online communities based on the ecological cognition framework

Label	Typical Characteristics
Lurker	The lurker may experience a force, such as social, but will not act on it, resulting in them not fully taking part in the community.
Troll	Driven by chaos forces as a result of mental stimuli, would post provocative comments to incite a reaction.
Big Man	Driven by order forces as a result of mental stimuli, will seek to take control of conflict, correcting inaccuracies and keeping discussions on topic.
Flirt	Driven by social forces as a result of social stimuli, will seek to keep discussions going and post constructive comments.
Snert	Driven by anti-social forces as a result of social stimuli, will seek to offend their target because of something they said.
E-venger	Driven by vengeance forces as a result of emotional stimuli, will seek to get personal justice for the actions of others that wronged them.
MHBFY Jenny	Driven by forgiveness forces, as a result of experiencing emotional stimuli. As managers they will seek harmony among other members.
Chat Room Bob	Driven by existential forces as a result of experiencing gross stimuli, will seek more intimate encounters with other actors.
Ripper	Driven by thanatotic forces as a result of experiencing gross stimuli, seeks advice and confidence to cause self-harm.
Wizard	Driven by creative forces as a result of experiencing action stimuli, will seek to use online tools and artifacts to produce creative works.
Iconoclast	Driven by destructive forces as a result of experiencing action stiumli, will seek to destroy content that others have produced.

author receiving formal training in this method. Yang's approach, while allowing the gathering of diverse and varied information, would not allow the research to experience the completeness of Kim's (2000) membership lifecycle, or be able to fully evaluate the character theory and whether the characters in it can be identified.

Location and Participants

An online community was selected for study, this one serving Wales and those with an interest in the geographical locations of Pontypridd and the Rhondda and Cynon Valleys in South Wales. Its members consist of workers, business owners, elected members, and expatriates of the area the online community serves. This online community, known to its members as 'Ponty Town', with 'Ponty' being the shortened term for Pontypridd, was chosen by the author due to his cognitive interest in the Pontypridd constituency and his belief that he would be a representative member of the community and fit in due to holding similar personal interests to the members. This is in line with Figallo (1998), who argues that similar interests is what convinces some members of online communities to form and maintain an Internet presence. The members of the community each had their own user ID and username that they used to either portray or mask their identity. They ranged from actual names, such as 'Mike Powell' and 'Karen Roberts' that were used by elected representatives, names from popular culture, such as 'Pussy Galore', to location-based and gendered names, such as 'Ponty Girl', 'Ponty Boy' and 'Kigali Ken.'

Equipment and Materials

A Web browser was used to view and engage with the online community, and a word processor used to record data from the community.

Procedure

The author joined the online community under investigation and interacted with the other members. The community members did not know the author personally, however, he utilized his real name. Even though the author could have posted under a pseudonym it would have made the study less ecologically valid and more difficult for the author to assess the reaction of the participants. The author carried out activities in the online community by following the membership lifecycle stages, which manifested in not posting any responses, posting a few responses on specific topics to regularly posting as an active member of the community. Additionally, data collected by Livingstone and Bober (2006) was used to understand the results.

Results

Undertaking the ethnography proved to be time consuming, though revealing about the nature of online communities and the characteristics of the actors that use them. Of the eleven characters identified in the proposed character theory, eight were found in the investigated online community.

Lurkers could be identified by looking at the member list, where it was possible to find that 45 of the 369 members were lurkers in that they did not post any messages.

The Troll

The troll was easily identified as an actor that went by the pseudonym Pussy Galore, who even managed to provoke the author.

This Bishop baiting is so good I'm sure there will soon be a debate in the Commons that will advocate abolishing it. – Pussy Galore, Female, Pontypridd

Some of the troll's comments may be flames, but their intention is not to cause offence, but to present an alternative and sometimes intentionally provocative viewpoint, often taking on the role of devil's advocate by presenting a position that goes against the grain of the community.

There is some evidence of the troll existing in the data collected by Livingstone and Bober (2006), as out of a total of 996 responses, 10.9% (164) of those interviewed agreed that it is 'fun to be rude or silly on the internet.'

The Big Man

Evidence of actors being driven by order forces was also apparent, as demonstrated by the following comment from a big man.

I don't think so. Why should the actions of (elected member) attacking me unprovoked, and making remarks about my disability lead to ME getting banned? I am the victim of a hate crime here. – The Victim, Male, Trefforest

The example above clearly demonstrates the role of the big man as absorbing the conflicts of the community and having to take responsibility for the actions of others. While the big man may appear similar to the snert by challenging the actions of others, their intention is to promote their own worldview, rather than to flame and offend another person. The big man may resemble the troll by continually presenting alternative viewpoints, but their intention is not to provoke a flame war based on a viewpoint they do not have, but to justify a position they do have.

The importance of the big man to the community was confirmed approximately a year after the ethnography was completed when during additional observations, a particular actor, ValleyBoy, appeared to take over the role from the big man that was banned, and another member, Stinky, called for banned members to be allowed to return, suggesting the community was lacking

strong and persistent characters, such as the big man.

The Flirt

In the studied online community, there was one remarkable member who posted mostly constructive posts in response to others' messages, known by her pseudonym Ponty Girl who was clearly identifiable as a flirt. Her comments as a whole appear to promote sociability as she responds constructively to others' posts. The flirt's approach to dealing with others appears to differ from the big man who absorbs conflict as it seems to resonate with the constructive sides of actors leading them to be less antagonistic towards the flirt than they would be the big man.

Yes, I've seen him at the train station on quite a few occasions," "A friend of mine in work was really upset when she had results from a feedback request from our team - I'd refused to reply on the principle that she is my friend and I would not judge her, but a lot of the comments said that she was rude, unsympathetic and aloof. She came to me to ask why people thought so badly of her. – Ponty Girl, Female, Graig

The Snert

There were a significant number of members of the community that responded to posts in an anti-social manner, characteristic of snerts. While members like Stinkybolthole frequently posted flames, the online community studied had one very noticeable snert, who went by the name of JH, whom from a sample of ten of his posts, posted six flames, meaning 60% of his posts were flames.

Nobody gives a shit what you want to talk to yourself about. Get a life," "I'm getting the picture. 'Fruitcake Becomes Councilor' is such an overused newspaper headline these days," "Sounds like you've won the lottery and haven't told us. Either that or your husband is a lawyer, accountant or drug dealer,, "The sooner we start to re-colonize ooga-booga land the better, then we'll see Britains (sic) prosperity grow. Bloody pc wimps, they need to get laid. – JH, Male, Trallwn

The existence of the snert is evident. The data collected by Livingstone and Bober (2006) reveals that from a sample of 1,511, 8.5% (128) of individuals have received nasty comments by e-mail and 7% (105) have received nasty comments in a chat room. Of the 406 that had received nasty comments across different media, 156 (38.42%) deleted them straight away. A total of 124 (30.54%) tried to block messages from the snert, 84 (20.69%) told a relative, 107 (26.35%) told a friend, 74 (18.23%) replied to ask the snert to stop their comments, and 113 (27.83%) engaged in a flame war with the snert.

The eVenger

Evident in the online community investigated was the masked e-venger, who in the case of this particular community was an actor who signed up with the pseudonym elected member, claiming to be an elected representative on the local council, who the members quickly identified to be someone who had been banned from the community in the past. This user appeared to have similar ways of posting to the snert, posting flames and harassing other members. The difference between the e-venger and the snert is that the former is driven by wanting to get even for mistreatment in the past whereas the latter responds unconstructive to the present situation.

Poor sad boy, have you met him? He's so incompitent (sic). His dissabilty (sic) is not medical it's laughable. The lad has no idea about public perception," "Don't give me this sh#t, she and they cost a fortune to the taxpayer, you and I pay her huge salary. This is an ex-education Cabinet Member who was thrown out by the party, un-

elected at the next election and you STILL pay her wages!", "I'll see you at the Standards meeting [Mellow Pike]. 'Sponsor me to put forward a motion'! Bring it on. – Elected Member

The member appeared to be driven by emotional stimuli activating vengeance forces, seeking to disrupt the community and even making personal attacks on the members including the author. As outlined above, the data collected by Livingstone and Bober (2006) reveals that 27.83% of people that are flamed will seek revenge by posting a flame back.

The MHBFY Jenny

Sometimes the remarks of members such as flirts and big men are accepted, which can lead other actors to experience emotional stimuli activating forgiveness forces as was the case with Dave, the investigated online community's MHBFY Jenny.

Mind you it was funny getting you to sign up again as 'The Victim'. – Dave, Male, Pontypridd

While many of the MHBFY Jenny's comments are constructive like the flirt, they differ because the former responds to their internal dialogue as was the case with Dave above, whereas the flirt responds to external dialogue from other actors, as Ponty Girl clearly does.

The Chat Room Bob

The online community investigated, like many, had its own chat room Bob. The actor taking on this role in the investigated online community went by the name of Kigali Ken, and his contributions make one wonder whether he would say the same things in a real-world community.

Any smart women in Ponty these days? Or any on this message board? I've been out of the country for a while but now I'm back am looking for some uncomplicated sex. Needless to say I am absolutely lovely and have a massive... personality. Any women with an attitude need not apply. – Kigali Ken, Male, Pontypridd

While their action of seeking out others may appear to be flirting using the vernacular definition, the intention of the chat room Bob differs from the flirt who based on Heskell's (2001) definition is someone who feels great about themselves and resonates this to the world to make others feel good, as they will make pro-social comments about others and in response to others. The chat room Bob on the other hand, appears to be only after their own gratification, responding to their physical wants.

The existence of the chat room Bob is evident. The data collected by Livingstone and Bober (2006) reveals that 394 people from a sample of 1,511 have reported that they have received sexual comments from other users. Of these 238 (60.4%) deleted the comment straight away, 170 (43.14%) attempted to block the other person, 49 (12.44%) told a relative, 77 (19.54%) told a friend, and 75 (19.04%) responded to the message. This suggests that the chat room Bob is an unwanted character in online communities whose contributions people will want to delete and whom they may try to block.

The Ripper, Wizard, and Iconoclast

Despite studying the online community for over a year, there was no evidence of there being a ripper, a wizard or an iconoclast in the community, beyond the administrators of the site posting and deleting content and adding new features, such as polls. The closest an actor came to being a ripper was an actor called choppy, who faked a suicide and then claimed a friend had hijacked their account. Fortunately, it might be argued that a true ripper who was seeking to cause self-harm was not present, but the existence of this type of online community member should lead online

community managers to show concern for them, and members should not reply with comments such as "murder/suicide" when they ask for advice, as happens in some online communities.

While visual representations are often absent from bulletin board communities, actors will often make their first interpretations of others in virtual worlds when they book at another's avatar and evaluates them based on their worldview, which may provoke a relation leading to the actor developing an interest in the other actor. In the context of online communities, an avatar is a digital representative of an actor in a virtual environment that can be an icon of some kind or an animated object (Stevens, 2004). According to Aizpurua et al. (2004), the effective modeling of the appearance of an avatar is essential to catch a consumer's attention and making them believe that they are someone, with avatars being crucial to creating interaction between customers and vendors. According to Puccinelli (2006), many vendors understand that customers' decisions to engage in economic transactions are often influenced by their reactions to the person who sells or promotes it, which seems to suggest that the appearance of an avatar will affect the number of transactions other actors will have with it.

A STEREOTYPE THEORY FOR INTERPRETING AVATARS IN ONLINE COMMUNITIES

Technology-enhanced businesses led by business leaders of a black ethnicity have been some of the most innovative in the world, with companies like Golden State Mutual ending the 1950s with electronic data processing systems in place, $133 million of insurance in force and $16 million in assets (Ingham & Feldman, 1994). Representations of black actors have also been some of the most studied, with potential applications for studying avatars in online communities. Alvarado et al. (1987) argue that black actors fit into four social classifications: the exotic, the humourous, the dangerous and the pitied. Furthermore, Malik (2002) suggests that male black actors are stereotyped as patriarchal, timid, assiduous, and orthodox. Evidence for these can easily be found in contemporary print media, such as Arena magazine (Croghton, 2007) where an advertisement for an electronic gaming system displays a black individual as pitied. In the same publication Murray and Mobasser (2007) argue that the Internet is damaging relationships, where images of women are of those in 'perfect' bodies, and although they do not define what a perfect body is, it would be safe to assume they mean those depicted in the publication, such as Abigail Clancy and Lily Cole, the later of which described herself as 'hot stuff', and perhaps depictions of this sort could be iconographic of an exotic avatar. Alvarado et al. (1987) supported by Malik (2002) have argued that these stereotypes have been effective in generating revenue for advertisers and not-for-profit organizations. While these stereotypes may be useful for developing an understanding of avatars and how they can generate both social and economic capital for individuals, they need to be put into the context of a psychological understanding of how actors behave and interact with others.

Utilizing the ecological cognition framework (Bishop, 2007a; 2007c); it can be seen that the visual appearance of an actor's avatar could be based on the five binary-opposition forces, with some of the stereotypes identified earlier mapping on to these forces. The image of actors as orthodox and pariahs can be seen to map onto the forces occupied by the flirts and snerts, respectively; the assiduous and vanguard stereotypes appear to be in harmony with the forces occupied by the wizard and iconoclast, respectively, the dangerous and timid stereotypes are consistent with the forces connected with the e-venger and MHBFY Jenny, the exotic and pitied stereotypes can be seen to map on to the forces used by the chat room Bob and ripper, respectively, and the patriarchal and humourous stereotypes appear to be consistent

with the forces, respectively, used by the big man and troll. The stereotype theory provides a useful basis for investigating the role of avatars in online communities and the effect they have on social and economic transactions.

Location and Participants

A study was carried out in the second life virtual world and involved analyzing the avatars used to create a visual representation of the actor and profile pages displaying their personal details and avatar of 189 users, known as residents, of the community who met the criteria of having given at least one rating to another actor, a feature that has since been discontinued in the system despite it showing how popular a particular actor was.

Equipment and Materials

The Second Life application was used to view and engage with the online community, and a word processor and spreadsheet was used to record data from the community in the form of the number of times a person had received a gift or response from another.

Procedure

The author became a member of the online community under investigation and interacted with the other members over a period of three months. The members of the community did not know the author, especially as a pseudonym was adopted, as is the norm with Second Life. The author carried out activities in the online community by following the membership lifecycle stages that each individual member of an online community goes through as discussed earlier in the chapter. A search was done for actors and possible locations and groups of specific avatars identified. After an avatar was categorized, data from their profile was recorded and the return on investment (ROI) calculated. According to Stoehr (2002), calculating the ROI is a way of expressing the benefit-cost ratio of a set of transactions, and can be used to justify an e-commerce proposal.

Results

The results, as summarized in Table 3, reveal that the avatar with greatest return on investment was the patriarchal stereotype with a 428.43% return and the one with the least ROI was the assiduous with a 12.63% loss. The most common avatar was the humorous, followed by exotic and pariah. The least common avatar was the vanguard, followed by the dangerous and pitied.

An independent samples test using the Mann-Whitey method was carried out on one of the highest ROI avatars, the patriarchal, with one of the lowest, the pariah. It revealed, as expected, a significant difference in the return on investment (Z=-3.21, p<0.002). Also interesting was the difference between the specific attributes rated. The mean appearance rating for the patriarchal stereotype was 29.24 compared to 17.27 for the pariah (Z=-3.10, p<0.003), the mean building rating for the patriarchal was 29.03 compared to 17.40 for the pariah (Z=-3.06, p<0.003), and the mean behavior rating was 30.62 for the patriarchal stereotype and 16.37 for the pariah (Z=-3.68, p<0.001). This would seem to suggest that as well as not getting as high a return on investment, other actors will not judge the more antisocial-looking pariah as well as they judge the more sophisticated-looking patriarchal avatar. Examples of the avatars are presented in Figure 1. Studies

Table 3. Mean (M) dollars ($) given and received by actors of specific avatars and their ROI (%)

Stereotype	Character	N	M Given $	M Received $	M ROI %
Exotic	Chat Room Bob	30	1171.67	1731.67	237.26
Pitied	Ripper	11	1743.18	2534.09	141.19
Humourous	Troll	48	362.5	446.35	120.69
Patriarchal	Big Man	17	4500	4588.24	428.43
Orthodox	Flirt	16	4393.75	4575	49.48
Pariah	Snert	26	149.04	107.69	-1.24
Assiduous	Wizard	16	267.19	159.38	-12.63
Vanguard	Iconoclast	4	75	162.5	233.33
Dangerous	E-venger	6	2587.5	2095.83	0.62
Timid	MHBFY Jenn	15	6150	4395	101.82

Figure 1. Examples of avatars in order top-bottom, left-right as Table 3

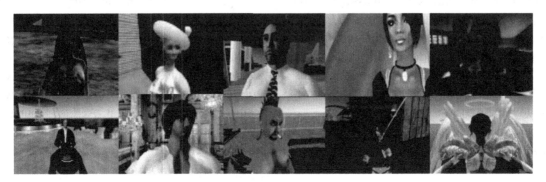

such as those by Zajonc (1962) and Goldstein (1964) have demonstrated that actors will seek to avoid the uncovering of beliefs and other thoughts that come about when an actor experiences threatening behavior from others or uncomfortable emotions. This being the case, it could be that when an actor is presented with an avatar that causes them discomfort or 'dissonance', then they will seek to resolve the conflict created by avoiding that particular avatar. This would explain why the pariah stereotype produces a limited number of economic transactions and has the one of the worst returns on investment, which would seem to support the findings of Eagly et al. (1991) that people that appear less discomforting are more popular with peers and receive preferential treatment from others.

THE FUTURE OF SOCIAL NETWORKING COMMUNITIES

In science fiction, the future is often portrayed as utopian or dystopian, where possible future outcomes of social trends or changes that are the result of scientific discoveries are depicted and the implications of them assessed (Csicsery-Ronay, 2003). In the cyberpunk genre of science fiction, the dystopian future is often made up of corporations, who ruthlessly corrupt, corrode, exploit and destroy (Braidotti, 2003). Social networking communities are quickly being subsumed into corporate structures. In July 2005, News Corporation bought Myspace.com, which is a social networking service that integrates message boards with personal homepages and utilizes the Circle of Friends social networking technique, and in December of that year, the British broadcaster ITV bought the old school tie-based Friends Reunited social networking service (BBC, 2005; Scott-Joynt, 2005).

The ecological cognition framework has the potential to radically transform minor Web sites into highly persuasive and engaging communities where relationships between vendors and customers can be enhanced and the goals of each can be met. While there is also the possibility that a corporation that understands online communities can manipulate its members in such a way that it can easily exploit them, the model could be used by vendors with more of an interest in helping customers meet their goals to market their products and services effectively. Vendors that understand the stage of Kim's (2000) lifecycle they are at and the stage the consumer is at can more effectively target their messages in such a way that they are persuasive. Using the model, vendors can design avatars that provoke the particular responses they want from customers and continue that initial appeal by adopting the appropriate character type. This works well in some media texts where according to Kress (2004) media producers can use the appearance of their characters to convey that character's personality and build on that throughout the text.

DISCUSSION

The rise of online communities in Internet environments has set in motion an unprecedented shift in power from vendors of goods and services to the customers who buy them, with those vendors who understand this transfer of power and choose to capitalize on it by organizing online communities and being richly rewarded with both peerless customer loyalty and impressive economic returns. A type of online community, the virtual world, could radically alter the way people work, learn, grow consume, and entertain. Understanding the exchange of social and economic capital in virtual worlds could involve looking at what causes actors to spend their scarce resources on improving someone else's reputation. Actors' reputations may affect how willing others are to trade with them or even give them gifts, and their reputation is in part influenced by their appearance and how they interact with other actors and often feedback from other actors are displayed on their profile.

The ecological cognition framework provides a theoretical model for developing a character theory for online communities based on bulletin board and chat room models. The five forces and their opposites can be used to develop the types, and the judgments of ignorance and temperance can be used to explain the behavior of those that do not participate, namely lurkers, which were accounted for in the investigated online community where it was possible to find that 45 of the 369 members were lurkers for the reason that they did not post any messages. The ECF would suggest that chaos forces drive trolls, as they attempt to provoke other members into responding as a result of experiencing mental stimuli. The troll was easily identified in the investigated online community as an actor that went by the pseudonym Pussy Galore, who even managed to provoke the author. Order forces can be seen to drive the big man and was represented in the investigated online community by the victim. Those actors who are plainly obnoxious and offend or harass other actors through posting flames are known as snerts, who were most obviously represented in the investigated online community by a user called JH. Flirts are members that respond to the text posted by other members as social stimuli, and will respond to it after activating their social forces and in the studied online community there was one remarkable member who posted mostly constructive posts in response to others' messages, know by her pseudonym Ponty Girl. There are actors driven by their vengeance forces, which could be labeled as e-vengers, represented in the investigated online community by elected member and those actors driven by forgiveness forces could be called MBHFY Jenny, represented in the studied online community by Dave. An actor in an online community that is driven by existential forces, known to many as the chat room Bob, who seeks out members who will share nude pictures or

engage in sexual relations with them, was apparent in the investigated online community using the name Kigali Ken. There is evidence for an online community member driven by thanatotic forces, who could be called a ripper, but this member was not found in the investigated online community beyond an actor called Choppy. There are also theoretically two more characters in online communities, driven by action stimuli that results in them experiencing creative or destructive forces, with the one driven by creative forces being the wizard, and the opposite of iconoclast being the one that seeks to destroy content in online communities.

These character types are particularly evidenced in bulletin board communities, but in the virtual world it is likely that an actor's avatar will have some effect on how others perceive them before they are spoken to. The extent to which an actor is able to sustain an appeal to another could be analyzed as seduction. An actor's avatar forms an important part of the intimacy stage of seduction, as the visual appearance of an actor could possibly have an impact on how others perceived them, and an actor may construct an image based on their identity or the image they want to project and the relationship between an actor's avatar and their identity can be understood as elastic as even the best and strongest elastic can break, with there being the possibility that avatars can develop to the point where connection between them and the identities of the actors using them can be stretched so far that they cease to exist. There has been a debate over whether identity is unitary or multiple with psychoanalytic theory playing a complicated role in the debate. If there is a lifecycle to an actor's membership in an online community, then it is likely that they will develop different cognitions, such as beliefs and values at different stages that may become 'joindered'. This would mean that an actor's behavior would be affected by the beliefs and values they developed when joining the community when they are at a more advanced stage in their membership of the community. Utilizing the ecological cognition framework, it can be seen that the visual appearance of an actor's avatar could be based on the five binary-opposition forces, with some of the stereotypes identified earlier mapping on to these forces. The investigation found that the avatar with the greatest return on investment was the patriarchal stereotype with a 428.43% return and the one with the least ROI was the assiduous with a 12.63% loss. The most common avatar was the humorous, followed by exotic and pariah. The least common avatar was the vanguard, followed by the dangerous and pitied. An independent samples test revealed, as expected, a significant differences between the pariah and the patriarchal stereotype with the later having a greater return on investment, and higher ratings on appearance, building and behavior, suggesting that as not getting as high a return on investment, other actors will not judge the more antisocial-looking pariah as well as they judge the more sophisticated-looking patriarchal avatar.

The research methods used in this study were an ethnographical observation and document analysis. These methods seem particularly suited to online communities, where behavior can be observed through participation and further information can be gained through analyzing user profiles and community forums. The study has demonstrated that online communities, in particular virtual worlds, can be viewed as a type of media, and traditional approaches to media, such as investigating stereotypes, can be applied to Internet-based environments.

ACKNOWLEDGMENT

The author would like to acknowledge all the reviewers that provided feedback on earlier drafts of this chapter. In addition the author would like to thank S. Livingstone and M. Bober from the Department of Media and Communications at the London School of Economics and Political Science for collecting some of the data used in this study, which was sponsored by a grant from the Economic and Social Research Council. The Centre for Research into Online Communities and E-Learning Systems is part of Glamorgan Blended Learning Ltd., which is a Knowledge Transfer Initiative, supported by the University of Glamorgan through the GTi Business Network of which it is a member.

REFERENCES

Aizpurua, I., Ortix, A., Oyarzum, D., Arizkuren, I., Ansrés, A., Posada, J., & Iurgel, I. (2004). Adaption of mesh morphing techniques for avatars used in Web applications. In: F. Perales & B. Draper (Eds.), *Articulated motion and deformable objects: Third international workshop.* London: Springer.

Anon. (2003). *Net grief for online 'suicide'.* Retrieved from http://news.bbc.co.uk/1/hi/technology/2724819.stm

Armstrong, A., & Hagel, J. (1997). *Net gain: Expanding markets through virtual communities.* Boston, MA: Harvard Business School Press.

BBC. (2005). *ITV buys Friends Reunited Web site.* London: BBC Online. Retrieved from http://news.bbc.co.uk/1/hi/business/4502550.stm

Beaty, J., Hunter, P., & Bain, C. (1998). *The Norton introduction to literature.* New York, NY: W.W. Norton & Company.

Berne, E. (1961). *Transactional analysis in psychotherapy.* New York, NY: Evergreen.

Berne, E. (1964). *Games people play: The psychology of human relationships.* New York, NY: Deutsch.

Bernstein, T., & Wagner, J. (1976). *Reverse dictionary.* London: Routledge.

Bishop, J. (2002). *Development and evaluation of a virtual community.* Unpublished dissertation. http://www.jonathanbishop.com/ publications/ display.aspx?Item=1.

Bishop, J. (2003). Factors shaping the form of and participation in online communities. *Digital Matrix, 85,* 22–24.

Bishop, J. (2005). The role of mediating artefacts in the design of persuasive e-learning systems. In: *Proceedings of the Internet Technology & Applications 2005 Conference.* Wrexham: North East Wales Institute of Higher Education.

Bishop, J. (2006). Social change in organic and virtual communities: An exploratory study of bishop desires. *Paper presented to the Faith, Spirituality and Social Change Conference.* University of Winchester.

Bishop, J. (2007a). The psychology of how Christ created faith and social change: Implications for the design of e-learning systems. *Paper presented to the 2nd International Conference on Faith, Spirituality, and Social Change.* University of Winchester.

Bishop, J. (2007b). Increasing participation in online communities: A framework for human-computer interaction. *Computers in Human Behavior, 23,* 1881–1893. doi:10.1016/j.chb.2005.11.004

Bishop, J. (2007c). Ecological cognition: A new dynamic for human computer interaction. In: B. Wallace, A. Ross, J. Davies, & T. Anderson (Eds.), *The mind, the body and the world* (pp. 327-345). Exeter: Imprint Academic.

Braidotti, R. (2003). Cyberteratologies: Female monsters negotiate the other's participation in humanity's far future. In: M. Barr (Ed.), *Envisioning the future: Science fiction and the next millennium.* Middletown, CT: Wesleyan University Press.

Bressler, S., & Grantham, C. (2000). *Communities of commerce.* New York, NY: McGraw-Hill.

Campbell, J., Fletcher, G., & Greenhil, A. (2002). Tribalism, conflict and shape-shifting identities in online communities. *Proceedings of the 13th Australasia Conference on Information Systems.* Melbourne, Australia.

Chak, A. (2003). *Submit now: Designing persuasive Web sites.* London: New Riders Publishing.

Chan, T.-S. (1999). *Consumer behavior in Asia.* New York, NY: Haworth Press.

Croughton, P. (2007). *Arena: The original men's style magazine.* London: Arena International.

Csicsery-Ronay, I. (2003). Marxist theory and science fiction. In: E. James & F. Mendlesohn (Eds.), *The Cambridge companion to science fiction.* Cambridge: Cambridge University Press.

Eagly, A., Ashmore, R., Makhijani, M., & Longo, L. (1991). What is beautiful is good, but…: A meta-analytic review of research on the physical attractiveness stereotype. *Psychological Bulletin, 110*(1), 109–128. doi:10.1037/0033-2909.110.1.109

Feller, J. (2005). *Perspectives on free and open source software.* Cambridge, MA: The MIT Press.

Figallo, C. (1998). *Hosting Web communities: Building relationships, increasing customer loyalty and maintaining a competitive edge.* Chichester: John Wiley & Sons.

Freud, S. (1933). *New introductory lectures on psycho-analysis.* New York, NY: W.W. Norton & Company, Inc.

Givens, D. (1978). The non-verbal basis of attraction: Flirtation, courtship and seduction. *Journal for the Study of Interpersonal Processes, 41*(4), 346–359.

Goffman, E. (1959). *The presentation of self in everyday life.* Garden City, NY: Doubleday.

Goldstein, M. (1964). Perceptual reactions to threat under varying conditions of measurement. *Journal of Abnormal and Social Psychology*, *69*(5), 563–567. doi:10.1037/h0043955

Harvey, L., & Myers, M. (1995). Scholarship and practice: The contribution of ethnographic research methods to bridging the gap. *Information Technology & People*, *8*(3), 13–27. doi:10.1108/09593849510098244

Heskell, P. *Flirt Coach: Communication tips for friendship, love and professional success.* London: Harper Collins Publishers Limited.

Ingham, J., & Feldman, L. (1994). *African-American business leaders: A biographical dictionary.* Westport, CT: Greenwood Press.

Jansen, E. (2002). *Netlingo: The Internet dictionary.* Ojai, CA: Independent Publishers Group.

Jordan, T. (1999). *Cyberpower: An introduction to the politics of cyberspace.* London: Routledge.

Kiesler, S., & Sproull, L. (1992). Group decision making and communication technology. *Organizational Behavior and Human Decision Processes*, *52*(1), 96–123. doi:10.1016/0749-5978(92)90047-B

Kim, A. (2000). *Community building on the Web: Secret strategies for successful online communities.* Berkeley, CA: Peachpit Press.

Kress, N. (2004). *Dynamic characters: How to create personalities that keep readers captivated.* Cincinnati, OH: Writer's Digest Books.

Kyttä, M. (2003). *Children in outdoor contexts: Affordances and independent mobility in the assessment of environmental child friendliness.* Doctoral dissertation presented at Helsinki University of Technology, Espoo, Finland.

Li, D., Li, J., & Lin, Z. (2007). Online consumer-to-consumer market in China—a comparative study of Taobao and eBay. *Electronic Commerce Research and Applications.* doi:.doi:10.1016/j.elerap.2007.02.010

Lindahl, C., & Blount, E. (2003). Weblogs: Simplifying Web publishing. *IEEE Computer*, *36*(11), 114–116.

Livingstone, S., & Bober, M. (2006). *Children go online, 2003-2005.* Colchester, Essex: UK Data Archive.

Malik, S. (2002). *Representing black Britain: A history of black and Asian images on British television.* London: Sage Publications.

Mann, C., & Stewart, F. (2000). *Internet communication and qualitative research: A handbook for Research Online.* London: Sage Publications.

Mantovani, F. (2001). Networked seduction: A test-bed for the study of strategic communication on the Internet. *Cyberpsychology & Behavior*, *4*(1), 147–154. doi:10.1089/10949310151088532

Mena, J. (1999). *Data mining your Web site.* Oxford: Digital Press.

Mitchell, W. (2005). *What do pictures want?: The lives and loves of images.* Chicago, IL: University of Chicago Press.

Murray, S., & Mobasser, A. (2007). Is the Internet killing everything? In: P. Croughton (Ed.), *Arena: The original men's style magazine.* London: Arena International.

Orczy, E. (1905). *The scarlet pimpernel.* Binding Unknown.

Orengo Castellá, V., Zornoza Abad, A., Prieto Alonso, F., & Peiró Silla, J. (2000). The influence of familiarity among group members, group atmosphere and assertiveness on uninhibited behavior through three different communication media. *Computers in Human Behavior*, *16*, 141–159. doi:10.1016/S0747-5632(00)00012-1

Propp, V. (1969). *Morphology of the folk tale*. Austin, TX: University of Texas Press.

Puccinelli, N. (2006). Putting your best face forward: The impact of customer mood on salesperson evaluation. *Journal of Consumer Psychology*, *16*(2), 156–162. doi:10.1207/s15327663jcp1602_6

Rhiengold, H. (2000). *The virtual community: Homesteading on the electronic frontier*. London: The MIT Press.

Scott-Joynt, J. (2005). *What Myspace means to Murdoch*. London: BBC Online. Retrieved from http://news.bbc.co.uk/1/hi/business/4697671.stm.

Shen, X., Radakrishnan, T., & Georganas, N. (2002). vCOM: Electronic commerce in a collaborative virtual worlds. *Electronic Commerce Research and Applications*, *1*, 281–300. doi:10.1016/S1567-4223(02)00021-2

Smith, C. (2000). Content analysis and narrative analysis. In: H. Reis & C. Judd (Eds.), *Handbook of research methods in social and personal psychology*. Cambridge: Cambridge University Press.

Sternberg, R. (1986). A triangular theory of love. *Psychological Review*, *93*(2), 119–135. doi:10.1037/0033-295X.93.2.119

Stevens, V. (2004). Webhead communities: Writing tasks interleaved with synchronous online communication and Web page development. In: J. Willis & B. Leaver (Eds.), *Task-based instruction in foreign language education: Practices and programs*. Georgetown, VA: Georgetown University Press.

Stoehr, T. (2002). *Managing e-business projects: 99 key success factors*. London: Springer.

Turkle. (1997). *Life on the screen: Identity in the age of the Internet*. New York, NY: Touchstone.

Vogel, D. (1999). *Financial investigations: A financial approach to detecting and resolving crimes*. London: DIANE Publishing.

Wallace, P. (2001). *The psychology of the Internet*. Cambridge: Cambridge University Press.

Werry, C. (2001). Imagined electronic community: Representations of online community in business texts. In: C. Werry & M. Mowbray (Eds.), *Online communities: Commerce, community action and the virtual university*. Upper Saddle River, NJ: Prentice Hall.

Yang, G. (2003). The Internet and the rise of a transnational Chinese cultural sphere. *Media Culture & Society*, *25*, 469–490. doi:10.1177/01634437030254003

Zajonc, R. (1962). Response suppression in perceptual defense. *Journal of Experimental Psychology*, *64*, 206–214. doi:10.1037/h0047568

This work was previously published in Social Networking Communities and E-Dating Services: Concepts and Implications, edited by Celia Romm-Livermore and Kristina Setzekorn, pp. 60-77, copyright 2009 by Information Science Reference (an imprint of IGI Global).

Chapter 6

Moving from Cyber–Bullying to Cyber–Kindness:
What do Students, Educators and Parents Say?

Wanda Cassidy
Simon Fraser University, Canada

Karen Brown
Simon Fraser University, Canada

Margaret Jackson
Simon Fraser University, Canada

ABSTRACT

The purpose of this chapter is to explore cyber-bullying from three different, but interrelated, perspectives: students, educators and parents. The authors also explore the opposite spectrum of online behaviour - that of "cyber-kindness" - and whether positive, supportive or caring online exchanges are occurring among youth, and how educators, parents and policy-makers can work collaboratively to foster a kinder online world rather than simply acting to curtail cyber-bullying. These proactive efforts tackle the deeper causes of why cyber-bullying occurs, provide students with tools for positive communication, open the door for discussion about longer term solutions, and get at the heart of the larger purposes of education – to foster a respectful and responsible citizenry and to further a more caring and compassionate society. In the course of this discussion, they highlight the findings from two studies they conducted in British Columbia, Canada, one on cyber-bullying and a later study, which addressed both cyber-bullying and cyber-kindness.

DOI: 10.4018/978-1-4666-2803-8.ch006

INTRODUCTION

The proliferation of electronic media in recent years has allowed children and adolescents to take schoolyard bullying to an entirely new level – into the realm of cyber-bullying. Cyber-bullying in the school context involves using emails, websites, text messaging, camera phones, blogs, YouTube, Facebook and other forms of social networking technology to spread hurtful, nasty, derogatory, vulgar or untrue messages to or about other students, teachers or acquaintances. Cyber-bullying typically threatens the reputation, wellbeing, security and/or safety of the targeted victim. Unlike face-to-face bullying, which typically happens at a given location, in a particular moment in time, and by a perpetrator that is known or seen by the victim, messages posted online can be spread globally, can exist in perpetuity, and the cyber-bully can hide his/her identity by using an avatar. Cyber-bullying can impact the victim in serious ways: low self-esteem; inability to concentrate on schoolwork; anger; anxiety; depression; and even suicide (Brown, Jackson & Cassidy, 2006; Gradinger, Strohmeier & Spiel, 2009; Willard, 2006; Ybarra and Mitchell, 2008).

Unfortunately, cyber-bullying has expanded into a global phenomenon (Kowalski et al, 2008), while educators, parents, and policymakers struggle to develop effective solutions (Campbell, 2005; Shariff, 2005; Belsey, 2006; Brown et al. 2006; Willard, 2006). While most of the attention in the current literature has been on trying to understand cyber-bullying, the cyber-bully and the impact on the victim (Beale & Hall, 2007; Brown et al. 2006; Li, 2007; Patchin & Hinduja, 2006; Shariff, 2008; Smith et al. 2008; Willard, 2007; Worthen, 2007; Ybarra, Diener-West and Leaf, 2007), little or no attention has been given to examining whether technology is also being used by youth to communicate thoughtful, kind and caring messages (Cassidy, Brown & Jackson, 2010; Cassidy, Jackson & Brown, 2009; Jackson, Cassidy & Brown, 2009a; Jackson, Cassidy &

Brown, 2009b). Further, the focus in schools primarily has been on punishing the cyber-bully and implementing anti-cyberbullying programs, rather than on developing holistic solutions that cultivate more respectful on-line exchanges and build a more caring school culture.

While it is important to understand why young people cyber-bully, the extent of cyber-bullying, and the characteristics of perpetrators and victims, it is equally as important to investigate whether cyberspace is also being used in positive ways. Do children and adolescents use technology in ways that communicate care and kindness to one another and foster peers' self-esteem? Can young people's negative exchanges be re-directed in more positive ways that assist parents, educators, school counselors and other professionals in diminishing the potential risks associated with hurtful cyber-bullying victimization?

Green and Hannon (2007) suggest that digital technologies offer a "third space" (p. 60) between official and informal contexts, where young people can "create portfolios of digital media, engage in peer teaching and develop their confidence and voice" (Sharples, Graber, Harrison & Logan, 2009: 72). Ybarra et al. (2007) agree that online access can afford positive initiatives for youth, providing important information on a range of questions and concerns. Certainly children and adolescents are using the Internet as a positive and interactive venue for creating blogs (journaling), social dialoguing, sharing of ideas, and engaging in scholastic endeavours or innovative pursuits (Dowell, Burgess and Cavanaugh, 2009). As a result, the Internet offers potential for positive, helpful and caring communications between youth - or what we have termed "cyber-kindness."

This chapter examines the spectrum of students' on-line exchanges, from cyber-bullying to cyber-kindness, highlighting information from two research studies we conducted in British Columbia, Canada. In our first study (2005-2007), we examined 11-15 year old (Grades 6 to 9) students' perceptions and experiences with cyber-bullying and also interviewed fifteen teachers, vice-principals and principals. In our second study (2007-2009), we expanded our research to examine cyber-kindness among the same age/grade group of students, although at different schools, and also extended our research of school personnel to include school counselors and youth workers as well as teachers, vice-principals and principals (*n*=17). In the second study, we also surveyed parents, since cyber-bullying most often occurs using the home computer, and because parents play a critical role in preventing cyber-bullying and fostering cyber-kindness.[1] It is also important to solicit information from parents, as most parents underestimate the extent of cyber-bullying happening in their children's lives as well as the amount of time their children spend online each day (Bhat, 2008; Cassidy et al. 2010; Jackson et al. 2009a&b; Chou, Yu, Chen and Wu, 2009). While some researchers have found that close parental supervision is unrelated to youth engaging in risky online behaviour (Liau, Khoo & Ang, 2008), others find that lack of parental supervision may lead to increased chat room use (Sun et al. 2005). Further, parents tend to envision the potential risks of stranger online contact but often neglect to discuss with their children the dangers that can occur from friends and classmates (Bhat, 2008).

In the chapter we posit that school personnel need to work collaboratively with parents and with youth to develop policies and practices that proactively foster kindness, caring and respect towards one another rather than merely focusing on curtailing or punishing negative behaviour, such as cyber-bullying. This chapter, therefore, is situated within and contributes to two literatures: cyber-bullying, which to date primarily has focused on students and teachers and only recently on parents, and not at all on cyber-kindness (Brown et al. 2006; Cheever & Carrier, 2008; Cassidy et al. 2009; Cassidy et al. 2010; Jackson et al. 2009a&b; Kowalski, Limber & Agatston, 2008; Liau et al. 2008; Rosen, Cheever & Carrier, 2008;

Shariff, 2006); and the ethic of care literature, which stresses modeling, practice, dialogue and confirmation as a way to foster positive behaviour and which has not been sufficiently addressed in the cyber-world (Cassidy & Bates, 2005; Gregory, 2000; Noddings, 2002, 2005; Owen & Ennis, 2005; Rauner, 2000).

RESEARCH METHODS

Because our Study #1 is reported elsewhere in the literature (Cassidy et al. 2009; Jackson et al. 2009a; Jackson et al. 2009b), including the methods used to collect the data, we will not describe this study here, other than to note that it involved students in Grades six to nine (ages 11-15) (*n*=365) completing a fourteen-page questionnaire (192 variables), which included: closed-ended questions on cyber-bullying and open-ended questions asking respondents to provide examples of cyber-bullying; which students were most likely to be bullied online; solutions to the cyber-bullying problem; and other information on cyber-bullying they felt relevant. This study also included hour-long semi-structured interviews with fifteen teachers, vice-principals and principals. When discussing this study, we will cite published papers.

Study #2 on cyber-bullying and cyber-kindness was recently completed and the results have yet to be published (see Cassidy et al. 2010). In this study, 339 students in Grades six to nine (ages 11-15) were canvassed from a large metropolitan region of British Columbia, Canada. The researchers designed the student survey to accommodate youth of varying language abilities, meaning the wording used was relatively simple and the font was stylized for easy reading. Twenty background questions were asked of students; for example, their age, grade level, gender, ethnicity, home and first language, success with school, the number of computers in their home, the location of these computers, how often they were online, when they were online, and so on. The closed-ended questions (single-response questions, categorical-response items, rating scales), as well as the open-ended questions about cyber-bullying in the students' survey, inquired about the frequency and methods used in cyber-bullying and being cyber-bullied, the situation in which bullying took place, their reactions to being cyber-bullied, whom they talked to after they were cyber-bullied, information regarding cyber-bullying teachers and principals, and possible solutions to cyber-bullying.

Students were also asked a number of closed-ended and open-ended questions related to "cyber-kindness." This section included baseline statements to determine if respondents knew the differences between kind, thoughtful and caring messages and bullying types of comments, assessed the prevalence of, and methods used, in cyber-kindness, determined the situations in which cyber-kindness took place, asked for other examples of sent and received kind, thoughtful and caring messages, and canvassed students' opinions regarding what the school, students and parents could do to encourage kind and caring online behaviours.

In addition, in-depth, semi-structured interviews were conducted with 17 educators from these schools; respondents included two principals, four vice-principals, five teachers, four counselors and two youth workers. The 45-60 minute taped interviews included closed-category questions using a five-point Likert response scale ranging from 1 (extremely concerned) to 5 (not concerned at all) on the issue of cyberbullying, and another scale ranging from 1 (very familiar) to 5 (not familiar at all) on familiarity with technology (email, cellular phones/text messaging, Facebook, MSN/chat rooms, YouTube and Blogs). Another closed-category question using a five-point Likert response scale from 1 (not important at all) to 5 (extremely important) asked respondents how important is in the life of the school, to (a) prevent cyber-bullying and to (b) encourage cyber-kindness. Several open-ended questions probed

their experiences and suggestions regarding cyber-bullying and cyber-kindness, giving voice to their perspectives (Barron, 2000; Cook-Sather, 2002; Palys, 2003). Interview transcripts were reviewed and re-reviewed for emergent themes (and sub-themes), using a grounded theory approach and an open coding method (Miles & Huberman, 1994; McMillan & Schumacher, 1997).

Three hundred and fifteen parents of students at the participating schools were surveyed. The parents' survey included single-response questions, categorical-response items, rating scales and open-ended questions. The first section of the questionnaire assessed their knowledge of technology, concerns about cyber-bullying, supervision of their child's computer use, hours their child spends online, and knowledge of their child's cyber-bullying victimization and/or perpetration. The second section, comprised of open-ended questions, canvassed parents' knowledge of cyber-kindness and how educators and parents might help cultivate more considerate, respectful and caring online interactions among youth. Similar to the qualitative analysis undertaken with the educators' interviews, the themes which surfaced in the open-ended responses from parents, were determined through a process of review and re-review according to the frequency of responses and the strength of response (Miles & Huberman, 1994; McMillan & Schumacher, 1997).

WHAT STUDENTS, EDUCATORS AND PARENTS SAY ABOUT CYBER-BULLYING

Students

Where Does Cyber-Bullying Occur?

Since youth are very selective when engaging in popular online forums, determining their trends and behaviours and the media they choose for cyber-bullying, is important for policy-makers. In our first study (Cassidy et al. 2009), we discovered that the most common vehicle for cyber-bullying was chat rooms (53% of participants chose this option), while 37% said through emails or MSN, and only 7% said text messages. These results are similar to what Patchin and Hinduja (2006) found during the same time period; 384 respondents under age 18 reported that cyber-bullying most often occurred in chat rooms, but was less prevalent when texting or emailing. In our second study of 339 youth from the same age group, social networking sites like Facebook and MySpace had gained popularity among youth, such that this vehicle was the choice venue for cyber-bullying (52% of respondents), followed by email and MSN at 32%, and text messaging at 2.5%. Chat rooms were only recorded at 12%. This trend may change, of course, as new technology emerges and youth engage in new online agencies. This also underscores the importance of up-to-date research on this topic.

A disquieting conundrum among educators and parents is whether cyber-bullying starts at home or at school and the connection between the two domains. The findings from our first study (Cassidy et al. 2009) showed that 64% of youth claim that cyber-bullying is most likely to start at school and then continue at home, meaning that an incident at school precipitated the negative on-line exchange on the home computer. In this regard, findings from our second study are consistent with the first study, with 65% of respondents reporting that cyber-bullying starts at school. Schools typically place tighter controls and monitoring on computer use than do parents, although these controls are often sporadic and inconsistent, as demonstrated by the Valcke, Schellens, Van Keer and Gerarts (2007) study, which showed that approximately 50% of students found the controls to be sporadic, with youth in lower grades (Grade Three) admitting to significantly fewer computer controls than students in the higher grades (Grades Five and Six). If the research continues

to demonstrate that cyber-bullying on the home computer is a reaction to incidents that happened at school, this raises the issue of whether victims of schoolyard face-to-face bullying may become empowered on the home computer and become cyber-bullies. Conversely, it is also possible that students who are victims of face-to-face bullying at school will continue to be victimized in online ways through the home computer (Brown, et al., 2006) - a form of "double jeopardy" as described by Raskauskas and Stotz (2007).

Cyber-Bullying Victims

Online bullying can cause victims to experience a myriad of psychosocial effects ranging from depression, low self-esteem, anger, school absenteeism, poor grades, anxiety, and a tendency towards suicidal thoughts or suicide (Brown et al. 2006; Willard, 2006; Ybarra and Mitchell, 2008). The negative impact of the written word or a posted picture or video in cyberspace can be far-reaching and long-term, especially if the perpetrators are peers or someone they know (Ybarra and Mitchell, 2008). Victims often repeatedly re-visit the postings, causing continual re-victimization (Brown et al. 2006). Given the negative psychosocial effects of online victimization, educators and parents must be vigilant in reviewing possible academic and personal anomalies that may occur after an incident. School attendance and performance may suddenly drop or physical/mental ailments may occur (Bhat, 2008). Rigby (2005) noted that children who are cyber-bullied may be more prone to diminished levels of mental health. Bhat (2008) confirmed that victims tend to internalize feelings of anxiety, loneliness and depression, which can result in disengagement from school and peer relationships. One counselor, who was interviewed our second study, expressed concern about an adolescent male who was the victim of a lunch-hour school prank, conceived and videotaped by classmates and posted on YouTube. This student had totally disengaged from school activities and classmates after the incident, especially during the lunch hour. Ybarra and Mitchell (2008) reported a higher likelihood of youth who have been victimized online to bring a weapon (e.g. gun) to school.

A school youth worker from our second study, who teaches sex education and drug education, said in her interview that she finds Facebook scary, and gave the example of a fifteen-year-old youth attending a party, drinking alcohol or taking drugs, and then participating in sexual activity. The next day the sexual encounter is posted on Facebook because some reveler at the party photographed the event. In her view, this type of unfortunate exposure could ruin the adolescent's personal life as well as impact future academic opportunities (such as scholarships or university entrance), or, in the longer term, employment prospects. Dowell et al. (2009) confirmed that university personnel and potential employers have begun to review postings on the Internet (especially Facebook) for possible (mis)behaviours that might impact future employment or admissions of applicants.

Cyber-Bullying as Relational Aggression among Youth

Ybarra and Mitchell (2008) posit that youth who are cyber-bullied are often targeted by someone they know. In our first study, (Jackson et al. 2009b), we examined the cyber-bullying acquaintanceship issue through a theoretical relational aggression lens. That is, we sought to determine whether cyber-bullying could be considered a form of relational aggression among youth. Relational aggression is defined by Crick et al. (1999) as "behaviours that harm others through damage (or threat of damage) to relationships or feelings or acceptance, friendship, or group inclusion" (p. 177). Examples in the literature include spreading rumors with the intent to harm others or to socially exclude them through covert gossip or rumor

(Crick and Grotpeter, 1995). The technology of the Internet provides a perfect conduit to achieve these ends, and there are numerous examples cited in the cyber-bullying literature (Juvonen & Gross, 2008; Slonje & Smith, 2008).

Gender is an important variable to consider when examining cyber-bullying as relational aggression, given the observed preferences of girls to engage in covert gossip and rumor. Boys engage more in face-to-face bullying but girls employ covert tactics that affect social acceptance and friendships (Campbell, 2005). In analyzing the relationship between cyber-bullying and relational aggression (Jackson et al. 2009b), we found that slightly more girls (29%) than boys (21%) admit to engaging in cyber-bullying practices. This is consistent with the notion that girls will engage more often than boys in an activity, which can be defined as relational aggression. Further, almost 33 percent of the girls in our study reported that they have witnessed someone cyber-bullying online or on a cellular phone, as opposed to only 22 percent of the boys.

Living in fear or contemplating suicide after cyber-bullying are serious problems that educators, parents and clinicians must be prepared to handle. In our first study, (Cassidy et al. 2009), we found that 2.2 percent of boys and 1.6 percent of girls admit to being afraid as a result of cyber-bullying messages they had received, and another 4 percent confess to having suicidal thoughts ($n=365$). In the same study, approximately 40 percent of the respondents reported being a victim of cyber-bullying. Given the seriousness of the symptoms, medical personnel and school counselors/psychologists need to play a role in educating parents on how to reduce their children's risks for online victimization and to monitor their psychosocial health and online behaviour (Genuis and Genuis, 2005; Ybarra and Mitchell, 2008). Genuis and Genuis (2005) point out that health-care workers and clinicians must be prepared to become involved in the education of youth and their parents, because of their unique position that affords them the opportunity to interact with both groups. Similarly, Dowell et al. (2009) assert that school personnel need to be educated about cyberspace demeanors in order to better assess young people who may be at risk so they can intervene or refer youth to those who can help. In this vein, medical and psychologists needs to work collaboratively with schools in fostering better mentally and emotionally healthy young people.

Who are the Cyber-Bullies?

If cyber-bullying is to be effectively addressed, more needs to be known about who participates in this type of behaviour and whether there are some common characteristics among youth who cyber-bully. Ybarra and Mitchell (2004) recognized three cogent psychosocial antecedents: substance abuse issues; conventional bullying victimization; and delinquency. These researchers found that cyber-bullies are more likely to drink alcohol, smoke tobacco and engage in fights, and display significant academic under-achievement. Insubstantial caregiver-child relationships may also be a factor, with the same researchers demonstrating that 44 percent of cyber-bullies claim low emotional attachment to caregivers. They also suggested that frequent Internet use may lead to increased online abuse. Certainly the digital age is here to stay and most youth spend a significant amount of time online, primarily on home computers.

Although Ybarra and Mitchell (2004, 2008) found that these characteristics were present in those who cyber-bullied, in both our studies we found that this behaviour was far too prevalent in the population we surveyed to be only restricted to those with emotional, attachment, academic or psychosocial problems. We found that between one-quarter (first study) and one-third (second study) of students in Grades six to nine reported participating in cyber-bullying, and this behaviour

crossed socio-economic status, culture, gender and academic success. We did not, however, specifically collect data relating to attachment, psycho-social problems, delinquency, or substance abuse issues, but it is unlikely that these characteristics would be evident in all or even most of the youth we researched.

Bhat (2008) explained that young people may possess impulsive tendencies that make them inclined to entertain cyber actions in haste, without realizing the full ramifications of such behaviour. Thus, impulsivity in adolescence may underscore the lack of understanding or consideration for victims of their actions - a potential lack of insight into the full thrust of what they have done and the harm that befalls their victims. In our interviews with school personnel, they said that on a regular basis they were having to deal with students who had posted something online in an impetuous moment, only to be faced with the realization that the victim was devastated not only by the action itself but also by the magnitude of the audience that views it, be it strangers, friends or peers. The educators said that they reminded their students that their actions had short and long-term consequences; however, this admonition typically fell on deaf ears, and instances of cyber-bullying continued.

We found that age is also an important factor with cyber-bullying. In both studies, our findings indicate that cyber-bullying behaviour typically escalates between ages 13 and 14 and then diminishes as youth grow older. In our first study, between one-quarter and one-third of students aged 12 to 14 reported bullying others online versus those students aged eleven (17%) and fifteen (19%). The findings from our second study are consistent – 9 percent (aged 12), 34 percent (aged 13) and 35 percent (aged 14) occasionally cyber-bullied, with 2 percent (aged 13) and 7 percent (aged 14) often engaging in such behaviour. Younger and older students outside of this age range engaged less often.

Reporting Practices: To Whom do Youth Turn for Help?

If cyber-bullying occurs, to whom do the victims go to report the incident – parents, teachers, counselors, friends, police, or no one? Determining the answer to this question helps guide policymakers, educators and parents and sketches the possible isolation that many youth feel when targeted online. If youth are not confiding in adults or other professionals, then schools and parents maybe unaware of the extent of cyber-bullying, or downplay its impact, and victims may suffer in silence or confide only in peers who are not in a position to help with the long term effects of victimization. It also raises the important issue of bystander reporting, and how schools can work with peer groups in encouraging them to support their friends by informing adults about possible cyberspace infractions.

Studies show that youth are more likely to confide in friends rather than parents or educators (Patchin & Hinduja, 2006; Bhat, 2008). Our research (Cassidy et al. 2009) corroborated this, with 74% saying that would tell their friends, 57 percent would tell their parents, and only 47 percent would tell school officials. Almost no one would tell the police. We also found that students are more likely to report the cyber-bullying to school officials if they witnessed it than if they experienced it themselves. In our second study, approximately 27 percent of students would report cyber-bullying to school officials if they were victims as opposed to 40 percent who would report to a teacher, counselor or school administrator if they witnessed cyber-bullying taking place.

For those respondents who would not tell school personnel, 30 percent fear retribution from the cyber-bully (Jackson et al. 2009a). This robust response contravenes much of the current literature, which posits that youth are reluctant to report incidents to adults primarily out of fear that parents will limit or remove online computer time (Brown et al. 2006). Consonant with this finding

is Bhat's (2008) tragic recollection of Alex Teka, a young New Zealand girl who was the victim of cyber-bullying, who took her own life at the age of twelve because the cyber-bullying escalated once her mother complained to school authorities. Such findings impact the type of policies and practices educators need to develop in school to protect the victim from further retaliation if incidents are revealed to adults. Ybarra and Mitchell (2008) emphasize the need for funding initiatives such as online youth outreach programs and online mental health services which may assist youth who have been cyber-bullied.

Our study (Cassidy et al. 2009: 392) revealed other reasons why youth do not report cyber-bullying incidents to school personnel: the students see it as their problem not the responsibility of adults (29%); school personnel could not prohibit the bullying in any case (27%); friends could get into trouble if the cyber-bullying is revealed (26%); caregivers would restrict Internet access (24%), and other students would label them as "informers" (20%). As well, age is a paramount factor in reporting practices. Younger students are more likely to report incidents to school officials than those youth 14 years and older. Not surprisingly, as children enter puberty, confiding in adults (parents, school officials) is sometimes seen as the least viable option; they would rather turn to peers for support during their teenage years. Further, Williams and Guerra (2007) say that students tend to perceive adults as untrustworthy and schools as unsupportive. Unfortunately, in many cases, students may opt to remain silent, internally wrestling with the consequences of cyberspace victimization (Juvonen and Gross, 2008; Slonje and Smith, 2008).

Although there is a distinct gap in technological knowledge between youth and adults, youth need to be encouraged to discuss cyberspace problems with authorities, and adults need to reach out to youth to work collaboratively with them. Youth may fear that adults, coined digital immigrants by Prensky (2001), may not understand their digital culture and the philosophy of online life, so divulging confidential information may be hindered by this additional concern.

Students' Opinions about Cyber-Bullying

It is also vital to ascertain young people's opinions about cyberspace bullying in order to determine where they are coming from and any possible misconceptions they might have. Using a four-point Likert-type scale, we (Cassidy et al. 2009) asked our 365 respondents to rate certain statements, ranging from "strongly agree" to "strongly disagree". Responses to some questions were revealing. For example, almost 47 percent ("strongly agree", "agree") that freedom of expression is a right and online speech is border-less; this is despite Canadian legislation and the Canadian *Charter of Rights and Freedoms* that clearly defines limitations to freedom of expression (Shariff, 2006; Shariff and Gouin, 2005; Shariff and Johnny, 2007). Similar legislation exists in other democracies. Also of concern is that almost 50 percent ("strongly agree", "agree") that cyber-bullying is a normal part of the online world, and 32 percent ("strongly agree", "agree") that online bullying is less injurious than conventional face-to-face bullying, as cyber-bullying is merely "words in cyberspace" (Cassidy et al. 2009: 397). Also revealing is that almost three-quarters of the respondents indicate that cyber-bullying is more of a problem than prior years, suggesting that policy and practice initiatives undertaken by school districts and parents to counteract this growing problem are not working effectively to alleviate cyberspace transgressions. In this vein, more than 60 percent of respondents suggest that the solutions lie with young people who are far more familiar with technology than adults, again underscoring the tension between digital immigrants and digital natives (Brown et al. 2006; Prensky, 2001). Such admissions give credence to the view that policymakers, school officials and

parents should collaborate with youth on finding workable solutions.

Students' Solutions to Cyber-Bullying

When we asked our 11-15 year old participants (Cassidy et al. 2009) to choose the three best solutions for curtailing cyber-bullying behaviour from a list of ten options, the three most commonly selected solutions were (in the following ranked order):

1. Set up anonymous phone-in lines.
2. Develop programs to teach students about cyber-bullying and its effects.
3. Work on creating positive self-esteem in students.

In our second study, we asked a similar question. The results were as follows:

1. Develop programs to teach students about cyber-bullying and its effects (47%).
2. Set up anonymous phone-in lines (42%).
3. Focus on developing cyber-kindness instead of trying to stop cyber-bullying (40%).
4. Work on creating positive self-esteem in students (38%).

It is interesting that none of the most commonly selected choices had to do with punishment; rather students wanted to tell the authorities in confidence about what was happening, wanted curriculum on the topic, and saw the need for students to develop a better sense of self.

Educators: Their Views on Cyber-Bullying and Cyber-Kindness

In this digital age, it is important to probe educators' knowledge and experiences of cyber-bullying and their views on ways in which schools might promote, through the formal and informal curriculum, kinder, more caring and respectful online behaviours among youth. In our second study, we interviewed 17 educators to ascertain what is being done in their schools to address cyber-bullying and to foster "cyber-kindness." The results of these interviews are reported here.

Concerns about Cyber-Bullying

We used a recoded five-point Likert-style scale (1=not concerned at all; 5=extremely concerned), to determine the educators' level of concern about cyber-bullying. Surprisingly (given the level of cyber-bullying among youth), seven of out the seventeen educators (or about 42%) admitted that they were "not concerned at all" about the issue (M=3.6, SD=.9). However, when asked how important it was to prevent cyber-bullying compared to other competing priorities in the life of the school, almost 85 percent of these educators deemed it "extremely important", on a five-point scale between "not important at all" to "extremely important" (M=4.3, SD=.9). They stressed the need for teachers to learn new technology and proposed that the primary method for preventing cyber-bullying is education. It also was acknowledged that parents needed to be more involved in their children's online activities, particularly as many parents are not very knowledgeable about various forms of technology and, as Attewell, Suazo-Garcia and Battle (2003) point out, parents generally do not supervise their children's computer use. In our study, we found that almost 40 percent of youth have their computers in their bedrooms (Cassidy et al. 2010), thus preventing any real monitoring of behaviour.

Examples of Cyber-Bullying Provided by Educators

Each of the educators interviewed in our study was able to relay at least one story of a cyber-bullying incident that occurred among students at their school. One counselor confirmed that all of the bullying incidents at his school that year were cyber-oriented, not face-to-face, and that all involved girls – gossiping, backstabbing, excluding and criticizing. Lenhart, Madden, Macgill and Smith (2007) posited that girls are the primary bloggers, dominating the teen blogosphere, whereas boys are chiefly involved in video-sharing websites such as YouTube where they upload more so than girls who use these sites. Certainly the creation of blogs and Web pages are increasingly becoming more common among youth, as indicated in Chou et al.'s (2009) Taiwanese study; these sites could be used in positive ways instead of a vehicle for cyber-bullying. The examples of cyber-bullying among students reported by educators in our study showed both genders using Facebook to cyber-bully, with girls more likely to use words to communicate hurt and boys more likely to use visual media like pictures. However, as noted above, most of the cyber-bullying activity at their schools was happening under the radar of educators; only very few incidents were brought to their attention.

Here are three examples of cyber-bullying reported by educators:

1. A group of boys scattered spaghetti spray on another student, called the student names, videotaped the incident using a video camera and then posted it on Facebook. Once the video was uploaded to Facebook, other students made derogatory comments about the victim.

2. Educators at one secondary school were shocked when six or seven cyber-bullying incidents occurred within a month or so after school administrators conducted a school-wide seminar on cyber-bullying and the consequences of such behaviour. Administrators came to the conclusion that students were testing the school because one student on Facebook (who was bullying another student) admitted the idea originated

from the school presentation and he wanted to see if the school would actually follow through with its reprimand.

3. A video was posted on YouTube showing two students engaged in a physical fight off school property. The video also showed student bystanders watching the fight and cheering them on, some of whom were students in the school.

A number of educators in this study expressed concern that when parents were notified about their child's cyber-bullying adventures, several were defensive and showed surprise that their child was involved, even when it was clear that the negative behaviour occurred on the home computer. One vice-principal discussed the strong link between the home and school life vis-à-vis the Internet, and how cyber-bullying that starts at home negatively impacts the school milieu, and vice versa.

Teachers and Principals Being Cyber-Bullied

Several educators on our study expressed alarm at the growing propensity for teachers and principals to be the target of cyber-bullies. Indeed, in the student survey part of our study, 12 percent of the sample (*n*=334) admitted to personally participating in posting a mean, nasty, rude or vulgar message about a school official online. In addition to making reference to the "Rate Your Teacher" sites used by students, teachers provided examples such as the following:

1. Over the holiday season, a teacher was targeted on Facebook by a student for no apparent reason. It involved rude comments, swearing and name-calling. The teacher had set up a school Facebook account for a school challenge and the student created a fake Facebook account targeting the teacher.
2. Individual students created a group on Facebook to attack a teacher. This type of

behaviour then spread to different groups attacking other teachers on Facebook.

3. Students hid their cellular phones under their school desks and made fun of a teacher (hair, clothes, etc.) by secretly texting their friends. The teacher was oblivious to these interactions.

Indeed, media stories abound about educators being targeted online through Facebook, harrassed through polling sites or being blasted by pictures posted on YouTube (Froese-Germain, 2008). In a recent 2008 national poll conducted by the Canadian Teachers Federation, one in five teachers reported being victims of cyber-bullying (ATA News, 2009). This appears to be a growing phenomenon.

School Policies on Cyber-Bullying

Educators from the schools in both our studies confirmed that there was no specific cyber-bullying policy in place, and any problems that surfaced fell under the auspices of the face-to-face bullying policy, which tend to be district wide and described in the school students' handbook. One principal hoped that any policy concerning cyber-bullying would not be too specific so as to limit the options the principal could choose, and which could be tailored to the specific context.

Although it is important for adults to understand what youth are saying online, they also need to frame this understanding within youth culture and their styles of communication. Communication among youth is very different than that of the older generation. Wright and Lawson (2004) espouse the benefits of using social networking sites and other online tools as pedagogy in the classroom, as this is the form of communication youth use and this knowledge can be re-directed in positive ways in the classroom. In a Flemish study, however, 78 principals from primary schools were asked to describe the way technology was used in their classrooms. All but one principal listed rules, restrictions and monitoring placed on the use of computers to provide safe Internet use. Only one principal used technology as curriculum, incorporating a project that collaborated with teachers and students to focus on safe Internet use (Valcke et al. 2007).

Further, parents are often left out of the loop when it comes to developing policy involving social communication tools and Internet use. This restricts the ways in which they might help their children develop more positive ways to communicate online. In the end, parents and teachers, who are typically cyberspace foreigners, must strengthen their knowledge base when it comes to online forums and technology, as children and adolescents are fast becoming the experts in this area. In doing so, parents and teachers will be able to better understand the attitudes and philosophies that guide young people's Internet use (Chou et al. 2009). As it stands now, children in households are the computer experts, which Stahl and Fritz (2002) suggest, "disrupts the guiding role of parents" (p. 9). Ybarra and Mitchell (2008) suggest that schools re-focus their efforts from restricting online access in schools to employing mental health interventions for susceptible youth and cyberspace safety education that teaches youth positive and caring online communications.

Parents: Views, Roles, Supervision

Parents are key to preventing cyber-bullying and to addressing it when their children are victims or perpetrators, yet very little research has been done on parents' perceptions, understandings, or experiences. This is why we surveyed parents (n=315) in our second study, and linked these findings with what students were saying in their survey and what educators were saying in their interviews. We uncovered five important findings.

First, parents seriously underestimate the amount of time their children spend on line each day. From our parents' questionnaire (see Table 1), just under one-quarter of parents believe their child spends less than one hour online (*n*=303). Only 6 percent believe their child is online five hours or more. Alternatively, as shown in Table 2, the significance of variation between parents' estimation and actual usage by their child is evident, as 11 percent of students admit spending more than five hours a day online. Approximately 52 percent of parents estimate their children spend one to two hours per day online, which is fairly constant with the students' self-reports of 48 percent. The same can be said of Internet usage of three to four hours in which 15.5 percent of parents versus 16.8 of students self-report such involvement. However,

Table 1. (Parents) (recode) How many hours per day (on average) do you think your child spends online?

		Frequency	Percent	Valid Percent	Cumulative Percent
Valid	1-2 hours	163	51.7	53.8	53.8
	3-4 hours	49	15.6	16.2	70.0
	5-6 hours	13	4.1	4.3	74.3
	More than 6 hours	6	1.9	2.0	76.2
	Less than 1 hour	67	21.3	22.1	98.3
	None	5	1.6	1.7	100.0
	Total	303	96.2	100.0	
Missing	System	12	3.8		
Total		315	100.0		

Table 2. (Children) How often are you online?

		Frequency	Percent	Valid Percent	Cumulative Percent
Valid	not every day	107	31.6	31.6	31.6
	usually 1-2 hours per day	138	40.7	40.7	72.3
	usually 3-4 hours per day	57	16.8	16.8	89.1
	usually 5-6 hours per day	19	5.6	5.6	94.7
	usually more than 6 hours per day	18	5.3	5.3	100.0
Total		339	100.0	100.0	

at this point an inverse pattern begins to emerge (3 hours or more) in which the parents' estimation and student self-reports reverse, with far more parents underestimating the actual hours their child spends online. As the literature reveals, many parents may miscalculate the amount of time their child spends on the Internet, or are simply unaware of their child's computer usage (Liau et al. 2008).

Bjørnstad and Ellingsen (2004) also found the same result as did Liau et al. (2008). In some cases, a youth may spend hours online after the parent(s) go to bed, particularly in some immigrant communities where he or she is communicating with friends or family from a different time zone, or where the youth is involved in blogs, chat rooms, or global games. This miscalculation of computer use time is more serious when parents are not aware of what their children are doing on the computer. In our study, approximately 40 percent of parents either do not or only minimally supervise their child's time on the computer. This corroborates Lenhart and Madden's (2007) study, where they found that although some parents do invoke computer rules governing their child's Internet use, others feel disinclined to do so because they feel their child is safe or mature enough to handle specific circumstances. Even in cases where there is some degree of supervision or parameters around computer use in the home, this does not mean that youth adhere to the rules. It may be that rules are ignored or circumvented by youth. In any case, the supervisory techniques identified by parents in our study were broad in scope – from walking back and forth occasionally or walking into the room at regular intervals, installing filters/trackers, or checking the online history. A few parents used more invasive strategies, like prohibiting social networking and chat sites, removing Internet access from the home, or strictly monitoring computer time.

Second, parents' familiarity with technology fluctuated widely. Most parents (77%) were very familiar with older technology such as email and cellular phones but were ambiguous about weblogs (69%) and Facebook (65%). Overall, we found no significant relationship between the parents' degree of technological knowledge and their level of concern about cyber-bullying, but correlation analysis was significantly related ONLY with parents' knowledge of older technology (emails) ($p=0.025$; $p<.05$). Accordingly, there was no link between those parents who were more familiar with new technology such as weblogs and Facebook and those who expressed more concern about cyber-bullying and the potential dangers of online communications. As set out in Figure 1, approximately 20 percent of parents indicated that they were "not too concerned" about cyber-bullying while 44 percent were "quite concerned" ($n=276$, M=3.5, SD=1.19).

Third, we found no significant differences in the level of supervision and the level of concern about cyber-bullying ($x^2=8.145$, $p>.05$); the extent of supervision had no correlation with the level of concern parents had about cyber-bullying.

Figure 1.

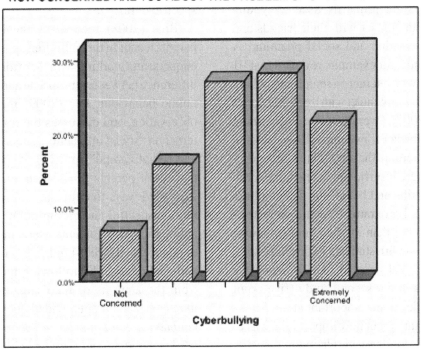

Fourth, we noted a disparity between what parents said about the extent to which their children were involved in cyber-bullying, and what the youth on the surveys reported. Approximately 10 percent of parents reported their child had been involved in some form of cyber-bullying, while in the student survey, 35 percent admitted to cyber-bullying and 32 percent of students reported being victimized. Seemingly parents are unsuspecting of their children's cyber-bullying experiences, or their children have kept them out of the knowledge loop.

Lastly, when we asked parents about cyber-bullying prevention, five themes emerged: tighten controls and restrict behaviour; focus on prevention, education and positive communication; unfamiliarity with cyber-bullying and possible solutions; holding the school responsible; and cyber-bullying is inevitable. Approximately 41 percent of respondents (*n*=108) said that the best way to prevent cyber-bullying was to impose strict

controls and punish if breached. However, 36 percent (*n*=95) offered a more holistic, contextual approach to solving cyber-bullying problems (discussed in the next section as this connects to the discussion on cyber-kindness). Ten parents reported that they were unfamiliar with the term "cyber-bullying" and its consequences, and therefore had no suggestions for solutions. Three parents felt the onus should be on schools to take the lead in thwarting cyber-bullying, while another three cynical parents admitted that cyber-bullying was inevitable and prevention was impossible.

CYBER-KINDNESS: CARING AND RESPECTFUL ONLINE RELATIONSHIPS

Electronic arenas are not all negative and can be a haven for positive discourse where youth seek a safe, nurturing environment for behaviours that promote social responsibility and encourage caring and respectful interactions. For example, social networking sites are primarily used by young people to keep in contact with their friends and engage in conversation and social planning. A Pew Internet study, for example, revealed that 91 percent of teens who interact on social networking sites do so to maintain contact with frequent friends (Lenhart et al. 2007). Overall, online engagement can create and further friendships (Kowalski and Limber, 2007). Some of the school-based interventions, however, are insufficient in teaching youth the necessary skills and behaviour when online, with many school programs focusing on restrictive methods rather than instructional strategies and collaboration with students to develop skills and knowledge (Valcke et al. 2007). When we asked educators in our second study (five-point Likert-style scale, 1=not important at all, 5=extremely important) about how important it was to encourage cyber-kindness in relation to the various competing priorities in the school, 65 percent ($n=17$, SD=1.03, M=4) claimed that this issue is an important one that needs to be developed and practiced in the classroom and in personal communications. Only one youth worker in the school did not feel that teaching about cyber-kindness was important at all.

Our findings from the parent survey revealed similar results. While most parents use the control method as a way to monitor their children's online behaviour, and seek to restrict or punish those who cross the line, many acknowledge that schools and families need to take a preventative approach and encourage and model positive online behaviour. Indeed, students in our earlier study (Cassidy et al. 2009), suggest that the solutions to cyber-bullying lie with addressing students' self-esteem and peer relationships and that education and modeling appropriate behaviours in the family and in the school are the most effective strategies. Ybarra and Mitchell (2008) claim that educators and other professionals should concentrate on

youths' behaviour as the best method for reducing online victimization rather than governing specific online sites.

Bhat (2008) recognizes the importance of parents teaching their children appropriate online etiquette and working with them to distinguish the difference between appropriate and inappropriate online behaviour. Lee (2009) makes the same observation, and discusses the impact of the Internet on "social interaction, social relationships, and social capital" (p. 510). Approximately 71 percent of parents ($n=224$) in our second study suggested ways to cultivate a more considerate, kind, respectful and caring online environment, although fifteen parents were mystified about the concept of cyber-kindness – its authenticity and usefulness. The following themes emerged from the analysis of their suggestions, offered in response to an open-ended question: *If cyber-kindness is the opposite of cyber-bullying, how might educators and parents help cultivate more considerate, kind, respectful and caring online interactions between youth today?"* - the importance of adults' modeling the right behaviour; the importance of the home environment; the need for classroom instruction and rewards in this area; and the importance of providing online examples to show what cyber-kindness might look like. Each of these suggestions is discussed in turn.

Modeling

The importance of modeling was by far the most prominent suggestion (55%, $n=124$). Parents were aware that all adults, including educators, needed to model acceptable behaviour in their own lives so that young people will see this behaviour and emulate it. For example, if adults are not acting kindly and respectfully in their own personal online communications, how can youth be expected to do so? Practicing and modeling the right values is a powerful moral educator; in fact, typically what is taught in the informal interactions of life are often more effective than what is said or admonished (Jackson, Hansen and Boostrom, 1995). Modeling and practice are key components for fostering the ethic of care, as articulated by philosopher and educator, Nel Noddings (2002, 2005) and others (Cassidy and Bates, 2005; Gregory, 2000; Rauner, 2000). Parents in our study understood that they needed to "practice what they preached", so their words were not empty admonitions. One parent stressed the need to make kindness and caring among youth the "cool" thing to do, so others would follow in this trendsetting behaviour. Another parent explained that if kindness and respect were prevalent in the real world, then perhaps such behaviour would spill over into the virtual world.

Indeed, the cyberspace realm offers much potential for positive initiatives. For instance, just over 50 percent of the four million (and growing) blogs are created by youth between the ages of 13 and 19, and this is an arena for youth and adults alike to share stories, voice their opinions, and record journals (Chou et al. 2009). Tynes (2007) explained that youth, especially those who may not have siblings or parents to help them with homework, often ask online friends for help with homework or advice about particular classes. Facebook, in particular, can be an invaluable resource of information and consultation.

Home Environment

In our study, parents emphasized that learning begins at home, and that parents had to assume more responsibility to teach and model respect for others. As one parent said: "it all starts in the home. If children do not see kindness and respect from siblings, friends and parents, the war is nearly lost." Tynes (2007) argued for parents to discuss the positive aspects of the Internet, rather than focus on the negative aspects. Problems that are inadequately addressed at home may predispose youth to vulnerable online activities. Unfortunately, as we discussed earlier, parents often lack the skills and knowledge about online technology;

this fact was acknowledged, however, by only two parents in our study who stated the need to become more technologically savvy in order to help their children. This lack of knowledge, however, should not stop parents from discussing and modeling respectful behaviours, so youth can recognize and maintain such actions in their online world.

Classroom Instruction and Resources

Parents presented the idea of schools developing homework assignments based on random acts of cyber-kindness, in collaboration with parents and peer groups at the school and at home. Other parents suggested that educators should provide examples of positive language that youth could practice online with a friend or a senior citizen such as a grandparent. In a similar vein, Bhat (2008) stressed the need for school counselors to take an active role in offering training to students and parents about "netiquette" standards of behaviour since many youth may not understand what are acceptable online expressions. Other parents suggested in-class activities such as role-playing, a cyber communication course and basic instructions on the ethics of Internet use and respect.

The website Net-Detectives (www.net-detectives.org) uses a role-play model that could be adapted for cyber-bullying challenges and as a way for students to identify with different perspectives using the role-play strategy (Wishart, Oades and Morris, 2007). An Internet safety advice website, Kidsmart (www.kidsmart.org.uk), is helpful for schools, parents and students and provides lesson plans and slides for parents. Bhat (2008) outlines a number of online programs and resources that educators might want to use or adapt for their classroom: for example,

i-SAFE (http://www.i-safe.org), Bullying, No Way! (http://www.bullyingnoway.com.au/), SAMHSA (http://www.samhsa.gov/index.aspx), Steps to Respect: A Bullying Prevention Program (http://www.cfchildren.org/programs/str/overview/), and Media Awareness Network (http://www.media-awareness.ca/english/index.cfm), have some specific resources on cyber-bullying for classroom use.

Contests and Rewards/ Online Examples

Some parents in our study thought that if students regularly saw examples of positive messages online they might be more likely to do the same. For example, each week the school could install a new screensaver on the school computers showing examples of positive online exchanges submitted by students in the school. Others thought rewards or contests might foster positive initiatives – rewarding students with I-tune dollars or celebrate kindness through an award system at school. Overall parents expressed their concern about how cyber-bullying seemed part of the wider culture of how youth related to one another, and because of this it required a deep and sustained effort on the parts of parents and educators.

TEACHERS: OPINIONS ABOUT CYBER-KINDNESS

In our second study, the 17 educators in their interviews were asked to give examples of times when students used technology to communicate positive, supportive, respective and kind messages to each other or to school personnel. They were also queried on ways in which they might incorporate cyber-kindness into the curriculum. When asked on a five-point *Likert-style scale* how important it is to teach cyber-kindness in relation to competing priorities, 65 percent ($n=11$, SD=1.03, M=4), stated that cyber-kindness is "important" or "very important."

Educators, though, had few examples of cyber-kindness to share, and even fewer sugges-

tions for fostering cyber-kindness. One counselor speculated that the majority of the interactions between youth online are positive. Another teacher offered an example of a student who initiated an act of kindness towards a classmate, by hooking up a camera in class to MSN and connecting with the student who was ill at home so he could follow the lesson. This same teacher involved his students in two charities, which operate online and provide resources and support to the needy. A vice-principal mentioned that one teacher was using a Wiki site for students to post positive annotations about their class discussions. The administrators at this school also used Wiki for teachers to share curriculum ideas and readings. This school also used Facebook to raise $75,000 for juvenile diabetes.

Another counselor who had worked in the school system for 29 years viewed cyber-kindness as being practiced in the school environment when students helped each other with homework online or initiated correspondence with students who were new to the school. Several educators mentioned kind or considerate thank you emails or examples of students bolstering others' self-esteem by saying flattering comments about their appearance.

ENCOURAGING CYBER-KINDNESS

Suggestions from educators and parents for ways to encourage a kinder online world are in line with the current literature: educating teachers about social networking sites (Worthen, 2007); engaging parents and teachers in collaborative solutions (Beale and Hall, 2007); designing effective curriculum (Sharples et al. 2009); modeling appropriate values and behaviours in the school and home (Noddings, 2002, 2005); and building trusting relationships with youth so that open and respectful dialogue can occur (Cassidy & Bates,

2005; Palmer, 1998). Our study of cyber-bullying and cyber-kindness, however, raises questions for further research.

For example, because technology is changing so rapidly and young people are on the cutting edge of the new media, what will be the impact of this increasing gap in knowledge between the adult generation and youth? Should young people be given the responsibility for setting up their own ways of monitoring online behaviour and also play a role in developing school curriculum for addressing the deeper issues as to why they and their peers cyber-bully? If cyber-bullying is so widespread among the youth culture, then researchers need to investigate why this is so. Is it because, as our study suggests, that youth see negative online exchanges as just a regular part of the online world and something to be tolerated? If so, then how can this perception be addressed? Further investigation also needs to be undertaken with respect to gender differences in cyber-bullying (Jackson et al. 2009b), and since there are differences in how girls verses boys cyber-bully perhaps there needs to be differential approaches according to gender for curtailing cyber-bullying. Further, additional studies need to be undertaken to assess whether there are differences in perceptions and experiences depending on culture, or social-economic status or other characteristics. Also, the new field of how to foster cyber-kindness is an area full of opportunity for further research.

CONCLUSION AND IMPLICATIONS

It is obvious that there is a generational gap between educators and youth regarding familiarity with technology. This restricts the ways in which teachers might creatively address issues of cyber-bullying and work towards cultivating a kinder online world. Educators' priorities seem more geared to curtailing cyber-bullying than to building a positive school culture of cyber-kindness. It is

apparent that few educators can provide examples of cyber-kindness and even fewer can suggest ways of fostering a more positive online world with their students. It is also apparent from the literature that schools do not provide sufficient opportunity for discussing cyber-bullying among stakeholders and for engaging youth in collaborative efforts to develop policies and practices for countering this growing problem. Nor are parents and other related professionals (medical staff, psychologists, counselors, police) invited to participate in the dialogue and to work with schools to develop effective measures and to deal with the aftermath of victimization (Tynes, 2007).

It is also evident that much of the cyber-bullying activity is happening under the radar of school staff and parents. Educators who participated in our two studies were only able to provide a few examples of cyber-bullying incidents that happened in their schools, yet up to 32 percent of students in these schools had been cyber-bullied in the last year and up to 36 percent had participated in cyber-bullying activities towards others. Also, 12 percent admitted to posting mean, nasty, rude or vulgar messages about a school official online, another aspect of cyber-bullying that is worrisome. Similarly parents were generally unaware of the extent of cyber-bullying being experienced or perpetrated by their children. Only thirty-three parents (approximately 10%) were aware that their child had been involved in various types of cyber-bullying activities, either as a victim or perpetrator, indicating a glaring discrepancy between what the parents know about and what is actually happening in their child's life. Parents, then, are generally uninformed about the extent of cyber-bullying being experiencing by their children. It is apparent that concerted and attentive dialogue about cyber-bullying involving all the major players in the lives of youth should occur and that stakeholders should work collaboratively to form long-term solutions that encourage kindness and caring in the online world.

REFERENCES

Attewell, P., Suazo-Garcia, B., & Battle, J. (2003). Computers and Young Children: Social Benefit or Social Problem? *Social Forces, 82*(1), 277–296. doi:10.1353/sof.2003.0075

Barron, C. (2000). *Giving Youth a Voice: A Basis for Rethinking Adolescent Violence.* Halifax, Nova Scotia: Fernwood Publishing.

Beale, A., & Hall, K. (2007). Cyberbullying: What school administrators (and parents) can do. *Clearing House (Menasha, Wis.), 81*(1), 8–12. doi:10.3200/TCHS.81.1.8-12

Bhat, C. (2008). Cyber Bullying: Overview and Strategies for School Counsellors Guidance Officers and All School Personnel. *Australian Journal of Guidance & Counselling, 18*(1), 53–66. doi:10.1375/ajgc.18.1.53

Bjørnstad and Ellingsen. (2004). *Onliners: A report about youth and the Internet.* Norwegian Board of Film Classification.

Brown, K., Jackson, M., & Cassidy, W. (2006). Cyber-bullying: Developing policy to direct responses that are equitable and effective in addressing this special form of bullying. *Canadian Journal of Educational Administration and Policy, 57.*

Campbell, M. (2005). Cyberbullying: An older problem in a new guise? *Australian Journal of Guidance & Counselling, 15*(1), 68–76. doi:10.1375/ajgc.15.1.68

Cassidy, W., & Bates, A. (2005). "Drop-outs" and "push-outs": finding hope at a school that actualizes the ethic of care. *American Journal of Education, 112*(1), 66–102. doi:10.1086/444524

Cassidy, W., Brown, K., & Jackson, M. (2010). *Redirecting students from cyber-bullying to cyber-kindness.* Paper presented at the 2010 Hawaii International Conference on Education, Honolulu, Hawaii, January 7 to 10, 2010.

Cassidy, W., Jackson, M., & Brown, K. (2009). Sticks and stones can break my bones, but how can pixels hurt me? Students' experiences with cyber-bullying. *School Psychology International, 30*(4), 383–402. doi:10.1177/0143034309106948

Chou, C., Yu, S., Chen, C., & Wu, H. (2009). Tool, Toy, Telephone, Territory, or Treasure of Information: Elementary school students' attitudes toward the Internet. *Computers & Education, 53*(2), 308–316. doi:10.1016/j.compedu.2009.02.003

Cook-Sather, A. (2002). Authorizing students' perspectives: Toward trust, dialogue, change in Education. *Educational Researcher, 24*(June-July), 12–17.

Crick, N., & Grotpeter, J. (1995). Relational aggression, gender, and social-psychological adjustment. *Child Development, 66*, 710–722. doi:10.2307/1131945

Crick, N., Werner, N., Casas, J., O'Brien, K., Nelson, D., Grotpeter, J., & Markon, K. (1999). Childhood aggression and gender: A new look at an old problem. In Bernstein, D. (Ed.), *Nebraska symposium on motivation* (pp. 75–141). Lincoln, NE: University of Nebraska Press.

Dowell, E., Burgess, A., & Cavanaugh, D. (2009). Clustering of Internet Risk Behaviors in a Middle School Student Population. *The Journal of School Health, 79*(11), 547–553. doi:10.1111/j.1746-1561.2009.00447.x

Froese-Germain, B. (2008). Bullying in the Digital Age: Using Technology to Harass Students and Teachers. *Our Schools. Our Selves, 17*(4), 45.

Genuis, S., & Genuis, S. (2005). Implications for Cyberspace Communication: A Role for Physicians. *Southern Medical Journal, 98*(4), 451–455. doi:10.1097/01.SMJ.0000152885.90154.89

Gradinger, P., Strohmeier, D., & Spiel, C. (2009). Traditional Bullying and Cyberbullying: Identification of Risk Groups for Adjustment Problems. *The Journal of Psychology, 217*(4), 205–213.

Green, H., & Hannon, C. (2007). *TheirSpace: Education for a digital generation.* Demos, London. Retrieved July 6, 2009 from: http://www.demos.co.uk/publications/theirspace

Gregory, M. (2000). Care as a goal of democratic education. *Journal of Moral Education, 29*(4), 445–461. doi:10.1080/713679392

Jackson, M., Cassidy, W., & Brown, K. (2009a). Out of the mouths of babes: Students "voice" their opinions on cyber-bullying. *Long Island Education Review, 8*(2), 24–30.

Jackson, M., Cassidy, W., & Brown, K. (2009b). *"You were born ugly and youl die ugly too": Cyberbullying as relational aggression. In Education (Special Issue on Technology and Social Media (Part 1),* 15(2), December 2009.http://www.ineducation.ca/article/you-were-born-ugly-and-youl-die-ugly-too-cyber bullying-relational-aggression

Jackson, P. W., Hansen, D., & Boomstrom, R. (1993). *The moral life of schools.* San Francisco: Jossey-Bass Publishers.

Juvonen, J., & Gross, E. F. (2008). Extending the school grounds? Bullying experiences in cyberspace. *The Journal of School Health, 78*(9), 496–505. doi:10.1111/j.1746-1561.2008.00335.x

Kowalski, R., & Limber, S. (2007). Electronic Bullying Among Middle School Students. *The Journal of Adolescent Health, 41*(6), S22–S30. doi:10.1016/j.jadohealth.2007.08.017

Kowalski, R. M., Limber, S. E., & Agatston, P. W. (2008). *Cyber bullying: Bullying in the digital age.* Malden, MA: Blackwell Publishers. doi:10.1002/9780470694176

Lee, S. (2009). Online Communication and Adolescent Social Ties: Who benefits more from Internet use? *Journal of Computer-Mediated Communication, 14*(3), 509–531. doi:10.1111/j.1083-6101.2009.01451.x

Lenhart, A., & Madden, M. (2007). Teens, Privacy & Online Social Networks: How teens manage their online identities and personal information in the age of MySpace. *Pew Internet & American Life Project.* Retrieved from: http://www.pewinternet.org

Lenhart, A., Madden, M., & Macgill, A. (2007). Teens and Social Media: The use of social media gains a greater foothold in teen life as they embrace the conversational nature of interactive online media. *Pew Internet & American Life Project.* Retrieved from: http://www.pewinternet.org

Li, Q. (2007). New bottle but old wine: A research of cyberbullying in schools. *Computers in Human Behavior, 23*(4), 1777–1791. doi:10.1016/j.chb.2005.10.005

Liau, A., Khoo, A., & Ang, P. (2008). Parental awareness and monitoring of adolescent Internet use. *Current Psychology (New Brunswick, N.J.), 27*(4), 217–233. doi:10.1007/s12144-008-9038-6

ATA News (2009). *RCMP and CTF join forces to fight cyberbullying, 43*(10), 6.

Noddings, N. (2002). *Educating moral people: A caring alternative to character education.* New York: Teachers College Press.

Noddings, N. (2005). *The challenge to care in schools: An alternative approach to education.* New York: Teachers College Press.

Owen, L., & Ennis, C. (2005). The ethic of care in teaching: An overview of supportive literature. *Quest, 57*(4), 392–425.

Palmer, P. J. (1998). *The courage to teach: Exploring the inner landscape of a teacher's Life.* New York: Jossey-Bass.

Palys, T. (2003). *Research Decisions: Quantitative and Qualitative Perspectives* (3rd ed.). New York: Thomson Nelson.

Patchin, J. W., & Hinduja, S. (2006). Bullies move beyond the school yard: A preliminary look at cyber bullying. *Youth Violence and Juvenile Justice, 4*(2), 148–169. doi:10.1177/1541204006286288

Prensky, M. (2001). Digital natives, digital immigrants. *Horizon, 9*(5), 1–6. doi:10.1108/10748120110424816

Raskauskas, J., & Stoltz, A. D. (2007). Involvement in traditional and electronic bullying among adolescents. *Developmental Psychology, 43*(3), 564–575. doi:10.1037/0012-1649.43.3.564

Rauner, D. (2000). *They still pick me up when I fall: The role of caring in youth development and community life*. New York: Columbia University Press.

Rigby, K. (2005). What children tell us about bullying in schools. *Children Australia Journal of Guidance and Counselling, 15*(2), 195–208. doi:10.1375/ajgc.15.2.195

Rosen, L., Cheever, N., & Carrier, L. (2008). The association of parenting style and child age with parental limit setting and adolescent MySpace behavior. *Journal of Applied Developmental Psychology, 29*(6), 459–471. doi:10.1016/j.appdev.2008.07.005

Shariff, S. (2006). Cyber-Dilemmas: Balancing Free Expression and Learning in a Virtual School Environment. *International Journal of Learning, 12*(4), 269–278.

Shariff, S. (2008). *Cyber-bullying: Issues and solutions for the school, the classroom and the home*. New York: Routledge.

Shariff, S., & Gouin, R. (2005). *Cyber-Dilemmas: Gendered Hierarchies, Free Expression and Cyber-Safety in Schools*.paper presented at Oxford Internet Institute, Oxford University, U.K. International Conference on Cyber-Safety. Paper available at: www.oii.ox.ac.uk/cybersafety

Shariff, S., & Johnny, L. (2007). Cyber-Libel and Cyber-bullying: Can Schools Protect Student Reputations and Free Expression in Virtual Environments? *Education Law Journal, 16*(3), 307.

Sharples, M., Graber, R., Harrison, C., & Logan, K. (2009). E-safety and Web 2.0 for children aged 11-16. *Journal of Computer Assisted Learning, 25*(1), 70–84. doi:10.1111/j.1365-2729.2008.00304.x

Slonje, R., & Smith, P. (2008). Cyberbullying: Another main type of bullying? *Scandinavian Journal of Psychology, 49*(2), 147–154. doi:10.1111/j.1467-9450.2007.00611.x

Smith, P., Mahdavi, J., Carvalho, M., Fisher, S., Russell, S., & Tippett, N. (2008). Cyberbullying: its nature and impact on secondary school pupils. *Journal of Child Psychology and Psychiatry, and Allied Disciplines, 49*(4), 376–385. doi:10.1111/j.1469-7610.2007.01846.x

Stahl, C., & Fritz, N. (2002). Internet safety: adolescents' self-report. *The Journal of Adolescent Health, 31*(1), 7–10. doi:10.1016/S1054-139X(02)00369-5

Sun, P., Unger, J., Palmer, P., Gallaher, P., Chou, P., & Baezconde-Garbanati, L. (2005). Internet accessibility and usage among urban adolescents in southern California: Implications for web-based health research. *Cyberpsychology & Behavior, 8*(5), 441–453. doi:10.1089/cpb.2005.8.441

Tynes, B. (2007). Internet safety gone wild? Sacrificing the educational and psychosocial benefits of online social environments. *Journal of Adolescent Research, 22*(6), 575–584. doi:10.1177/0743558407303979

Valcke, M., Schellens, T., Van Keer, H., & Gerarts, M. (2007). Primary school children's safe and unsafe use of the Internet at home and at school: An exploratory study. *Computers in Human Behavior, 23*(6), 2838–2850. doi:10.1016/j.chb.2006.05.008

Willard, N. (2006). Flame Retardant. *School Library Journal, 52*(4), 55–56.

Willard, N. (2007). The authority and responsibility of school officials in responding to cyberbullying. *The Journal of Adolescent Health, 41*(6), 64–65. doi:10.1016/j.jadohealth.2007.08.013

Williams, K., & Guerra, N. (2007). Prevalence and Predictors of Internet Bullying. *The Journal of Adolescent Health, 41*(6), S14–S21. doi:10.1016/j.jadohealth.2007.08.018

Wishart, J., Oades, C., & Morris, M. (2007). Using online role play to teach internet safety awareness. *Computers & Education, 48*(3), 460–473. doi:10.1016/j.compedu.2005.03.003

Worthen, M. (2007). Education policy implications from the expert panel on electronic media and youth violence. *The Journal of Adolescent Health, 41*(6), 61–62. doi:10.1016/j.jadohealth.2007.09.009

Wright, E. R., & Lawson, A. H. (2004). *Computer-mediated Communication and Student Learning in Large Introductory Sociology Courses*. Paper presented at the Annual Meeting of the American Sociological Association, Hilton San Francisco & Renaissance Parc 55 Hotel, San Francisco, CA. Retrieved from: http://www.allacademic.com/meta/p108968_index.html

Ybarra, M., Diener-West, M., & Leaf, P. (2007). Examining the overlap in Internet harassment and school bullying: Implications for school intervention. *The Journal of Adolescent Health, 41*(6), 42–50. doi:10.1016/j.jadohealth.2007.09.004

Ybarra, M., & Mitchell, K. (2004). Youth engaging in online harassment: associations with caregiver-child relationships, Internet use and personal characteristics. *Journal of Adolescence, 27*(3), 319–336. doi:10.1016/j.adolescence.2004.03.007

Ybarra, M., & Mitchell, K. (2008). How Risky are Social Networking Sites? A comparison of places online where youth sexual solicitation and harassment occurs. *Pediatrics, 121*(2), 350–357. doi:10.1542/peds.2007-0693

ENDNOTES

[1] Parents and other primary caregivers such as grandparents, legal guardians, etc. will be referred to here collectively as "parents."

Chapter 7
Online Empathy

Niki Lambropoulos
Intelligenesis Consultancy Group, UK

BACKGROUND AND HYPOTHESIS

Scientific research on empathy started in the early 20th century. Only in 1992 did the development of cognitive neuroscience help di Pellegrino, Fadiga, Fogassi, Gallese and Rizzolatti to identify the mirror neurons related to representations of an Object from a Subject, verifying Lipps' (1903) and McDugall's (1908) suggestions on empathy. Primary empathy is related to the automatic matching of the feelings of the other person (Fischer, 1980). An example is the relationship newborns have between each other on their first days on the earth. Another verification of mirror neurons was made by Rizzolati and Arbib (1998), as well as identification of the areas where the mirror neurons are located, interacting with areas in both hemispheres (Broca area 44 and PE/PC).

Hiltz and Turrof (1978) referred to members' comments on their closest friends, whom seldom or never see each other face to face. Preece and Ghozati (2001) made the first serious attempt to search and analyze empathy in online communities as well as understand it better towards sociability and usability. They used the process-based model of knowing, feeling and responding compassionately for *distress* by Levenson and Ruef

(1992), and the results showed that empathy is widespread in communities. In 2002, Preston and de Waal presented their Perception-Action Model (PAM), a process-based suggestion on empathy. PAM states that:

attended perception activates subject's representations of the state, situation and object, and that activation of this representation automatically primes or generates the associated autonomic and somatic responses, unless inhibited (p. 4).

PAM also relates empathy to the levels of awareness, reconciliation and vicarious learning as well as effortful information processing. The latter, and the mirror neurons discovery, suggest that empathy can be taught and learned as it creates symmetries between the Subject and the Object, activating the primary empathy in human perception. Self-awareness leads to self-directed behaviour, then empathy arousal and, as such, arousal of shared intentions, feelings and thoughts for common goals, desires and beliefs for community building. Eslinger, Moll and de Oliveira-Souza (2002) are among the first neuroscientists to search for Subjects' empathy from written text. They found that text judging showed

DOI: 10.4018/978-1-4666-2803-8.ch007

different human brain pattern activation, strongly influenced by emotional experience of the text due to reasoning and judgment.

As such, the hypothesis was the following: If empathic members are sensitive organs who have the ability to simulate members' common visions, needs and suggestions (Goleman, McKee & Boyatzis, 2002), they could be detected on the Internet, form a group and be mediators or messengers between authorities and the public.

THE STUDIES

In the first study, conducted in 2003, 13 individuals accepted to participate in a discussion forum on Peace and War before, during and after the invasion in Iraq (March 10 through August 23, 2003; the invasion was conducted on March 20). The subjects were from 16 to 48 years old and came from Canada, India, Greece and the U.K. They had to read the messages in the forum and keep notes simultaneously for three weeks in a self-observatory way. Then, they had to answer semi-structured interviews and hand in their notes. The results suggested that all members were initially open to other members. The second week, two respondents developed empathy (15.3%), as they reported identification of other members' profiles, writing styles and similarities in feelings derived from the text. The same members decided to reply to the forum, which indicated that 100% of the members who developed empathy were activated.

In the second study, conducted July 2004, a focus group of 28 online community managers discussed active participation and groupz-ware. The Social Network Analysis (SNA) on social behaviour and interaction using Netminer as the research software and Content Analysis on textual communication using ATLAS.ti revealed that the most empathic members gave important insights to the discussion. In addition, specific message structures appeared from empathic members who followed PAM.

DISCUSSION

The mechanism of reading and understanding others' meaning from a text leads to emotional contagion and motoric responses exhibited as online engagement. PAM showed that empathy has social consequences on community building, relating activation as a motoric response to empathy. Familiarity and similarity increase the levels of interpersonal trust (Feng, Lazar, & Preece, 2004), which is very important for rising and, as such, predictability of members' actions could facilitate towards goals and common visions. The levels of empathy are increased when based on empathic members' accessing and assessing members' states, which direct them to initiate or terminate actions for the sake of the community. Interpersonal trust is the key for developing empathy in order to cross the red line of inactive participation to energetic engagement. SNA research could bring the actors on stage and suggest the individuals who are able to help the community.

CONCLUSION

The results indicated that online empathy exists; in addition, it helps members to construct roles as in Community of Practice. The members identify and define roles for themselves and roles for the other members in a self-organized, organic way. If we use the Internet and online communities to reach our targets as citizens in an active society, then we incorporate the basic qualities of eDemocracy. We suggest the wide use of focus groups in order to gather and assess online communities' members suggestions and identification of people who are able to help our communities, either as volunteers or, even better, as our representatives in public affairs.

REFERENCES

di Pellegrino, G., Fadiga, L., Fogassi, L., Gallese, V., & Rizzolatti, G. (1992). Understanding motor events: A neurophysiological study. *Experimental Brain Research*, *91*, 176–180.

Eslinger, P. J., Moll, J., & de Oliveira-Souza, R. (2002). Emotional and cognitive processing in empathy and social behaviour. *The Behavioral and Brain Sciences*, *25*, 34–35.

Feng, J., Lazar, J., & Preece, J. (2004). (in press). Empathic and predictable communication influences online interpersonal trust. *Behaviour & Information Technology*.

Fischer, K. W. (1980). A theory of cognitive development: The control and construction of hierarchies of skills. *Psychological Review*, *87*, 477–531. doi:10.1037/0033-295X.87.6.477

Goleman, D., McKee, A., & Boyatzis, R. E. (2002). *Primal leadership: Realizing the power of emotional intelligence*. Harvard, MA: Harvard Business School Press.

Hiltz, S. R., & Turoff, M. (1978). *The network nation: Human communication via computer*. Reading, MA: Addison Wesley.

Levenson, R. W., & Ruef, A. M. (1992). Empathy: A physiological substrate. *Journal of Personality and Social Psychology*, *63*(2), 234–246. doi:10.1037/0022-3514.63.2.234

Lipps, T. (1903). Einfühlung, Innere Nachahmung und Organempfindung. *Archiv für die Gesamte Psychologie*, *1*, 465–519.

McDougall, W. (1908). *An introduction to social psychology*. London: Methuen. doi:10.1037/12261-000

Peace and War in Taking IT Global. (2004). Retrieved November 5, 2004, from www.takingitglobal.org/

Preece, J. (2000). *Online communities: Designing usability, supporting sociability*. Chichester, UK: John Wiley & Sons.

Preece, J., & Ghozati, K. (2001). Observations and explorations of empathy online. In Rice, R. R., & Katz, J. E. (Eds.), *The Internet and health communication: Experience and expectations* (pp. 237–260). Thousand Oaks, CA: Sage Publications. doi:10.4135/9781452233277.n11

Preston, S. D., & de Waal, B. M. (2002). Empathy: Its ultimate and proximate bases. *The Behavioral and Brain Sciences*, *25*, 1–72.

Rizzolatti, G., & Arbib, M. A. (1998). Language within our grasp. *Trends in Neurosciences*, *21*, 188–194. doi:10.1016/S0166-2236(98)01260-0

Wenger, E. (1998). *Communities of practice: learning, meaning and identity*. Cambridge: Cambridge University Press.

KEY TERMS AND DEFINITIONS

Empathy: Matching other persons' feelings. There is a distinction between Empathy and projection; the direction of matching the feelings is opposite. In empathy, the Subject moves towards the Object of observation; whereas in projection; the Subject projects his/her own feelings to the Object, acquiring a false image of the Object.

Energetic Engagement: Active participation that refers to members' ability to suggest changes on the policies, the structure and the environment/system.

Online Communities of Interest: Online groups that grow from a common interest in a subject. They develop norms based on shared values and meanings.

Online Communities of Practice: Online groups that grow from common professional and specific practices in a subject. The newcomers engage in CoP via legitimate peripheral participation based on the old "master and disciple" relationship. Observation is the means of acquiring knowledge, as the community is based on specific practices such as mimicry, demonstration and collaborative work (Wenger, 1998).

Sociability: Sociability is concerned with the collective purpose of a community, the goals and roles of the individuals in a community and policies generated to shape social interaction (Preece, 2000).

Social Network Analysis: Depicts the communication and relationships between people and/or groups through diagrams based on social relationships between a set of actors.

Usability: A measure of quality of a user's experience when interacting with a product or a system. It is described by the ease of learning, efficiency of use, memorability, error frequency and severity, and subjective satisfaction.

This work was previously published in Encyclopedia of Virtual Communities and Technologies, edited by Subhashish Dasgupta, pp. 346-348, copyright 2006 by Idea Group Reference (an imprint of IGI Global).

Chapter 8
Politeness as a Social Computing Requirement

Brian Whitworth
Massey University, New Zealand

Tong Liu
Massey University, New Zealand

ABSTRACT

This chapter describes how social politeness is relevant to computer system design. As the Internet becomes more social, computers now mediate social interactions, act as social agents, and serve as information assistants. To succeed in these roles computers must learn a new skill—politeness. Yet selfish software is currently a widespread problem and politeness remains a software design "blind spot." Using an informational definition of politeness, as the giving of social choice, suggests four aspects: 1. respect, 2. openness, 3. helpfulness, and 4. remembering. Examples are given to suggest how polite computing could make human-computer interactions more pleasant and increase software usage. In contrast, if software rudeness makes the Internet an unpleasant place to be, usage may minimize. For the Internet to recognize its social potential, software must be not only useful and usable, but also polite.

INTRODUCTION

Social Computing

Computers today are no longer just tools that respond passively to directions or input. Computers are just as mechanical as cars, but while a car inertly reflects its driver's intentions, computers now ask questions, request information, suggest actions, and give advice. Perhaps this is why people often react to computers as they would to a person, even though they know it is not (Reeves & Nass, 1996). Miller notes that if I accidentally hit my thumb with a hammer, I blame myself not the hammer, yet people may blame an equally mechanical computer for errors they initiate (Miller, 2004). Software it seems, with its ability to make choices, has crossed the threshold from inert machine to interaction participant as the term human-computer interaction (HCI) implies. Nor are computers mediating a social interaction, like e-mail, simply passive, as the software, like a facilitator, affects the so-

DOI: 10.4018/978-1-4666-2803-8.ch008

cial interaction possibilities (Lessig, 1999). As computers evolve, people increasingly find them active collaborators and participators rather than passive appliances or media. In these new social roles, as agent, assistant, or facilitator, software has a new requirement—to be polite.

To treat machines as people seems foolish, like talking to an empty car, but words seemingly addressed to cars on the road are actually to their drivers. While the cars are indeed machines, their drivers are people. Likewise, while a computer is a machine, people "drive" the programs interacted with. Hence, people show significantly more relational behaviours when the other party in computer mediated communication is clearly human than when it is not (Shectman & Horowitz, 2003), and studies find that people do not treat computers as people outside the mediation context (Goldstein, Alsio, & Werdenhoff, 2002)—just as people do not usually talk to empty cars. Reacting to a software installation program as if to a person is not unreasonable if the program has a social source. Social questions like: "Do I trust you?" and "What is your attitude to me?" now apply. If computers have achieved the status of semi-intelligent agents, it is natural for people to treat them socially, and thus expect politeness.

A social agent is taken as an interacting entity that represents another social entity in an interaction, either person or group, for example, if an installation program represents a company (a social entity), the installation program is a social agent, if it interacts with the customer on behalf of the company. The interaction is social even if the social agent is a computer, and an install creates a social contract even though the software is not a social entity itself. In the special case where a software agent is working for the party it is interacting with, it is a software assistant, working both for the user and to the user. In such cases of human-computer interaction (HCI), social concepts like politeness apply.

If software can be social it should be designed accordingly. A company would not let a socially ignorant person represent it to important clients. Yet, often, today's software interrupts, overwrites, nags, changes, connects, downloads, and installs in ways that annoy and offend users (Cooper, 1999). Such behaviour is probably not illegal, but it is certainly impolite.

Selfish Software

The contrast to polite software is "selfish software." Like a selfish person who acts as if only he or she exists, so selfish software acts as if it were the only application on your computer. It typically runs itself at every opportunity, loading at start-up and running continuously in the background. It feels free to interrupt you any time, to demand what it wants, or announce what it is doing, for example, after installing new modem software, it then loaded itself on every start-up and regularly interrupted me to go online to check for updates to itself. It never found any, even after many days, so finally after yet another pointless "Searching for upgrades" message I (first author) decided to uninstall it. As in *The Apprentice* TV show, one reaction to assistants that do not do what you want is: "You're fired!"

Selfish software is why after 2-3 years Windows becomes "old." With computer use, the Windows taskbar soon fills with icons, each an application that finds itself important enough to load at start-up and run continuously. Such applications always load, even if you never use them, for example, I never use Windows messenger but it always loads itself onto my taskbar. When many applications do this, it slows down the computer considerably, and taskbar icon growth is just the tip of the iceberg of what is happening to the entire computer. Because selfish programs put files wherever they like, uninstalled applications are not removed cleanly, and over time Windows accretes an ever increasing "residue" of files and registry records left-over from previous installs. Giving selfish applications too much freedom degrades performance until eventually only reinstalling the entire operating system can recover system performance.

Polite Computing

Polite computing is about how software design can support HCI politeness. It is not about how people should be polite to people online, which various "online etiquette" guides cover. This chapter aims to define, specify, and illustrate an information vision of polite computing.

Politeness is distinct from both usefulness and usability requirements. Usefulness addresses a system's functionality, while usability concerns how people use that functionality. The first focuses on what the computer does, and the second on how the user gets the computer to do that. Polite computing, however, is not about what the computer does, nor how one can better get it to do it. It is about social relations rather than computer power or cognitive ease. It enables software that "plays well" in a social setting and encourages users to do the same. It addresses the requirements for social interaction, enabling better social collaboration, rather than better tool use. The contexts differ, so software could be easy to use yet rude, or polite but hard to use. While usability reduces training and documentation costs, only politeness lets a software agent work with a competent user without frustration. Both usability and politeness, however, fall under the rubric human-centred design.

BACKGROUND

The Oxford English Dictionary (http://dictionary. oed.com) defines politeness as:

… behaviour that is respectful or considerate to others

Considering and respecting others, a critical success factor in physical society, is equally relevant to online society. The predicted effect of polite computing is better human-computer interactions. While one may mistrust a polite door-to-door salesman as much as an impolite one, the polite one will get more "air time" because interacting with them is more pleasant. If politeness makes social interaction more pleasant, a polite society is a nicer place to be than an impolite one, and its people will be more willing to interact beneficially with others. Polite computing can contribute to computing by:

1. Increasing legitimate interactions.
2. Reducing anti-social attacks.
3. Increasing synergistic trade.
4. Increasing software use.

There is nothing to stop programmers faking politeness, just as nothing stops people in the physical world from doing so, but when people behave politely, cognitive dissonance theory finds they also tend to feel more polite (Festinger, 1957). Likewise, if programmers design for politeness, the overall effect will be positive, even though some may pretend.

Politeness Supports Legitimate Interactions

Legitimate interactions, defined as those that are both fair and in the common good, have been proposed as the complex social source of civilized prosperity (Whitworth & deMoor, 2003) and a core requirement for any prosperous and enduring community (Fukuyama, 1992). Conversely, societies where win-lose corruption and conflicts still reign are among the poorest in the world (Transparency-International, 2001). Legitimate interactions offer all parties a fair choice and are in the public good, while anti-social interactions, like theft or murder, give the "victim" little choice and harm society overall. In contrast, polite acts are more than fair. To do as the law requires is not politeness precisely because it is required, for example, one does not thank a driver who stops at a red light, yet one thanks the driver who stops to let you into a line of traffic. While laws specify what citizens should do, politeness is about what they could do. If politeness involves offering more choices in an interaction than the law requires then it begins where fixed laws end. If criminal acts fall below the law, then polite acts rise above it, and polite, legitimate, and anti-social acts can be ordered by the degree of choice offered to the other party or parties (Figure 1). In this view politeness increases social "health," just as criminality poisons it.

Politeness Reduces Anti-Social Attacks

Polite computing may have value, but should not it take a back seat to security issues? Is politeness relevant if we are under attack? Yet upgrading security every time an attack exploits another loophole is a never-ending cycle. An alternative is to develop strategies to reduce motivation to attack (Rose, Khoo, & Straub, 1999). Politeness can help one common source of attacks—resentment or anger against a system where the powerful are perceived to predate the weak (Power, 2000). Often hacking is vengeance against a person, a company, or the capitalist society in general (Forester & Morrison, 1994). Politeness contradicts the view that since everyone takes what they can, so can I. That some people are polite and give choice to others, may cause those neutral to society to copy, or those against society to become neutral. Politeness and security seem two sides of the

same coin of social health. By analogy, a gardener defends his or her crops from weeds but does not wait for every weed to be killed before fertilizing. If politeness grows a better society, one should not wait to use it until every threat is purged. If security reduces anti-social acts, and politeness encourages social acts, they are complementary not mutually exclusive functions.

Politeness Increases Prosperity

Over thousands of years, as physical society became more "civilized," this has created enormous prosperity, so for the first time in history some economies now produce more food than their people can eat (as their obesity epidemics testify). The bloody history of humanity seems to represent a social evolution from zero-sum (win-lose) interactions, such as war, to non-zero sum (win-win) interactions, such as trade (Wright, 2001). Scientific research illustrates this social synergy, as for researchers to freely give their hard earned knowledge to all seems at first foolish, but when a critical mass do this, people gain more than they could have by working alone. Synergy means that when many people give to each other, they gain more than is possible by selfish activity. The success of the open source software (OSS) movement illustrates this, as open source products like Linux now compete with commercial products like Windows. The mathematics of social synergy are that while individual gains increase linearly with group size, synergy gains increase geometrically, as they depend on the number of interactions not the number of group members. The Internet illustrates social synergy, as we each only "sow" a small part of it, but from it can "reap" the world's knowledge interactions.

Politeness Increases Software Use

A study of reactions to a computerized Chinese word-guessing game found that when the software apologized after a wrong answer by saying "We

Figure 1. The social choice dimension

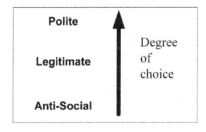

are sorry that the clues were not helpful to you" the game was rated more enjoyable than when the computer simply said "This is not correct" (Tzeng, 2004). Brusque and often incomprehensible error messages like the "HTTP 404—File not Found" response to an unavailable Web page can imply a user fault, while a message like: "Sorry I could not find file xxxxx." does not. Accusatory error messages can rub users the wrong way, especially if it is a software error in the first place.

In general, politeness improves the social interactions of a society, which makes it a nicer place to be. The reader can judge for him or herself whether the world wide Web is currently a nice place to be or whether its "dark side" which includes spam, spyware, viruses, hackers, pop-up ads, nagware, identity theft, solicitations, pornography, spoofers, and worms (Power, 2000), means it could benefit from polite computing. If software were more polite, people might be more willing to use it and less willing to abuse it.

AN INFORMATION DEFINITION OF POLITENESS

Reinventing Politeness Online

To apply politeness to computer programming, it must be defined in information terms. If politeness is "considering others," then since different societies "consider" differently, what is polite in one culture can be rude in another. Given no universal "polite behaviour," there seems no basis to apply politeness to the logic of programming. Yet while different countries have different laws, the goal of fairness that underlies the law can be attributed to every society (Rawls, 2001). Likewise, different cultures could have different "etiquettes" but a common goal of politeness. Figure 2 distinguishes the goals of Figure 1 from their specific implementations. In this view, while each society may implement a different etiquette, politeness remains the common design goal, just

Figure 2. Social goal vs. implementation

Goal	Implementation
Politeness	Etiquette
Legitimacy	Laws

as legitimacy is the spirit behind laws that vary in detail between societies.

If politeness can take different forms in different societies, to ask which implementation applies online is to ask the wrong question. The right question is how to "reinvent" politeness in each specific online case, whether for chat, wiki, e-mail, or other groupware. Just as each different physical society develop local etiquettes and laws, so different applications may need a different politeness implementation based on a general design "pattern," specifying politeness in information terms (Alexander, 1964).

Informational Politeness

If the person considered knows what is "considerate" for them, politeness can be defined abstractly as the giving of choice to another in a social interaction. Doing this is then always considerate if the other knows what is good for them, though the latter assumption may not always be true, for example, a young baby. In a conversation, where the locus of channel control passes back and forth between parties, it is polite to give control to the other party, for example, it is impolite to interrupt someone, as that removes their choice to speak, and polite to let them finish talking, as they then choose when to stop.

An information definition of politeness is:

... any unrequired support for situating the locus of choice control of a social interaction with another party to it, given that control is desired, rightful and optional. (Whitworth, 2005)

Unrequired means the choice given is more than required by the law, as a required choice is not politeness. Optional means the polite party has the ability to choose, as politeness must be voluntary. Desired by the receiver means giving choice is only polite if the other wants it. "After you" is not polite when facing a difficult task. Politeness means giving desired choices, not forcing the locus of control, with its burden of action, upon others. Finally, rightful means that consideration of someone acting illegally is not polite, for example, to considerately hand a gun to a serial killer about to kill, is not polite.

Other Definitions

Some define politeness as "being nice" to the other party (Nass, 2004) and argue that when another says "I think I'm a good teacher; what do you think?" polite people respond "You're great," even if they do not think so. In this view, agreeing with another's self praise is considered one of the "most fundamental rules of politeness" (Nass, 2004). Yet while agreeableness may often accompany politeness, it does not define it if one can be both agreeably impolite and politely disagreeable. One can politely refuse, beg to differ, respectfully object, and humbly criticize, that is, disagree but still be polite. Conversely, one can give charity to others yet be impolite, that is, be kind but rude.

Being polite is different from being kind, for example, kind parents may not give an infant many choices, but politeness does not apply to young children who are considered to not yet know what they really want. Do software creators consider software users to be like little children, unable yet to exercise choice properly? While inexperienced users may happily let software do as it thinks is best, when children grow up they want more choice (as teenagers illustrate). The view that "software knows best" is hard to justify for the majority of today's computer-literate users. Perhaps once computer users were child-like, but today they want respect and choices from their software.

Impolite Computing

Impolite computing has a long history. Spam, for example, fills inboxes with messages users do not want (Whitworth & Whitworth, 2004) and is impolite because it takes choice away from e-mail receivers. Pop-up windows are impolite, as they "hijack" the user's cursor or point of focus and take away the user choice of what they want to look at. Users do not like this, so many browsers prevent pop-ups. Impolite computer programs can:

1. Use your computer's services. Software can use your hard drive to store information cookies or your long distance phone service for downloads.
2. Change your computer settings. Like browser home page, e-mail preferences or file associations.
3. Spy on what you do online. Spyware, stealth-ware, or software back doors that gather information from your computer without your knowledge or record your mouse clicks as you surf the Web and, even worse, exchange your private information with others.

For example, Microsoft's Windows XP Media Player was reported to quietly record the DVDs it played and use the user's computer's connection to "phone home," that is, send data back to Microsoft (Editor, 2002). Such problems differ from security threats, where hackers or viruses break in to damage information. This problem concerns those invited into our information home, not those who break in, for example, "software bundling," where users choose to install one product but are forced to get many:

When we downloaded the beta version of Triton [AOL's latest instant messenger software], we also got AOL Explorer—an Internet Explorer shell that opens full screen, to AOL's AIM Today home page when you launch the IM client—as well as Plaxo Helper, an application that ties in with the Plaxo social-networking service. Triton also installed two programs that ran silently in the background even after we quit AIM and AOL Explorer (Larkin, 2005).

Likewise, Yahoo's "typical" installation of their IM also downloads their Search Toolbar, anti-spyware and anti-pop-up software, desktop and system tray shortcuts, as well as Yahoo Extras, which inserts Yahoo links on your browser. It also alters the users' home page and auto-search functions to point to Yahoo by default. Even Yahoo employee Jeremy Zawodny dislikes this:

I don't know which company started using this tactic, but it is becoming the standard procedure for lots of software out there. And it sucks. Leave my settings, preferences and desktop alone (http://jeremy.zawodny.com/blog/archives/005121.html).

A similar scheme is to use security updates to install new products, for example:

Microsoft used the January 2007 security update to induce users to try Internet Explorer 7.0 whether they wanted to or not. But after discovering they had been involuntarily upgraded to the new browser, they next found that application incompatibility effectively cut them off from the Internet (Pallatto, 2007).

Security cannot defend against people one invites in, especially if it is the security system taking advantage! However, in a connected and free society, social influence can be very powerful. In physical society the withering looks given to the impolite are not toothless, as what others think of you affects how they behave towards you. In old societies banishment was often considered worse than a death sentence. Likewise, what online

users think of a company that creates a software agent can directly impact sales. A reputation for riding roughshod over computer user's rights is not good for business.

SPECIFYING SOFTWARE AGENT POLITENESS

The widespread problem of software that is rude, inconsiderate, or selfish is a general software design "blind spot" (Cooper, 1999). The specification of politeness in information terms is in its infancy, but previous work (Whitworth, 2005) suggests polite software should:

1. Respect the other's rights. Polite software respects the user, does not pre-empt user choices, and does not act on or copy information without its owner's permission.
2. Openly declare itself. Polite software does not sneak or change things in secret, but openly declares what it does, who it represents, and how they can be contacted.
3. Help the other party. Polite software helps users make informed choices, giving useful and understandable information when needed.
4. Remember the interaction. Polite software remembers past user choices in future interactions.

Respectful

Respect includes not taking another's rightful choices. If two parties jointly share a resource, one party's choices can deny the other's; for example, if I delete a shared file, you can no longer print it. Polite software should not preempt rightful user information choices regarding common HCI resources like the desktop, registry, hard drive, task bar, file associations, quick launch, and other user configurable settings. Pre-emptive acts, like changing a browser home page without asking, act unilaterally on a mutual resource and so are impolite.

Information choice cases are rarely simple, for example, a purchaser can use the software but not edit, copy, or distribute it. Such rights can be specified as privileges, in terms of specified information actors, methods, and objects (Table 1). To apply politeness in such cases requires a legitimacy baseline, for example, a software provider has no right to unilaterally upgrade a computer the user owns (though the Microsoft Windows Vista End User License Agreement (EULA) seems to imply this). Likewise, users have no right to unilaterally upgrade, as this edits the product source code. In such cases politeness applies, for example, the software suggests an update and the user agrees, or the user requests an update and the software agrees (for the provider). Similarly, while a company that creates a browser owns it, the same logic means users own data they create with the browser, for example, a cookie. Hence, software cookies require user permission, and users should be able to view, edit, or delete "their" cookies.

A respectful assistant does not interrupt unnecessarily, while selfish software, like a spoilt child, repeatedly does, for example, Windows Update advises me when it starts, as it progresses, and when it finishes its update. Its modal window interrupts what I am doing, seizes the cursor, and loses my current typing. Since each

time Update only needs me to press OK, this is like being repeatedly interrupted to pat a small child on the head. The lesson of Mr. Clippy, that software serves the user not the other way around, seems still unlearned at Microsoft.

It is hard for selfish software to keep appropriately quiet, for example, Word can generate a table of contents from a document's headings. However, if one sends the first chapter of a book to someone, with the book's table of contents (to show its scope), every table of contents heading line without a page number loudly declares: "ERROR! BOOKMARK NOT DEFINED." This, of course, completely spoils the sample document impression, and even worse, this is not apparent until the document is received. Why could the software not just quietly put a blank instead of a page number? Why must it announce its needs so rudely? What counts is not what the software needs, but what the user needs.

Open

Part of a polite greeting in most cultures is to introduce oneself and state one's business. Holding out an open hand, to shake hands, shows that the hand has no weapon and that nothing is hidden. Conversely, to act secretly behind another's back, to sneak or to hide ones actions, for any reason, is impolite. Secrecy in an interaction is impolite because the other has no choice regarding things

Table 1. Socio-technical actors, objects, and methods

Actors	Objects	Methods
People	*Persona* (represent people)	*Create/Delete/Undelete*
Groups	*Containers* (contain objects)	*Edit/Revert*
Agents	*Items* (convey meaning)	*Archive/Un-archive*
	- *Comments* (dependent meaning)	*View/Hide*
	- *Mail* (transmit meaning)	*Move/Undo*
	- *Votes* (choice meaning)	*Display/Reject*
		Join/Resign
		Include/Exclude

they do not know about. Hiding your identity reduces my choices, as hidden parties are untouchable and unaccountable for their actions. When polite people interact, they declare who they are and what they are doing.

If polite people do this, polite software should do the same. Users should see who is doing what on their computer. However, when Windows Task Manager shows a cryptic process like CTSysVol. exe, attributed to the user, it could be system critical process or one left over from a long uninstalled product.

An operating system Source Registry could link all online technical processes to their social sources, giving contact and other details. "Source" could be a property of every desktop icon, context menu item, taskbar icon, hard drive file, or any other resource. A user could delete all resources allocated by a given source without concern that they were system critical. Windows messages could also state their source so users know who a message is from. Source data could be optional, making it backward compatible. Applications need not disclose themselves, but users will prefer sources that do. Letting users know the actions of their computer's inhabitants could help the marketplace create more polite software.

Helpful

A third politeness property is to help the user by offering understandable choices, as a user cannot properly choose from options they do not understood. Offering options that confuse us inconsiderate and impolite; for example, a course text Web site offers these choices:

- OneKey Course Compass
- Content Tour
- Companion Web site
- Help Downloading
- Instructor Resource Centre

It is unclear how the "Course Compass" differs from the "Companion Web site," and why both seem to exclude "Instructor Resources" and "Help Downloading." Clicking on these choices, as is typical for such sites, leads only to further confusing menu choices. The impolite assumption is that users enjoy clicking links to see where they go. Yet information overload is a serious problem for Web users, who have no time for hyperlink merry-go-rounds.

Yet to not offer choices at all, on the grounds that users cannot understand them, is also impolite. Installing software can be complex, but so is installing satellite TV technology. In both cases users expect to hear their choices in an understandable way. Complex installations are simplified by choice dependency analysis of how choices are linked, as Linux's installer does. Letting a user choose to install an application they want minus a critical system component is not a choice but a trap. Application-critical components are part of the higher choice to install or not, for example, a user's permission to install may imply access to hard drive, registry, and start menu, but not to desktop, system tray, favourites, or file associations.

Remember

Finally, it is not enough to give choices now but forget them later. If previous responses are forgotten, the user must redo them, which is inconsiderate. Hence, software that actually listens and remembers past user choices is a wonderful thing. Polite people remember previous encounters, yet each time I open Explorer it fills its preferred directory with files I do not want to see, then returns the cursor to me to select the directory I want to look at, which is never the one displayed. Each time, Explorer acts as if it were the first time I had used it, yet I am the only person it has ever known. Why can it not remember where I was last time and return me there? The answer is simply that it is impolite by design.

Such "amnesia" is a trademark of impolite software. Any document processing software could automatically open the user's last document and put the cursor where they left off, or at least give that option (Raskin, 2000). The user logic is simple: "If I close the file I am finished, but if not, put me back where I was last time." Yet most software cannot even remember what we were doing last time we met. Even within an application, like Outlook's e-mail, if one moves from inbox to outbox and back, it "forgets" the original inbox message and one must scroll back to it.

If a choice repeats, to ask the same question over and over, for the same reply, is to pester or nag like the "Are we there yet?" of children on a car trip. This forces the other party to again and again give the same choice reply, for example, uploading a batch of files creates a series of overwrite questions, and software that continually asks "Overwrite Y/N?" forces the user to continuously reply "Yes." Hence, most copy software also offers the "Yes to All" meta-choice that remembers for the choice set. Offering choices about choices (meta-choices) reduces information overload, as users need only set repeated access permissions once, for example:

1. Always accept.
2. Always reject.
3. Let me choose.

A general meta-choice console (GMCC) would give users a common place to see or set all meta-choices (Whitworth, 2005).

IMPLEMENTATION CASES

The Impolite Effect

In HCI interactions, impoliteness can cause a social failure every bit as damaging as a logic failure, for example, the first author's new 2006 computer came with McAfee Spamkiller, which when activated overwrote my Outlook Express mail server account name and password with its own values. When checking why I could no longer receive mail, I retyped in my mail server account details and fixed the problem. However, next time the system rebooted, McAfee rewrote over my mail account details again. The McAfee help person explained that Spamkiller was protecting me by taking control and routing all my e-mail through itself. To get my mail I had to go into McAfee and tell it my specific e-mail account details. That this did not work is less the issue than why this well known software:

1. Felt entitled to overwrite the e-mail account details a user had typed in.
2. Could not copy my account details, which it wrote over, to create its own account.

This same software also "took charge" whenever Outlook started, forcing me to wait as it did a slow foreground check for e-mail spam. Yet in 2 weeks of use, it never found any spam at all! I (first author) concluded it was selfish software, and uninstalled it.

Interaction Situations

Other human computer interactions where politeness applies include:

1. **Errors:** Polite error messages say we have an error rather than you have an error. While computers tend to take charge when things go well, when they go wrong, software seems to universally agree that the user is in fact "in charge." To ask what "we" (rather than you) want to do about an error implies the computer should also suggest solution options. Studies of users in human-computer tutorials show significant differences based on how politely the computer addresses the user, that is, users respond differently to "Click the Enter button" vs. "Let's click

the Enter button" (Mayer, Johnson, Shaw, & Sandhu, 2006).

2. **Advice and Notifications:** To interrupt impolitely disturbs the user's train of thought. For complex work, like programming, even short interruptions can cause a mental "core dump," as the user drops one thing to attend to another. The real interruption effect is then not just the interruption time, but also the user recovery time (Jenkins, 2006), for example, if a user takes three minutes to refocus after an interruption, a 1 second interruption every 3 minutes can reduce productivity to zero. Mr. Clippy, Office '97's paper clip assistant, had this problem, since as one user noted: "It wouldn't go away when you wanted it to. It interrupted rudely and broke your train of thought." (Pratley, 2004). Searching the Internet for "Mr. Clippy" gives comments like "Die, Clippy, Die!" (Gauze, 2003), yet its Microsoft designer wonders: "If you think the Assistant idea was bad, why exactly?" (Pratley, 2004). To answer simply, he was impolite, and in XP, is replaced by polite smart tags.

3. **Action Requests:** Asking permission is polite because it gives the other choice and does not pre-emptively act on a common resource, such as a zip extract product that puts the files it extracted as icons onto the desktop, without asking! Such software tends to be used only once.

4. **Information Requests:** If software asks for and gets choices from a user, it should remember them. Polite people do not ask "What is your name?" every time they meet, yet software often has no interaction memory whatsoever, for example, when reviewing e-mail offline in Windows XP, actions like using Explorer trigger a "Do you want to connect?" request every few minutes. No matter how often one says "No!" it keeps asking, because the software has no interaction memory.

5. **Installations:** Installation programs are notorious for pre-emptive acts, for example, the Real-One Player adds desktop icons and browser links, installs itself in the system tray, and can commandeer all video and sound file associations. Customers resent such invasions, which while not illegal, are impolite. An installation program changing your PC settings is like furniture deliverers rearranging your house because they happen to be in it. Software upgrades continue the tradition, for example, Internet Explorer upgrades that make MSN your browser home page without asking. Polite software does not do this.

Online Learning

Online learning software, like WebCT or Blackboard, illustrates how politeness issues vary with channel type. While channel richness (rich vs. lean) was once thought the main property of computer-mediated communication (Daft & Lengel, 1986), channel properties like linkage (one-to-one, one-to-few or one-to-many) and interactivity (one-way or two-way) now also seem relevant (Whitworth, Gallupe, & McQueen, 2001). For example, instructor-student online communications, like e-mail, text messaging, chat, podcasts, cell phone, or video-computer interaction are usually one-to-one and two-way. In contrast, instructor-class communications are one-to-many and one-way. The rich-lean dimension is orthogonal to this distinction, for example, an instructor can post lean text assignments, graphical lecture slides, or rich video-lessons. E-mail still plays a major role in online learning, though it remains largely plain text, because it is interactive. Online learning system's e-mail and chat functions unnecessarily duplicate existing e-mail services, like Hotmail. Having a separate e-mail for each class taken or taught requires students or instructors to check each class e-mail, in addition to their normal e-mail. For online learning systems to create normal e-mail lists would be much more user considerate, as then students would only have to check their normal e-mail.

In 1:1 two-way communications, like e-mail, "the conversation channel" is the shared resource. Yet while physical society recognizes the joint ownership of communication channels and offers everyone the right not to interact (e.g., to remain silent, to not receive junk mail, to not answer the phone, etc.), the core e-mail system gives all senders the right to put any message into any receiver's inbox. This unfairly gives all rights to the sender and none to the receiver and enables the ongoing spam epidemic that plagues us all.

A more fundamental problem with e-mail in online learning is that one-to-one teacher-student interactions do not scale well (Berners-Lee, 2000). While one can as easily post lessons to a large class as to a small one, handling e-mails for classes over 50 can be difficult. The legitimacy baseline is that students have paid for class tuition, not one-to-one on-demand tuition. Experienced instructors often restrict the use of e-mail to personal requirements, like arranging meetings. They discourage its use for course content, for example, "Sorry I could not make the last class, what did I miss?" is a real student e-mail that I discouraged. Politeness in an interaction works two-ways, so training students to be e-mail polite is a valid learning goal, for example, polite e-mails are:

1. **Signed:** Give your name clearly—e-mails from nicknames like "fly-with-wind" are often unanswered.
2. **Understandable:** Give course/class number in the e-mail title so the instructor knows the context.
3. **Personal:** Use personal e-mail for personal issues, not issues that affect the entire class, for example, an online instructor may paste a "When is the exam?" e-mail into an online discussion board and answer it there, so other students can see the answer.

Class to instructor interactions, like an online assignment submission box, illustrate many-to-one one-way communication. For multi-choice quizzes, the computer can also grade the submissions and give student feedback. This is scalable as the computer can handle any class size, and can remember previous tests, telling the student if he/she is improving or not. However, while online exams do not need politeness, as students must take them, voluntarily online learning is a different matter. The distinction is:

*   **Formal Testing Quizzes:** Usually begin and end at a fixed time, shuffle questions and options to prevent cheating, and give little content feedback. Being mandatory, politeness applies only minimally.
*   **Informal Learning Quizzes:** Offer choices like pausing to restart later, optional tips, answer feedback, and choice of difficulty level. Being voluntary, politeness can help involve the student in the learning process.

If learning means changing one's own processing, a case can be made that all learning is voluntary. If so, polite interaction may help engage students in voluntarily online learning. The difference between a forced online quiz and an online learning experience may be politeness and respect. Online quizzes can support face-to-face lessons, for example, if students answer online questions on a textbook chapter the week before lecture. This questioning encourages them to actively find information from the textbook, and prepares them for the weekly face-to-face class. Unlike a testing quiz, which is given after the class, and is graded by percentage correct, a "learning participation" quiz occurs before the taught class, and any reasonable participation (e.g., 30+%) gets full points. However, the quiz must be done in the week stated, and there are no "resits" for weekly participations. The quiz answers are not released until the week finishes, and students can do or redo the quiz any time in the given week. In practice, those who do poorly in testing quizzes also tend to omit the learning quizzes. However, the good students find them an excellent way to learn.

Most online learning systems seem designed to give information to teachers rather than students, who get learning feedback only with difficulty, for example, Figure 3a shows a "View Scores" button, which when clicked gives Figure 3b, that shows a score. Few students then realize that clicking the underlined "1" gives feedback on the right answers. While online teachers can "see" everything, like when and for how long students are online, students struggle to see what could help them learn in online software.

Class-to-class FAQ boards, where students answer each other's questions, are many-to-many, two-way interactions that scale well to all class sizes. Respecting and using class member knowledge is not only popular with students but for fast changing subjects, like Web-programming, almost essential. If young people learn mainly from their peers, involving their peers in online learning

seems sensible, and polite computing could enable this.

Polite computing suggests voluntary choice is a new online learning dimension. Its application however, requires a complete redesign of current teacher focused systems like WebCT. The online classroom must move from what is essentially a software supported dictatorship to a system that invites voluntary student participation, based on a balance of rights and choices.

FUTURE TRENDS

Polite computing suggests computers will increasingly:

1. Remember interaction data rather than object data.
2. Become human assistants or agents rather than independent actors.
3. Support politeness rather than selfishness in online interaction.

Figure 3. Getting quiz feedback in WebCT

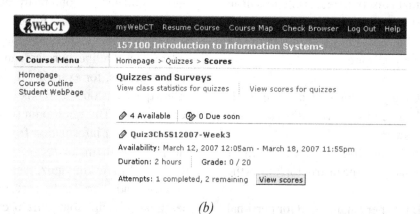

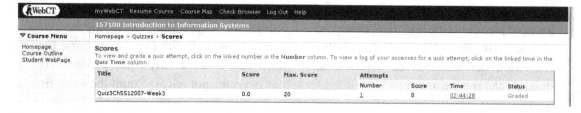

Remember the Interaction

It is astounding that major software manufacturers like Microsoft gather endless data on users, but seem oblivious to data on how their software interacts with the user. Like Peter Sellers in the film "Being There," such software "likes to watch," but cannot relate to people. To spy on users at every opportunity is not a user relationship, For example, Mr. Clippy watched your document actions but could not see his interactions with you, and so was oblivious to the rejection and scorn he evoked. Most software today is in the same category, and modern airport toilets seem more aware of their users than the average personal computer. Hopefully tomorrow's software will make HCI memory its business, as its primary role will be to work for people, not for itself.

Computers as Assistants or Agents

There are several reasons why people should control computers, not the reverse. Firstly, while computers manage vast amounts of data with ease, they handle context changes poorly, and outside their fixed parameters can seem very stupid. So-called "smart" computing (Kurzweil, 1999) usually needs a human "minder." Secondly, computers are not accountable for what they do, as they have no "self" to bear any loss. If society makes people accountable for what computers do, as it does, people need control over computer choices. Thirdly, the resistance of people to computer domination is predictable. Software designers should not underestimate the importance of user choice. In human history, freedom and choice are the stuff of revolutions, and a grassroots Internet movement against impolite software is not inconceivable.

Fortunately, the future of computers probably lies not in becoming so clever or powerful that people are obsolete, nor in being passive human tools, but in contributing to a human-computer combination that performs better than either people or computers alone. The runaway IT successes of the last decade (cell-phones, Internet, e-mail, chat, bulletin boards, etc.) all support people rather than supplant them. As computers develop this co-participant role, politeness will be a critical success factor. These arguments suggest that if the role of computers is to assist, they should learn to be polite.

Online Politeness Will Grow

Today, many users feel at war with their software: removing things they did not want added, resetting changes they did not want changed, closing windows they did not want opened, blocking e-mails they did not want to receive, and so forth. User weapons in this unnecessary war include third party blockers, cleaners, and filters of various sorts, whose main aim is to put users back in charge of their computer estate. Such applications are the most frequent accesses at Internet download sites. Like all wars, if software declares war on user choice, everyone will lose in the long run. If the Internet is a battlefield, no-one will want to go there. Some compare the Internet to the U.S. Wild West, and others talk of the "hunter-gatherers of the information age" (Meyrowitz, 1985). Yet the Stone Age and the U.S. Wild West evolved into civil society, and so perhaps it is time to introduce civility to the Internet. What took physical society thousands of years may occur online in only a few years for example, Wikipedia began with few rules and one leader, but now to combat "trolls" who trash data, has many rules (including copyright) and many roles, like "Steward," "Bureaucrat," and "Sysop" (Whitworth, Aldo de Moor, & Liu, 2006). Yet the real force behind Wikipedia is the majority's enjoyment of working together considerately, not its ability to deal with the anti-social minority.

Many successful online traders find politeness profitable. EBay's customer reputation feedback gives users optional access to valued information relevant to their purchase choice, which by the previous definition is polite. Amazon gives customers information on the books similar buyers buy, not by pop-up ads but as a view option below. Rather than a demand to buy, it is a polite reminder of same-time purchases that could save the customer postage. Politeness is not about selling but improving the customer relationship that leads to sales. By giving customers choice, polite companies win business because customers given choices come back. Perhaps one reason the Google search engine swept all before it was that its simple white interface, without annoying flashing or pop-up ads, made it pleasant to interact with. Google ads sit quietly at screen right, as options not demands. Yet while many online companies know that politeness pays, for others the lesson is still being learned, and for still others, hit-and-run rudeness is an online way of life.

FUTURE RESEARCH

The users of modern software increasingly choose whether to use it or not, for example, President Bush's 2001 decision not to use e-mail because he did not trust its privacy. The ability of software to hold users hostage to its power may be declining. Where customers choose their software, a simple prediction is made: Polite software will be used more and deleted or disabled less than impolite software.

An experimental test of polite computing value requires a comparison of polite versus impolite applications on measures like willingness to use, attitude to the software, willingness to purchase, and user satisfaction. Politeness here is defined to apply not just to language, conversations, or people, but also to human-computer interactions. Research can show if computer users really value politeness in HCI interactions like application installations, user help, online learning, e-mail, messaging, and bulletin boards, to mention a few. This politeness is not just the words used, but also the software actions taken. The relative value of the proposed politeness sub-aspects (respect, open-

ness, helpfulness, and remembering) can also be compared. Correlational studies could compare rated application politeness with market success. Longitudinal studies could determine if successful applications become more polite over time. Ethnographic studies could explore how users perceive polite and impolite software.

The scope of online politeness also bears investigation. The definition implies that young or inexperienced users will tolerate impolite agents like Mr. Clippy more than experienced users. Also, it has been proposed that for interactions mandated by law, or other coerced acts, politeness will apply less. Other individual differences including gender, age, and culture, may also mediate the user reaction to impolite software. Cultural differences in polite computing raise highly complex issues of roles and social structures and may affect the boundary between what is required and what is polite.

CONCLUSION

Polite software asks before it allocates computer resources, openly declares itself and its acts, does not unnecessarily interrupt or draw attention to itself, offers understandable choices, and remembers past interactions. Conversely, impolite software acts without asking, does things secretly, interrupts unnecessarily, offers confusing choices, and has no recall of its past interactions with you.

If polite software attracts users, impolite software can drive them away. This implies a new type of IS error—social error. A program syntax error fails to support the needs of the computer technology. A software usability error fails to support the psychological needs of the computer user. However, a social error means the software fails to support the equally critical needs of human social interaction. While users misunderstand systems designed with poor usability, they understand impolite software all too well, and that is why they walk away from the interaction. Whether a system fails because the computer cannot run it, the user cannot run it, or the user will not run it, makes no difference. The end effect is still that the application does not run. A software social error gives the same outcome as a software crash or user failure. Indeed, social errors may be even worse, as it is in the nature of people to actively seek retribution against those who wrong others in social interactions.

A future is envisaged where software politeness is a critical requirement for socio-technical system success, especially where user willingness to participate counts. Polite computing could be taught in system design classes, along with other system requirements. A "politeness seal" could credit applications that give rather than take user choice. If physical society in general sees the value of politeness, online society should follow that lead. As software becomes not only useful and usable but also polite, the Internet may at last recognize its social potential.

ACKNOWLEDGMENT

Many thanks to Guy Kloss, Massey University, for his very useful comments and insights.

REFERENCES

Alexander, C. (1964). Notes on the synthesis of form. Cambridge, Ma: Harvard University Press.

Berners-Lee, T. (2000). Weaving the Web: The original design and ultimate destiny of the world wide Web. New York: Harper-Collins.

Cooper, A. (1999). The inmates are running the asylum—Why high tech products drive us crazy and how to restore the sanity. USA.

Daft, R. L., & Lengel, R. H. (1986). Organizational information requirements, media richness and structural design. Management Science, 32(5, May), 554-571.

Festinger, L. (1957). A theory of cognitive dissonance. Stanford University Press.

Forester, T., & Morrison, P. (1994). Computer ethics. London: MIT Press.

Gauze, C. F. (2003). I see you're writing an article. INK19. Retrieved on http://www.ink19.com/issues/march2003/webReviews/iSeeYoureWritingAn.html

Goldstein, M., Alsio, G., & Werdenhoff, J. (2002). The media equation does not always apply: People are not polite to small computers. *Personal and Ubiquitous Computing*, 6, 87–96. doi:10.1007/s007790200008

Jenkins, S. (2006). Concerning interruptions. *Computer*, (November): 114–116.

Kurzweil, R. (1999). The age of spiritual machines. Toronto: Penguin Books.

Larkin, E. (2005). PC World, December, 28.

Lessig, L. (1999). Code and other laws of cyberspace. New York: Basic Books.

Mayer, R. E., Johnson, W. L., Shaw, E., & Sandhu, S. (2006). Constructing computer-based tutors that are socially sensitive: Politeness in educational software. *International Journal of Human-Computer Studies*, 64(1), 36–42. doi:10.1016/j.ijhcs.2005.07.001

Meyrowitz, J. (1985). No sense of place: The impact of electronic media on social behavior. New York: Oxford University Press.

Miller, C. A. (2004). Human-computer etiquette: Managing expectations with intentional agents. *Communications of the ACM*, 47(4), 31–34. doi:10.1145/975817.975840

Nass, C. (2004). Etiquette equality: Exhibitions and expectations of computer politeness. *Communications of the ACM*, 47(4), 35–37. doi:10.1145/975817.975841

Pallatto, J. (2007, January 22). Monthly Microsoft patch hides tricky IE 7 download. Retrieved on http://www.eweek.com/article2/0,1895,2086423,00.asp

PCMagazine. (2001). 20th anniversary of the PC survey results. Retrieved on http://www.pcmag.com/article2/0,1759,57454,00.asp.

Power, R. (2000). Tangled Web: Tales of digital crime from the shadows of cyberspace. Indianapolis: QUE Corporation.

Pratley, C. (2004). Chris_Pratley's OneNote WebLog. Retrieved http://weblogs.asp.net/chris_pratley/archive/2004/05/05/126888.aspx

Raskin, J. (2000). The humane interface. Boston: Addison-Wesley.

Rawls, J. (2001). Justice as fairness. Cambridge, MA: Harvard University Press.

Reeves, B., & Nass, C. (1996). The media equation: How people treat computers, television, and new media like real people and places. New York: Cambridge University Press/ICSLI.

Rose, G., Khoo, H., & Straub, D. (1999). Current technological impediments to business-to-consumer electronic commerce. Communications of the AIS, I(5).

Shectman, N., & Horowitz, L. M. (2003). Media inequality in conversation: How people behave differently when interacting with computers and people. Paper presented at the CHI (Computer Human Interaction) 2003, Ft Lauderdale, Florida.

Technology threats to privacy. (2002, February 24). New York Times, Section 4, p. 12.

Transparency-International. (2001). Corruption perceptions. Retrieved on www.transparency.org

Tzeng, J. (2004). Toward a more civilized design: studying the effects of computers that apologize. *International Journal of Human-Computer Studies, 61*(3), 319–345. doi:10.1016/j.ijhcs.2004.01.002

Whitworth, B. (2005). Polite computing. Behaviour & Information Technology, September 5, 353 – 363. Retrieved on http://brianwhitworth.com/polite05.pdf

Whitworth, B. (2006). Spam as a symptom of electronic communication technologies that ignore social requirements. In C. Ghaoui (Ed.), Encyclopaedia of human computer interaction (pp. 559-566). London: Idea Group Reference.

Whitworth, B. Aldo de Moor, & Liu, T. (2006, Nov 2-3). Towards a theory of online social rights. Paper presented at the International Workshop on Community Informatics (COMINF'06), Montpellier, France.

Whitworth, B., & deMoor, A. (2003). Legitimate by design: Towards trusted virtual community environments. *Behaviour & Information Technology, 22*(1), 31–51. doi:10.1080/01449290301783

Whitworth, B., Gallupe, B., & McQueen, R. (2001). Generating agreement in computer-mediated groups. *Small Group Research, 32*(5), 621–661. doi:10.1177/104649640103200506

Whitworth, B., & Whitworth, E. (2004). Reducing spam by closing the social-technical gap. Retrieved on http://brianwhitworth.com/papers.html. *Computer*, (October): 38–45. doi:10.1109/MC.2004.177

Wright, R. (2001). Nonzero: The logic of human destiny. New York: Vintage Books.

Section 3
Trust and Participation Issues in Web 2.0 Systems at Risk of Internet Trolling

Chapter 9
The Psychology of Trolling and Lurking:
The Role of Defriending and Gamification for Increasing Participation in Online Communities Using Seductive Narratives

Jonathan Bishop

Centre for Research into Online Communities and E-Learning Systems, UK

ABSTRACT

The rise of social networking services have furthered the proliferation of online communities, transferring the power of controlling access to content from often one person who operates a system (sysop), which they would normally rely on, to them personally. With increased participation in social networking and services come new problems and issues, such as trolling, where unconstructive messages are posted to incite a reaction, and lurking, where persons refuse to participate. Methods of dealing with these abuses included defriending, which can include blocking strangers. The Gamified Flow of Persuasion model is proposed, building on work in ecological cognition and the participation continuum, the chapter shows how all of these models can collectively be used with gamification principles to increase participation in online communities through effective management of lurking, trolling, and defriending.

INTRODUCTION

The study of online communities has led to such colourful expressions as trolling, flaming, spamming, and flooding being developed in order to describe behaviours that benefit some people while disrupting others (Lampe & Resnick, 2004). Since the proliferation of technologies like the 'circle-

of-friends' (COF) for managing friends lists in online communities (Romm & Setzekom, 2008), the use of the Internet to build online communities, especially using social networking services has grown – but so has the amount of Internet abuse on these platforms. Facebook is currently one of the more popular COF-based websites (Davis, 2008). In addition to this, microblogging, such as Twit-

DOI: 10.4018/978-1-4666-2803-8.ch009

ter, have 'status updates', which are as important a part of social networks Facebook and Google+, as the circle of friends is. These technologies have made possible the instantaneous expression of and access to opinion into memes that others can access quickly, creating what is called, 'The public square' (Tapscott & Williams, 2010). The public square is the ability to publish and control editorial policy, and is currently available to all with access to and competency in using the Internet and online social networking services.

It is clear in today's age that there are a lot of demands on people's time, and they have to prioritise which social networking services, or other media or activity they use. This is often based on which is most gratifying and least discomforting. It has become apparent that introducing gaming elements into such environments, where they would not usually be – a concept called 'gamification' – can increase interest and retention in them. Such systems can promote positive activities by members and reduce the number of people not taking part, called 'lurkers' (Bishop, 2009c; Efimova, 2009). It can also promote activities like 'trolling' where content is created for the 'lulz' of it – that is for the fun of it. These can have upsides and downsides, but it is clear gamification can play a part in managing it.

The Problem of Lurking and Trolling Behaviour

Besides social software, gamification and consumerisation have been identified as the big themes for cloud applications (Kil, 2010). Gamification offers online community managers, also known as systems operators (sysops), the opportunity for a structured system that allows for equitable distribution of resources and fair treatment among members. Finding new ways to makes ones' website grow is a challenge for any sysop, so gamification may be the key. Often this is looked on in a technical way, where such platforms are encouraged to move from simple resource archives

toward adding new ways of communicating and functioning (Maxwell & Miller, 2008). It is known that if an online community has the right technology, the right policies, the right content, pays attention to the strata it seeks to attract, and knows its purpose and values then it can grow almost organically (Bishop, 2009c). A potential problem stalling the growth of an online community is lack of participation of members in posting content, as even with the right technology there is often still a large number of 'lurkers' who are not participating (Bishop, 2007b). Lurkers are defined as online community members who visit and use an online community but who do not post messages, who unlike posters, are not enhancing the community in any way in a give and take relationship and do not have any direct social interaction with the community (Beike & Wirth-Beaumont, 2005). Lurking is the normal behaviour of the most online community members and reflects the level of participation, either as no posting at all or as some minimal level of posting (Efimova, 2009). Lurkers may have once posted, but remain on the periphery due to a negative experience.

Indeed, it has been shown that lurkers are often less enthusiastic about the benefits of community membership (Howard, 2010). Lurkers may become socially isolated, where they isolate themselves from the peer group (i.e. social withdrawal), or are isolated by the peer group (i.e. social rejection) (Chen, Harper, Konstan, & Li, 2009). Trolling is known to amplify this type of social exclusion, as being a form of baiting, trolling often involved the Troller seeking out people who don't share a particular opinion and trying to irritate them into a response (Poor, 2005).

The Practice of Defriending in Online Communities

While the Circle of Friends allows the different techno-cultures that use online communities to add people as friends, it also gives them the power to remove or delete the person from their social network. This has been termed in the United States of America as 'unfriending' or in the United Kingdom as 'defriending'. Defriending is done for a number of reasons, from the innocent to the malicious to the necessary. For instance, a user can innocently suspend their account or want to 'tidy-up' their Circle of Friends, so that only people they actually know or speak to are in it. There can be malicious and ruthless acts of 'cutting someone dead' or permanently 'sending them to Coventry' so that they are no longer in one's network or able to communicate with oneself (Thelwall, 2009). And users can do it through a 'blocking' feature to cut out undesirable people who are flame trolling them so much that it impairs their ability to have a normal discourse. Being able to 'block' the people they don't want to associate with means that it is impossible for them to reconnect without 'unblocking'. Such practice on social networking sites can lead to users missing out on the context of discussions because they are not able to see hidden posts from the person they blocked or who blocked them, to them seeing ghost-like posts from people whose identities are hidden but whose comments are visible for the same reason. Any form of defriending, whether intended innocently or otherwise, can lead to the user that has been defriended feeling angry and violated, particularly if the rules for killing a community proposed by Powazek (2002) haven't been followed. This can turn the user into an E-Venger, where by the user will seek to get vengeance against the person that defriended them through all means possible. If they're a famous person then this could mean posting less than flattering content on their Wikipedia page or writing negative comments about them in other online communities. If they're a close friend whose personal details they have to hand, then it could mean adding their address to mailing lists, or sending them abusive emails.

Gamification

As of the end of 2010, the Facebook game, Farmville, had more than 60 million users worldwide, or 1 per cent of the world's population with an average of 70 minutes played weekly (Hurley, 2000). Concepts like "Gamification" which try to bring video game elements in non-gaming systems to improve user experience and user engagement (Yukawa, 2005) are therefore going to be an important part in current and future online communities in order to increase participation of constructive users and reduce that of unconstructive users. It seems however the gaming elements of online communities need not be 'designed' by the *sysops*, but developed independently by the users, in some cases unintentionally or unknowingly.

For instance, it has become a game on Twitter for celebrities to try and outdo one another by exploiting the 'trending' feature which was designed to tell users what was popular. Celebrities like the interviewing broadcaster, Piers Morgan, and reality TV personality Alan Sugar talked up in the press their programmes which went head to head, and Mr. Morgan claimed victory because he and his guest, Peter Andre, on his Life Stories programme appeared higher in the most mentioned topics on Twitter. Also, consumers joined in this activity which could be called 'ethno-gamification' by agreeing to prefix 'RIP' to various celebrities names in order to get that term to appear in the trending column. In the same way 'hypermiling' has become a term to describe ethno-gamification where people try to compete with one another on how can use the least amount of fuel in their vehicles, so this could be called 'hypertrending' as people seek to try to get certain terms to trend higher than others. Examples of both of these are in Figure 1.

So it seems that gaming is essential to the way humans use computer systems, and is something that needs to be exploited in order to increase participation in online communities, which may not have the membership or status of established

Figure 1. Piers Morgan's Twitter page and 'RIP Adele' search results showing 'ethno-gamification' in the form of 'hypertrending'

platforms like Facebook and Google+. Table 1 presents a restructuring of the extrinsic motivators and mechanical tasks in gamification identified by (Wilkinson, 2006) as interface cues, which are 'credibility markers' which act as mediating artefacts when attached to a user's cognitive artefacts (Bishop, 2005; Norman, 1991; Weiler, 2002). These are categorised according to whether they

are 'authority cues,' signalling expertise, or 'bandwagon cues,' which serve as 'social proof' by allowing someone to reply on their peers. These are followed by and inclusion of the UK health authority's guidance on communities and behaviour change (Esposito, 2010; Smith, 1996).

These stimuli and post types will need to be tailored to individuals dependent on their 'player

Table 1. Examples of interface cues and guidance for gamification use

Stimulus type (Post type)	Examples of interface cue	Guidance for use as mediating artefacts
Social (Snacking)	'group identity[1]', 'fun[2]', 'love[2]'	Users do perform snacking offer short bursts of content and consume a lot too. To take advantage of this, one should utilise local people's experiential knowledge to design or improve services, leading to more appropriate, effective, cost-effective and sustainable services. In other words allow the community to interact without fear of reprisals
Emotional (Mobiling)	'punishments[2]', 'rewards[2]'	Mobiling is where users use emotions to either become closer to others or make a distance from them. This can be taken advantage of to empower people, through for example, giving them the chance to increase participation, so as to also increase confidence, self-esteem and self-efficacy. This can be done through using leaders and elders to encourage newer members to take part.
Cognitive (Trolling)	'levels[1]', 'learning[2]', 'points[2]'	Trolling as a more generic pursuit seeks to provoke others, sometimes affect their kudos-points with others users. Such users should contribute to developing and sustaining social capital, in order that people see a material benefit of taking part.
Physical (Flooding)	'power[1]', 'mastery[2]'	Flooding is where users get heavily involved with others uses by intensive posting that aims to use the person for some form of gratification. Sysops should encourage health-enhancing attitudes and behaviour, such as encouraging members to abuse the influence they have.
Visual (Spamming)	'leader-boards[1]', 'badges[2]'	Spamming, often associated with unsolicited mail, is in general the practices of making available ones creative works or changing others to increase the success of meetings one's goals. Interventions to manage this should be based on a proper assessment of the target group, where they are located and the behaviour which is to be changed and that careful planning is the cornerstone of success. Designing visual incentives can be effective at reinforcing the message.
Relaxational (Lurking)	'meaning[2]', 'autonomy[1]'	Lurking is enacted by those on the periphery of a community. Their judgements for not taking part often relate to a lack of purpose or control. It is essential to build on the skills and knowledge that already exist in the community, for example, by encouraging networks of people who can support each other. Designing the community around allowing people to both see what others are up to, as well as allowing them to have a break from one another can build strong relationships. A 'd0 not bite the newbies' policy should be enforced.

type' and 'character type.' The dictionary NetLingo identified four types of player type used by trollers; playtime, tactical, strategic, and domination trollers (Leung, 2010). Playtime Trollers are actors who play a simple, short game. Such trollers are relatively easy to spot because their attack or provocation is fairly blatant, and the persona is fairly two-dimensional. Tactical Trollers are those who take trolling more seriously, creating a credible persona to gain confidence of others, and provokes strife in a subtle and invidious way. Strategic Trollers take trolling very seriously, and work on developing an overall strategy, which can take months or years to realise. It can also involve a number of people acting together in order to invade a list. Domination Trollers conversely extend their strategy to the creation and running of apparently bona-fide mailing lists.

UNDERSTANDING ONLINE COMMUNITY PARTICIPATION

Increasing participation in online communities is a concern of most sysops. In order to do this it is important they understand how the behaviour of those who take part in their community affects others' willingness to join and remain on their website.

The Lurker Profile

Lurkers often do not initially post to an online community for a variety of reasons, but it is clear that whatever the specifics of why a lurker is not participating the overall reason is because of the dissonance of their cognitions that they have experienced when presented with a hook into a conversation. Cognitions include goals, plans, values, beliefs and interests (Bishop, 2007b), and may also include 'detachments'. These may include that they think they don't need or shouldn't post or don't like the group dynamics (Preece, Nonnecke, & Andrews, 2004). In addition some of the plans of lurkers causing dissonance has been identified (Preece et al., 2004), including needing to find out more about the group before participating and usability difficulties. The cognitions of 'goals' and 'plans' could be considered to be stored in 'procedural memory, and the 'values' and beliefs could be considered to be stored in 'declarative memory'. The remaining cognitions, 'interest' and 'detachment' may exist in something which the author calls, 'dunbar memory', after Robin Dunbar, who hypothesised that people are only able to hold in memory 150 people at a time. It may be that lurkers don't construct other members as individuals, and don't therefore create an 'interest' causing their detachment cognitions to be dominant. The profile of a reluctant lurker therefore is that of a socially detached actor, fearing consequences of their actions, feeling socially isolated or excluded, trapped in a state of low flow but high involvement. Lurkers, it has been argued are no more "tied" to an online community than viewers of broadcast television are "tied" to the stations they view (Beenen et al., 2004). However, it can be seen that some more determined lurkers are engaged in a state of flow with low involvement in doubting non-participation. Some have

suggested lurkers lack commitment Building and sustaining community in asynchronous learning networks, but they are almost twice as likely to return to the site after an alert (Rashid et al., 2006). Indeed, lurkers belong to the community, and while they decide not to post in it, they are attracted to it for reasons similar to others (Heron, 2009). It has been argued that most lurkers are either shy, feel inadequate regarding a given topic, or are uncomfortable expressing their thoughts in written form (Jennings & Gersie, 1987), but others suggest lurking is not always an ability issue (Sherwin, 2006).

Some researchers characterised lurkers as against hasty conversation rather than a problem for the community (Woodfill, 2009). Often lurkers are afraid of flame wars and potential scrutinising of their comments (Zhang, Ma, Pan, Li, & Xie, 2010). Marked and excessive fear of social interactions or performance in which the person is exposed to potential scrutiny is a core feature of social phobia (Simmons & Clayton, 2010), which has similar facets to lurking (Bishop, 2009d). Perhaps one of the most effective means to change the beliefs of lurkers so that they become novices is for regulars, leaders and elders to nurture novices in the community (Bishop, 2007b). It is known that therapist intervention can help overcome social phobia (Scholing & Emmelkamp, 1993). It could be that through 'private messaging' features that a leader could speak to a registered member who is yet to post. After all, a community is a network of actors where their commonality is their dependence on one another, so feeling a need to be present is essential.

Feelings of uncertainty over the use of posted messages is common to lurkers All social situations carry some uncertainty, which people with social phobia find challenging (Waiton, 2009). Lurking can potentially lead to social isolation, such as not naming anyone outside of their home as a discussion partner (Pino-Silva & Mayora, 2010). Lurkers are less likely to report receiving social support and useful information and often have lower satisfaction levels with group participation sessions (Page, 1999). Leaders can post more messages to encourage all members to post messages (Liu, 2007). Uncertainty caused by poor usability leads to non-participation by lurkers (Preece et al., 2004), and this can be tackled by having the right technology and policies (Bishop, 2009c). Developing trust involves overcoming, particularly in trading communities (Mook, 1987). Such trust was evident in The WELL (Whole Earth 'Lectronic Link), where members use their real names rather than pseudonyms (Rheingold, 2000). Requiring actors to use their real names could help a lurker overcome their uncertainties about others' true intentions.

The Troller Profile

A generic definition of trolling by *'Trollers'* could be 'A phenomenon online where an individual baits and provokes other group members, often with the result of drawing them into fruitless argument and diverting attention from the stated purposes of the group' (Moran, 2007). As can be seen from Table 2, it is possible to map the types of character in online communities identified by (Bishop, 2009b) against different trolling practices. Also included is a set of hypnotised narrator types which affect the approach a particular character can take to influence the undesirable behaviour of others without resorting to defriending, which is explored in the empirical investigation later.

This makes it possibly to clearly see the difference between those who take part in trolling

Table 2. Troller character types and counter-trolling strengths as narrators

Troller Character Type	Hypothesised Narrator Types	Description
Lurker	Stranger	Silent calls by accident, etc., clicking on adverts or 'like' buttons, using 'referrer spoofers', modifying opinion polls or user kudos scores.
Elder	Catalyst	An elder is an out-bound member of the community, often engaging in 'trolling for newbies', where they wind up the newer members often without question from other members.
Troll	Cynic	A Troll takes part in trolling to entertain others and bring some entertainment to an online community.
Big Man	Sceptic	A Big Man does trolling by posting something pleasing to others in order to support their world view.
Flirt	Follower	A Flirt takes part in trolling to help others be sociable, including through light 'teasing'
Snert	Antagonist	A Snert takes part in trolling to harm others for their own sick entertainment
MHBFY Jenny	Pacifist	A MHBFY Jenny takes part in trolling to help people see the lighter side of life and to help others come to terms with their concerns
E-Venger	Fascist	An E-Venger does trolling in order to trip someone up so that their 'true colours' are revealed.
Wizard	Enthusiast	A wizard does trolling through making up and sharing content that has humorous effect.
Iconoclast	Detractor	An Iconoclast takes part in trolling to help others discover 'the truth', often by telling them things completely factual, but which may drive them into a state of consternation. They may post links to content that contradicts the worldview of their target.
Ripper	Rejector	A Ripper takes part in self-deprecating trolling in order to build a false sense of empathy from others.
Chatroom Bob	Striver	A chatroom bob takes part in trolling to gain the trust of others members in order to exploit them.

to harm, who could be called 'flame trollers' from those who post constructively to help others, called 'kudos trollers.' A flame is a nasty or insulting message that is directed at those in online communities (Leung, 2010). Message in this context could be seem to be any form of electronic communication, whether text based or based on rich media, providing in this case it is designed to harm or be disruptive. A 'kudos' on the hand can be seen to be a message that is posted in good faith, intended to be constructive.

The Effect of Gamification and Defriending on Online Community Participation

In 2007, as Facebook was emerging, (Bishop, 2007b) presented the ecological cognition framework (see Figure 2). The 'ECF' was able to show the different plans that actors make in online communities based on their different dispositional forces, which created 'neuro-responses' driving

them to act, such as 'desires.' Four years earlier in 2003, research was pointing out that there were unique characteristics among those people forming part of the *net generation* (i.e. those born between 1977 and 1997). These included having dispositional forces with preference for *surveillance* and *escape*, factors which were not part of the ECF.

These online social networking services have shown that the ties that used to bring people to form online communities are different than what they used to be prior to 2007. The personal homepage genre of online community (Bishop, 2009a) is now the most dominant model of online community enabled through these services. Through actors forming *profiles,* linked together with the circle of friends and microblogging content, they can control the visibility of objects such as actors (e.g. their friends) and artefacts (e.g. the content they want to see). They are in effect creating their own online community dedicated to the people they consider friends.

Figure 2. The ecological cognition framework

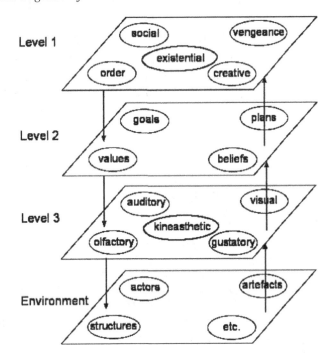

The Participation Continuum

One of the most important concepts in creating online communities that can harness gamification is the relationship between 'flow' and 'involvement.' When an actor is engaged in a state of flow their concentration is so intense that they forget about their fears and become fully immersed and completely involved in what they are doing (Csikszentmihalyi, 1990). Decision-making in such a state becomes more fluid and actors respond almost without thought for the consequences of their actions. In a high state of flow, Snerts will have low involvement cognitively and post flames with little restraint, often trolling for their own benefit, which then deters lurkers from becoming posters. A structure based on the ecological cognition framework for decision making in human-centred computer systems has been proposed (Bishop, 2007a), which introduced the concepts of deference, intemperance, reticence, temperance and ignorance. This was extended through the participation continuum, to suggest that these cognitive states will lead to empression, regidepression, depression, suppression and repression respectively in the case of the original five judgements (Bishop, 2011b). A six cognitive state, proposed in that paper, reflects the dilemma that lurkers go through, which is compression when they experience incongruence due to congruence

when trying to avoid cognitions which are not compatible with their ideal self. Decompression on the other hand is when they start to break this down. These concepts are presented in the model in Figure 3, called the participation continuum.

There appears to be a 'zone of participation dissonance' between the level at which an actor is currently participating and what they could achieve if there was greater support for usability and sociability. This distance between fully 'mediating' their transfer to enhancement of participation could be called the 'Preece Gap', after Jenny Preece, who set out how to design for usability and support sociability (Preece, 2001). As can be seen from the participation continuum in Figure 1, the higher the state of flow for a lurker, the more likely they are to be 'dismediating' from enhancement towards preservation by not to posting due to low involvement. Equally, the higher the state of flow for a poster the more likely they are to keep mediating towards enhancement and away from preservation within the community with little effort (i.e. involvement). The process in the middle resembles the visitor-novice barrier in the membership lifecycle (Kim, 2000). A lurker who has had bad experiences may be sucked into stagnation through rationalisation of non-participation, going from minimal posting (Efimova, 2009) to lurking (i.e. where they give up posting) and back out again after the intellectu-

Figure 3. The participation continuum

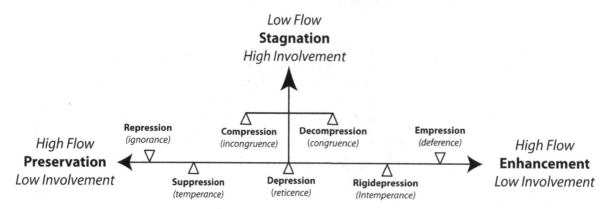

alisation process. This resembles a 'battering' cycle (Bishop, 2010), where the actor will be under a barrage of flaming abuse, then be told all is forgiven and they can come back as in (Bishop, 2009b).

AN INVESTIGATION INTO DEFRIENDING IN ONLINE COMMUNITIES

A study was designed to use a narrative analysis to analyse defriending activity and extend the understanding the ECF brings to online community research. Narrative analysis is a tool researchers can use to explore the intersection between the individual and society (Kil, 2010). Narrative analysis in Internet studies essentially uses both text and online "talk" to construct a holistic view of the online interactions, looking at cognition as well as affect (Yukawa, 2005). Narrative analysis is the most prevalent approach that has emphasized alternatives to categorising analysis, but much of narrative research, broadly defined, involves categorising as well as connecting analysis, and the distinction has not been clearly defined (Maxwell & Miller, 2008). Narratives were selected from Google's Blog Search by searching for the terms, "I deleted him as a friend", "he deleted me as a friend". "I deleted her as a friend" and "She deleted me as a friend". The ethnomethodological narrative analysis approach of (Bishop, 2011a) was then used to code the text in the blog posts to identify the different 'Methods', 'Memes', 'Amities', 'Rules' and 'Strategies' that impact on the decision to defriend someone or why someone was defriended.

Descriptives

The difficulties of a romantic relationship accounted for just over 2,700 (13.4%) of the cases where a female was defriended compared to less than 50 (0.47%) for men, suggesting that when a romantic relationship doesn't work out women are more likely to be defriended than men, or at least, people are more likely to disclose on a blog that they defriended a female because of relationship problems than they would males. Less than 20 males were defriended for a sex related issue compared to over 9,500 females. This may be because as Thelwall (2008) suggests, men use online social networking more for dating and women more for other forms of friendship. It became clear in the discourses there were often other people involved in the event leading to a person being defriended. In around 65 per cent of cases where males were defriended and 90 per cent where females were defriended there was another person involved. Over 3,000 females (16.4%) were defriended because someone was offended compared to only 4 males (0.08%) for the same reason (see Table 3).

Results

Analysing the data resulted in four key findings. Firstly, actors are provoked into responding to a state of disequilibrium, such as being defriended.

Table 3. Role of different factors in defriending narratives

Defriending Discourse Type	Males Defriended	Females Defriended
Effect of male on female friend	3,315	19,226
Effect of female on male friend	3,249	18,359
Employment mentioned	2,167	12,951
Sex	11	9,665
Break-ups and Dating	24	2,759
Offence	4	3,372
Little in common	3	1,835
Email related	25	1,386
Text message related	7	0
Application related	1	0
Total	*5,084*	*20,572*

Second, actors need to develop an awareness of the change in the environment before they are able to realise its impact on them. Thirdly, actors will first have a reaction to a state of disequilibrium before organising a response that causes them least dissonance. Fourthly and finally, actors will testify their experiences to others as a way of expressing their understanding in order to restore a state of equilibrium.

Finding 1: Actors are provoked into responding to a state of disequilibrium.

Understanding what drives actors to act is crucial to developing human-computer systems that adapt to and influence them. There has been extensive research into discovering what drives people, which has led to a number of theories, including psychoanalytic theory (Freud, 1933), hierarchical needs theory (Maslow, 1943), belief-desire-intention theory (Rao & Georgeff, 1998), which see desires as goals, and other desire-based theories, which see desires as instincts that have to be satisfied (Reiss, 2004). All of these theories suggest that actors are trying to satisfy some internal entity. This assumption ignores the role of the environment in shaping the behaviour of an actor and suggests that actors are selfish beings that only do things for shallow reasons.

There seemed from most of the narratives that there was something in the environment that provoked the actor to write about their defriending action. For instance, Era talking about a male she had known since the age of 12 who "made lots of sexual innuendos and jokes i.e. wolf whistles/ comments about my makeup, perfume etc." ended her narrative saying, "I told him goodbye and removed him as a friend on FB. I wished him all the best in his life. Then he replies and says he only likes me as a friend. He denied that he ever flirted with me and said I was crazy and that I over-analyse things," suggesting that recognition of her experience was important and writing in the blogosphere might be a way she saw to achieve it.

Finding 2: Actors need to develop an awareness of the change in the environment before they are able to realise its impact on them.

It was apparent in the data that those writing their narratives needed to gain an awareness of how the stimulus that provoked them affects them, so that they can understand its impact more appropriately. In one of the weblog narratives, a blogger, Julie, said; "I deleted her as a friend on Facebook because after waiting six months for her to have time to tell me why she was upset with me I got sick of seeing her constant updates (chronic posting I call it)". This supports the view accepted among many psychologists that perception and action are linked and that what is in the environment has an impact on an actor's behaviour. Perceptual psychologists have introduced a new dimension to the understanding of perception and action, which is that artefacts suggest action through offering affordances, which are visual properties of an artefact that determines or indicates how that artefact can be used and are independent of the perceiver (Gibson, 1986). This suggests that when an actor responds to a visual stimulus that they are doing so not as the result of an internal reflex, but because of what the artefact offers.

Finding 3: Actors will first have a reaction to a state of disequilibrium before organising a response that causes them least dissonance.

According to Festinger (1957) cognitive dissonance is what an actor experiences when their cognitions are not consonant with each other. For example if an actor had a plan to be social, but a belief that it would be inappropriate they would experience dissonance as a result of their plan not being consonant with their belief. Resolving this dissonance would achieve a state of consonance that would result in either temperance or intemperance. If this actor held a value that stated that they must never be social if it is inappropriate they could achieve consonance by abandoning the plan to be social which results in temperance. If the same actor had an interest in being social and a belief that it was more important to be social than not be social they might resolve to disregard their belief resulting in intemperance. If an actor experiences a desire without experiencing any dissonance they experience deference, as they will act out the desire immediately.

It became quite apparent early on in the analysis that those writing narratives would do to in such a way to cause least dissonance. For instance, one female blogger (Angie) when writing about a relationship breakdown with her friend, said, "I'm not sure if anything I write tonight will make any sense, but it's not as if anyone else reads these anyway so I guess it doesn't really matter how organized I keep it."

Finding 4: Actors will testify their experiences to others as a way of expressing their understanding in order to restore a state of equilibrium.

It became apparent from looking at the weblog entries that bloggers got some sort of closure from writing the narratives. For instance, closing one of her blogs, Angie said, "As you can see, my brain is a ridiculously tangled ball of yarn at the moment and my thoughts are all over the place. Maybe some good old REM's sleep will massage the knots out. Until next time." Psychological closure, it is argued, is influenced by the internal world of cognition as well as the external world of (finished or unfinished) actions and (challenging or unchallenging) life events. Weblogs, according to some, serve similar roles to that of papers on someone's office desk, for example allowing them to deal with emerging insights and difficult to categorise ideas, while at the same time creating opportunities for accidental feedback and impressing those who drop by (Efimova, 2009).

REVIEW OF FINDINGS

The findings when mapped on to the ECF suggest several things. The first is that in online communities a stimulus is presented that provokes an actor into realising that an opportunity exists to post. For instance, a person may read something on an online news website which they disagree with so much that it provokes them into blogging about it. The next stage of the ECF, the impetus is governed by understanding and at is at this stage the actor beings to gain an awareness of how the stimulus affects them. The next stage is the realisation of its relevance to them and where they gain the intention to respond to it. In reference to the earlier example, it may be that the news article is disparaging about a particular cultural group they belong to, and it reignites old memories of discrimination that they want to respond to. The next narrative stage is where the reaction to this knowledge, where they may form a plan to do something about giving them a sense of aspiration. The next stage of the ECF, Judgement, would be where the actor organises their responses to their reaction and weighs up the positives and negatives to acting on it. For example, their head may be flooded with emotions about how they responded to previous situations that were similar, which they may want to write down to contextualise the current situation. Once they have taken the bold step to write the post, they will then testify their opinions at the response stage and may cycle through their thoughts until they have given the response they are comfortable with. Table 4 presents the stages of the ECF and how these related to the findings of this study.

Towards the Gamification Flow of Persuasion Model

The constructivism proposed by Lev Vygotsky in *Mind in Society* (Vygotsky, 1930) says there is a gap between what people can achieve by themselves and what they can achieve with a more competent

Table 4. Description of stages of the ECF with reference to narrative stages

ECF Stage	Narrative Stage	Description
Stimuli	Provocation	There is a spark that makes someone want to post to an online community. This stimulus provokes an actor into seizing the opportunity to make a contribution.
Impetus	Awareness	Once someone has been given an incentive to post the next stage is to get an understanding of what they can do through gaining an awareness of what has happened.
Intent	Realisation	Once someone has an awareness of how an opportunity affects them the next stage is for them to realise how relevant it is to them to give them the intention to go further.
Neuroresponse	Reaction	Once someone has realised the relevance of a particular action to them they react to it without knowing the consequences giving them a feeling of aspiration.
Judgement	Organisation	Once someone has aspired to a particular course of action they may experience dissonance through organising the proposed action in line with their thoughts. They or their nervous system will then make the choice to take a particular action.
Response	Testimony	Once someone has made the judgement to take a particular action the next stage is to express that choice. In terms of narratives this is their testimony, which may encompass the various aspects of the previous stages.

peer. Vygotsky called this the zone of proximal development, and suggesting that through mediating with artefacts, which the author interprets to include signs such as language or tools such as software, an actor can have help to achieve their potential, in this case in learning. The preeminent *Oxford Dictionary of Law*, which defines a mediator as someone who assists two parties in resolving a conflict but has no decision-making powers, and the process and mediation, supports this conceptualisation proposed by Vygotsky, the author accepts. Equally the term 'dismediation' is the process where an actor, either through reflection or the intervention of another actor returns to a former state of preserving their original status quo. The example given in some texts on cognitive dissonance is where a consumer orders a car from a dealer and then experiences doubt over whether they made the right decision. It has been argued that a courtesy call can help an actor feel more confidence in their decision and reduce the experience, which I call reticence, as an intervention to create mediation towards enhancement, which in this case is the benefit from a new car, which acts as the 'seduction mechanism.' The seduction mechanism in this context refers to an intervention that stimulates substantial change in

an actor's goals, plans, values, beliefs, interests and detachments. An example, which can be found in the existing literature (Bishop, 2007c), is where someone who has been lurking is presented with a post that provokes them so much they feel compelled to reply. However, it is clear that not everyone reacts the same way to a seduction mechanism, as some may take longer to fully change their behaviour than others. A framework is therefore needed to explain these differences, and an extension to the participation continuum is presented in Figure 4.

DISCUSSION

Encouraging participation is one of the greatest challenges for any e-community provider. Attracting new members is often a concern of many small online communities, but in larger e-communities which are based on networks of practice, the concern is often retaining those members who make worthwhile contributions. These communities still have their 'classical lurkers' who have never participated, but they also appear to what could be called 'outbound lurkers', referred to as elders, who used to participate frequently, but now no longer do

Figure 4. The gamification flow of persuasion model

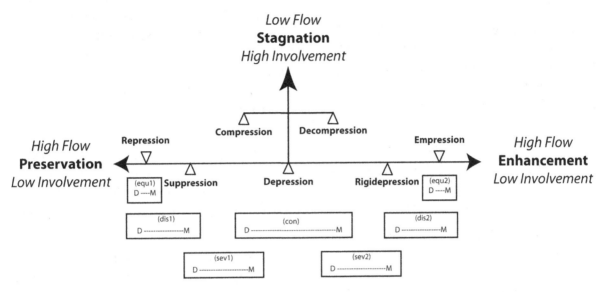

as much. One reason for this is that the actors have lost their ties through being 'defriended' by other actors in the network. Some of the reasons for this defriending behaviour has been explored in this chapter. They vary from issues in the workplace to difficulties in romantic relationships, whether romantic partners or strangers who take part in flame trolling. What is clear that defriending has an impact on those affected by them and are explained in the narratives they produce on weblogs. This suggests that while defriending can have an impact in one community, such as causing 'outbound lurking', it can increase participation in another. Actors will always have a desire to share their experiences, and as has been shown through this chapter they follow a clear six-part cycle in expressing themselves, and their narratives take on 10 different personas based on their individual differences. It could therefore be concluded that one online communities loss is another's gain, as participation in these environments has now become so pervasive that if a person is forced not to participate in them and therefore become an

'outbound lurker,' or elder, they can always find another to meet their desires to express themselves.

This chapter has argued that essential to ensuring 'responsible trolling' is the use of gamification techniques. Gamification introduces elements from video gaming, such as points and leader-boards in order to incentivise positive behaviours and disincentivise negative ones. A model, called the 'gamified flow of persuasion' is presented, which builds on the earlier participation continuum. This explains how gamification based systems can be designed so as to help users transfer from one level of participation to another.

ACKNOWLEDGMENT

The author would like to thank those who provided feedback to earlier versions of this paper. In particular he would like to thank Jean Bishop for the thorough proof reading and for suggesting the names of the narrator types.

REFERENCES

Beenen, G., Ling, K., Wang, X., Chang, K., Frankowski, D., Resnick, P., & Kraut, R. E. (2004). Using social psychology to motivate contributions to online communities. *Proceedings of the 2004 ACM Conference on Computer Supported Cooperative Work,* (p. 221).

Beike, D., & Wirth-Beaumont, E. (2005). Psychological closure as a memory phenomenon. *Memory (Hove, England), 13*(6), 574–593. doi:10.1080/09658210444000241

Bishop, J. (2005). The role of mediating artifacts in the design of persuasive e-learning systems. *Proceedings of the First International Conferences on Internet Technologies and Applications,* University of Wales, NEWI, Wrexham, (pp. 548-558).

Bishop, J. (2007a). Ecological cognition: A new dynamic for human-computer interaction. In Wallace, B., Ross, A., Davies, J., & Anderson, T. (Eds.), *The mind, the body and the world: Psychology after cognitivism* (pp. 327–345). Exeter, UK: Imprint Academic.

Bishop, J. (2007b). Increasing participation in online communities: A framework for human–computer interaction. *Computers in Human Behavior, 23*(4), 1881–1893. doi:10.1016/j.chb.2005.11.004

Bishop, J. (2007c). Increasing participation in online communities: A framework for human–computer interaction. *Computers in Human Behavior, 23*(4), 1881–1893. doi:10.1016/j.chb.2005.11.004

Bishop, J. (2009a). Enhancing the understanding of genres of web-based communities: The role of the ecological cognition framework. *International Journal of Web Based Communities, 5*(1), 4–17. doi:10.1504/IJWBC.2009.021558

Bishop, J. (2009b). Increasing capital revenue in social networking communities: Building social and economic relationships through avatars and characters. In Dasgupta, S. (Ed.), *Social computing: Concepts, methodologies, tools, and applications* (pp. 1987–2004). Hershey, PA: IGI Global. doi:10.4018/978-1-60566-984-7.ch131

Bishop, J. (2009c). Increasing membership in online communities: The five principles of managing virtual club economies. *Proceedings of the 3rd International Conference on Internet Technologies and Applications - ITA09,* Glyndwr University, Wrexham.

Bishop, J. (2009d). Increasing the economic sustainability of online communities: An empirical investigation. In Hindsworth, M. F., & Lang, T. B. (Eds.), *Community participation: Empowerment, diversity and sustainability.* New York, NY: Nova Science Publishers.

Bishop, J. (2010). *Multiculturalism in intergenerational contexts: Implications for the design of virtual worlds.* Paper Presented to the Reconstructing Multiculturalism Conference, Cardiff, UK.

Bishop, J. (2011a). *The equatrics of intergenerational knowledge transformation in technocultures: Towards a model for enhancing information management in virtual worlds. Unpublished MScEcon.* Aberystwyth, UK: Aberystwyth University.

Bishop, J. (2011b). Transforming lurkers into posters: The role of the participation continuum. *Proceedings of the Fourth International Conference on Internet Technologies and Applications (ITA11),* Glyndwr University.

Chen, Y., Harper, F. M., Konstan, J., & Li, S. X. (2009). Group identity and social preferences. *The American Economic Review, 99*(1). doi:10.1257/aer.99.1.431

Csikszentmihalyi, M. (1990). *Flow: The psychology of optimal experience.* New York, NY: Harper & Row.

Davis, S. (2008). With a little help from my online friends: The health benefits of internet community participation. *The Journal of Education, Community and Values, 8*(3).

Efimova, L. (2009). Weblog as a personal thinking space. *Proceedings of the 20th ACM Conference on Hypertext and Hypermedia,* (pp. 289-298).

Esposito, J. J. (2010). Creating a consolidated online catalogue for the university press community. *Journal of Scholarly Publishing, 41*(4), 385–427. doï:10.3138/jsp.41.4.385

Festinger, L. (1957). *A theory of cognitive dissonance.* Evanston, IL: Row, Peterson.

Freud, S. (1933). *New introductory lectures on psycho-analysis.* New York, NY: W.W. Norton & Company, Inc.

Gibson, J. J. (1986). *The ecological approach to visual perception.* Lawrence Erlbaum Associates.

Heron, S. (2009). Online privacy and browser security. *Network Security,* (6): 4–7. doi:10.1016/S1353-4858(09)70061-3

Howard, T. W. (2010). *Design to thrive: Creating social networks and online communities that last.* Morgan Kaufmann.

Hurley, P. J. (2000). *A concise introduction to logic.* Belmont, CA: Wadsworth.

Jennings, S., & Gersie, A. (1987). *Drama therapy with disturbed adolescents* (pp. 162–182).

Kil, S. H. (2010). Telling stories: The use of personal narratives in the social sciences and history. *Journal of Ethnic and Migration Studies, 36*(3), 539–540. doi:10.1080/13691831003651754

Kim, A. J. (2000). *Community building on the web: Secret strategies for successful online communities.* Berkeley, CA: Peachpit Press.

Lampe, C., & Resnick, P. (2004). Slash (dot) and burn: Distributed moderation in a large online conversation space. *Proceedings of the SIGCHI Conference on Human Factors in Computing Systems,* (pp. 543-550).

Leung, C. H. (2010). Critical factors of implementing knowledge management in school environment: A qualitative study in Hong Kong. *Research Journal of Information Technology, 2*(2), 66–80. doi:10.3923/rjit.2010.66.80

Maslow, A. H. (1943). A theory of motivation. *Psychological Review, 50*(4), 370–396. doi:10.1037/h0054346

Maxwell, J. A., & Miller, B. A. (2008). Categorizing and connecting strategies in qualitative data analysis. In Leavy, P., & Hesse-Biber, S. (Eds.), *Handbook of emergent methods* (pp. 461–477).

Mook, D. G. (1987). *Motivation: The organization of action.* London, UK: W.W. Norton & Company Ltd.

Moran, J. (2007). Generating more heat than light? Debates on civil liberties in the UK. *Policing, 1*(1), 80. doi:10.1093/police/pam009

Norman, D. A. (1991). Cognitive artifacts. In Carroll, J. M. (Ed.), *Designing interaction: Psychology at the human-computer interface* (pp. 17–38). New York, NY: Cambridge University Press.

Page, S. E. (1999). Computational models from A to Z. *Complexity, 5*(1), 35–41. doi:10.1002/(SICI)1099-0526(199909/10)5:1<35::AID-CPLX5>3.0.CO;2-B

Pino-Silva, J., & Mayora, C. A. (2010). English teachers' moderating and participating in OCPs. *System, 38*(2). doi:10.1016/j.system.2010.01.002

Poor, N. (2005). Mechanisms of an online public sphere: The website slashdot. *Journal of Computer-Mediated Communication, 10*(2).

Powazek, D. M. (2002). *Design for community: The art of connecting real people in virtual places.* New Riders.

Preece, J. (2001). *Online communities: Designing usability, supporting sociability.* Chichester, UK: John Wiley & Sons.

Preece, J., Nonnecke, B., & Andrews, D. (2004). The top 5 reasons for lurking: Improving community experiences for everyone. *Computers in Human Behavior, 2*(1), 42.

Rao, A. S., & Georgeff, M. P. (1998). Decision procedures for BDI logics. *Journal of Logic and Computation, 8*(3), 293. doi:10.1093/logcom/8.3.293

Rashid, A. M., Ling, K., Tassone, R. D., Resnick, P., Kraut, R., & Riedl, J. (2006). Motivating participation by displaying the value of contribution. *Proceedings of the SIGCHI Conference on Human Factors in Computing Systems,* (p. 958).

Reiss, S. (2004). Multifaceted nature of intrinsic motivation: The theory of 16 basic desires. *Review of General Psychology, 8*(3), 179–193. doi:10.1037/1089-2680.8.3.179

Rheingold, H. (2000). *The virtual community: Homesteading on the electronic frontier* (2nd ed.). London, UK: MIT Press.

Romm, C. T., & Setzekom, K. (2008). *Social network communities and E-dating services: Concepts and implications.* London, UK: Information Science Reference. doi:10.4018/978-1-60566-104-9

Scholing, A., & Emmelkamp, P. M. G. (1993). Exposure with and without cognitive therapy for generalized social phobia: Effects of individual and group treatment. *Behaviour Research and Therapy, 31*(7), 667–681. doi:10.1016/0005-7967(93)90120-J

Sherwin, A. (2006, April 3). A family of Welsh sheep - The new stars of Al-Jazeera. *Times (London, England),* (n.d), 7.

Simmons, L. L., & Clayton, R. W. (2010). The impact of small business B2B virtual community commitment on brand loyalty. *International Journal of Business and Systems Research, 4*(4), 451–468. doi:10.1504/IJBSR.2010.033423

Smith, G. J. H. (1996). Building the lawyer-proof web site. Paper presented at the *Aslib Proceedings, 48,* (pp. 161-168).

Tapscott, D., & Williams, A. D. (2010). *Macrowikinomics: Rebooting business and the world.* Canada: Penguin Group.

Thelwall, M. (2008). Social networks, gender, and friending: An analysis of MySpace member profiles. *Journal of the American Society for Information Science and Technology, 59*(8), 1321–1330. doi:10.1002/asi.20835

Thelwall, M. (2009). Social network sites: Users and uses. In Zelkowitz, M. (Ed.), *Advances in computers: Social networking and the web* (p. 19). London, UK: Academic Press. doi:10.1016/S0065-2458(09)01002-X

Vygotsky, L. S. (1930). *Mind in society.* Cambridge, MA: Waiton, S. (2009). Policing after the crisis: Crime, safety and the vulnerable public. *Punishment and Society, 11*(3), 359.

Weiler, J. H. H. (2002). A constitution for Europe? Some hard choices. *Journal of Common Market Studies, 40*(4), 563–580. doi:10.1111/1468-5965.00388

Wilkinson, G. (2006). Commercial breaks: An overview of corporate opportunities for commercializing education in US and English schools. *London Review of Education*, *4*(3), 253–269. doi:10.1080/14748460601043932

Woodfill, W. (2009, October 1). The transporters: Discover the world of emotions. *School Library Journal Reviews*, 59.

Yukawa, J. (2005). Story-lines: A case study of online learning using narrative analysis. *Proceedings of the 2005 Conference on Computer Support for Collaborative Learning: Learning 2005: The Next 10 Years!* (p. 736).

Zhang, P., Ma, X., Pan, Z., Li, X., & Xie, K. (2010). Multi-agent cooperative reinforcement learning in 3D virtual world. In *Advances in swarm intelligence* (pp. 731–739). London, UK: Springer. doi:10.1007/978-3-642-13495-1_90

This work was previously published in Virtual Community Participation and Motivation: Cross-Disciplinary Theories, edited by Honglei Li, pp. 160-176, copyright 2012 by Information Science Reference (an imprint of IGI Global).

Chapter 10

Negotiating Meaning in a Blog–Based Community:
Addressing Unique, Shared, and Community Problems

Vanessa P. Dennen
Florida State University, USA

ABSTRACT

This chapter addresses how members of a blog-based community share problems and support each other in the problem solving process, both sharing knowledge and offering support. Problems are divided into three categories, unique, shared, and community, each having its own particular norms for presentation, knowledge sharing, and resolution. Additionally, processes may differ based on discourse that centers on one individual blog versus discourse that spans multiple blogs. Findings show that intersubjectivity, norms, roles, and individual ownership of virtual space all are important elements contributing to the problem sharing and solving process.

INTRODUCTION

Virtual communities exist for a multitude of reasons. They may support work, learning, and personal development processes; provide a common arena for information sharing; or serve as entertainment and social spaces. In some cases they bring together people already organized as a community, providing additional means of communication. In other cases, the technology provides a platform for uniting people with mutual interests who might otherwise never meet. They may be closed and highly private (e.g. membership-only portals and listservs), they may flow and be defined through extended personal networks (e.g. accessible to people noted as friends of friends, such as is done on various social networking sites), or they may live out their existence in plain view (e.g. public web pages such as most blogs and Twitter feeds). Regardless of the specific context, interpersonal interaction is the underlying objective of participation in a virtual community.

Online information seeking is a fairly common activity with ever increasing numbers of people

DOI: 10.4018/978-1-4666-2803-8.ch010

turning to the web for answers to their questions and solutions to their problems ranging from simple and fact-based to complex and subjective. Trustworthiness of information becomes an issue for many of these people. Individuals can readily publish to the web so long as they have a computer and Internet access, with no vetting of content for accuracy or quality. A simple query in a search engine can yield pages of results: some accurate and some not; some empirical and some opinion-based; and some from reviewed or filtered sources and some not. Such is the beauty of the Internet and the simultaneous challenge of using it in an information seeking capacity.

If we marry these two concepts of virtual community and online information seeking, it becomes apparent that in the nexus is a possible solution for narrowing the expanse of the information landscape, increasing the pertinence and (perceived) reliability of one's online knowledge seeking and problem solving experiences, and encouraging a more personalized approach to the whole experience. The result, whether intentional and by design or merely a byproduct of the users' actions, is a knowledge sharing community.

In this chapter, I present a study of how a blogging community serves a knowledge sharing function for its members, with a particular focus on how members share their problems and then help each other by providing support, guidance, and information. The community is one of loosely networked sole-authored blogs written in a diaristic style. The bloggers generally do not know each other off-line, and many use pseudonyms which prevent actions like Googling each other, but they are bound by a shared profession: academe. In the blog world, they have sought interactions with like-minded others. Collectively, and in addition to socializing, these bloggers engage in knowledge sharing and problem solving processes on a regular basis. They do so informally, and without labeling their actions as such.

BACKGROUND

Virtual worlds do not exist in isolation from the physical world; what happens in one space readily influences the other. Bloggers, for example, frequently make connections between their online and offline worlds, which may represent actual friendships transcending the different media (Takhteyev & Hall, 2005) or may take the form of shared stories moving from one medium to another. Friendships that form online may not be as strong as those existing in the physical world (Cummings, Butler, & Kraut, 2002), but they can provide social support and contribute to a person's sense of well-being (Baker & Moore, 2008). Further, there are distinct purposes that those online relationships serve, such as providing support in contexts where face-to-face settings might be uncomfortable or unavailable.

The term "community" is one that has been used in an overly general sense, particularly with respect to online communities. Seemingly any collection of people drawn together by common interest or shared membership on a web site has been referred to as a community, a designation which becomes problematic given that community development requires more than just the ability to communicate in the same space (Kling & Courtright, 2003). In this paper, I use the term community to refer to collections of people who interact online on a regular basis, who have established online identities and are known to each other, and who exhibit collective characteristics such as trust, norms, and shared purposes. As such, communities are not mere message boards but are trending toward (if not actually representative of) what Wenger (1998) calls a Community of Practice.

Virtual communities exist based partly on the principle of reciprocity. Both socio-emotional and informational exchanges are likely to occur when a sense of virtual community has formed among participants (Ellonen, Kosonen, & Henttonen, 2007). The rationale for engaging in online communities for knowledge sharing purposes can both be for personal information gain as well as to support one's own social capital within a group (Kosonen, 2009).

Knowledge sharing in online communities is, essentially, a form of problem solving. Indeed, problem solving has been deemed a form of producing assets within a community (Wenger, White, & Smith, 2009). Although it is possible for individuals to simply share their knowledge for general purposes, many people are gravitating toward online communities to seek help in solving their own problems, to find others who share their problems and search for solutions together, and to share their own problem solving experiences in the hopes that others may find them helpful. Josefsson (2005) found that critically ill patients in Sweden used online communities for support purposes and pooled their information from online and offline sources in the interest of helping themselves and each other as they fought their illnesses. Similarly, Fox (2008) found in a survey of American patients that providing help and sharing knowledge about conditions and treatments is a common behavior. Chayko (2008) notes that people frequently cite helping others as a reason for participating in online communities, and telling stories that may help others learn from their experiences is a key motivation for some people to blog (Lee, Im, & Taylor, 2008; Lenhart & Fox, 2006; Ridings & Gefen, 2004). Although many examples of problem sharing in online communities focus on individuals with personal problems, as discussed above, the Internet also has become a platform for supporting larger scale community problem solving initiatives, such as civic engagement activities (Smith, Schlozman, Verba, & Brady, 2009).

Problem solving and knowledge sharing in online communities may be focused internally or externally. Two well known virtual ethnographies provide excellent contrasting examples: Rheingold (2000), in his book about a community called *The WELL*, discussed how community members came together in each others' times of need to provide support and assist in solving problems existing outside the community. Dibbell (1998), on the other hand, discussed how members of an online community called LambdaMOO approached a problem that existed entirely within the community, namely an offense committed by one community member against various others. Regardless of the locus of the problem – within or outside the community – the process of finding a solution in these as well as in other online communities relies on open communication and knowledge sharing among the membership.

A key element of knowledge sharing in online communities seems to be the attraction of like-minded people. Chayko (2002) notes that online relationships often help people feel less alone and more connected, not just on the level of having someone with whom they might communicate but at the level of being understood and sharing experience. Karlsson (2007) found that diary blogs tend to attract a fairly homogenous reader group who is similar also to the originating blogger. This assertion of homogeneity in blog-based networks is supported by other research (Kumar, Novak, Raghavan, & Tomkins, 2004). Further, bloggers have been found to have a particular audience in mind, even if that audience is comprised of unknown others (Qian & Scott, 2007). This assertion of like-mindedness does not mean that all participants must agree. Differences of opinion can readily be held within an online community, with trust as a critical factor in helping individual members assess the usefulness of information (Greenfield & Campbell, 2006). Still, individuals with differences treat each other respectfully to help maintain the sense of community (Dennen, 2009b). It is thus homogeneity of interest and not always of opinion that brings together a community.

Finally, it should be noted that interaction and community are not inherent to the practice of blogging, and not all bloggers desire feedback from an audience. In a sample of 260 bloggers,

Papacharissi (2006) found that self-expression was the key purpose for many bloggers and that while sharing information was of secondary importance, they were not heavily reliant on or anticipatory of feedback from readers. Further, many blogs exist in isolation, unconnected to others (Herring, et al., 2005). Among such bloggers, the likelihood of a knowledge community developing is slim, although there still might be knowledge to glean from reading these blogs. Thus, knowledge sharing and problem solving among bloggers is a deliberate activity sought out by particular individuals.

Research Context

In this chapter, I describe how problems are presented and discussed within a blogging community, a process that extends from initial sharing of the problem to (potentially) the indication and tale of resolution. In an attempt to organize, I developed a system for classifying problems according to factors that impact their presentation, discussion, and resolution. Specifically, I describe how this classification scheme was developed and then how interactions in a blogging community take place within each category.

When bloggers post about their problems they effectively initiate a knowledge sharing dialogue. That dialogue might include sympathetic or empathetic comments, shared experiences, support, and advice. Further, there are many potential turns and outcomes that can occur via that dialogue. It may present what ends up being the solution, being the catalyst for the blogger finding a solution, or it may lead nowhere at all. When multiple people respond, it may result in divergent or convergent thoughts, and as a result may bring people closer together or further apart, or otherwise alter the nature of their relationship.

We might view problem presentation and resolution within a community of bloggers as existing along two dimensions: number of bloggers posting about the problem and number of

bloggers affected by the problem. I use the term problem loosely; it is not necessarily indicative of a negative or conflict situation or experience, but rather of the need to address a situation, perhaps making a decision or taking action. An example of a non-conflict problem situation might be as follows: It's the end of the budget year and a faculty member has been told that she has $2500 in professional development funds to spend or forfeit in the next month. She is initially uncertain how to spend that money and seeks ideas.

For the sake of classification, I differentiate between those problems that are unique (affect one person), shared (affect multiple people in their own ways) or communal (particular to a community). Further, I note whether the problem is raised by one blogger or multiple bloggers. The classification system thus has 5 points:

1. Individual Unique
2. Individual Shared
3. Multiple Shared
4. Individual Community
5. Multiple Community

The reason for the classifications is that the nature of the interaction that surrounds a post about a problem varies based on these dimensions. Note that there is not a classification for Multiple Unique. Inherent in multiple people posting about a problem and there being a collective meaning in that act is some dimension of shared meaning and experience as well as a way to define the bounds of the phenomenon being discussed. Thus, although multiple bloggers can and do post about their unique problems, there is no clear rationale for grouping and discussing these otherwise discrete posts in a collective manner. Further, it is possible that multiple bloggers will post individually about one particular blogger's unique problem, seeking help or support for that problem from as wide of an audience as possible, but these requests tend to be focused more on garnering well wishes (e.g.

for someone's dissertation defense) or material support (e.g. for an ill person who lacks health insurance) and occur rather infrequently.

METHODS

Population and Sampling

For this study, I drew upon examples from a blogging community of academics. These bloggers tend to write in a diaristic manner, documenting elements of their personal and professional lives as well as the space where the two intersect. The community is informal and ill-defined, consisting of individual blogs that are interlinked to varying degrees and whose authors read and comment on the other blogs on a regular basis. For further discussion of what makes these blogs a community, see other works by Dennen and colleagues (Dennen, 2006; Dennen, 2009b; Dennen & Pashnyak, 2008).

Examples of each problem situation were selected using a quota sampling procedure. Essentially, at the time I choose to begin sampling, I conducted my regular daily observation of the blogs and each time I encountered an example of one of the problem situations I earmarked it until I had ten in each classification.

Examples to use as case discussions here were selected purposively in an attempt to represent a typical or mainstream case within that category. The determination of a case's suitability was determined not only based on its representativeness of common features in the 10 sampled blogs for each classification but also based on an overall fit noted from my extensive experience within this blogging community. By presenting the cases structurally and with secondary descriptions, each case effectively has been anonymized.

Procedure and Analysis

The procedure for this study involved three phases or levels of observation and analysis. In the first phase, general observations were culled from a year of field notes and archived blog posts. The field notes were related to a larger ethnographic study being done on this community and were not solely focused on documenting instances of problem situations on blogs, but nonetheless captured blogging about problems and their resolutions as a common narrative theme across the blog community. Through these observations, the classification scheme emerged. The classification scheme was then tested or triangulated by looking for evidence of each of these five classes of problem solving interactions in four additional online journaling communities in which I have either participated or researched in the last decade.

In the second phase, I returned to the academic blogging community and identified 10 instances of each problem class. For each of those 10 instances, I looked at trends and differences in how the problem sharing and solving process played out via the blog. The focus at this point was to identify the following elements:

- **Nature of the Problem Type:** What types of problems fit this category?
- **Typical Problem Presentation:** How did the blogger present the problem and what type of feedback, support, or interaction was being requested? In what way was feedback being requested?
- **Response and Knowledge Negotiation:** How did commenters respond to this type of problem? How did they negotiate their understanding of the problem, of what type of help the blogger was seeking, and of what would be a good solution?
- **Resolution:** Was problem resolution addressed on the blog? Did the commenters play a role in the resolution?

Note that one instance might span multiple blogs and multiple posts; only the simplest of problems were limited to discussion in a single post. From these instances, a description of typical phenomenon (or the range of typical) was developed. These descriptions then were triangulated against my field notes, which covered a far more extensive collection of problem-related posts although in less detail. Further, I returned to two other online communities from phase one to check that my descriptions would apply to problem situations in these communities, too.

In the third phase, I selected a case from each main category or problem type: unique, shared, and communal for further examination. For shared and communal problems I selected cases that demonstrate dialogues across multiple blogs. For the sake of simplicity in diagramming and discussing the cases, I purposely chose cases that, while typical, were relatively small in scope.

After selection, each case was analyzed for structural elements and major narrative events, and then I created a diagram demonstrating the key points of interaction that took place. Note that while these cases are not suitable for generalizing how many comments a problem post will receive or how quickly a problem will be resolved, they are useful for gaining a more vivid sense of how the blog-based problem solving process works.

Limitations

The major limitation of this study is its scope. It is a study of one community whose members are fairly homogenous in terms of career interests and level of education (although homogeneity of membership, as noted above, is not uncommon within virtual communities). It is the very nature of studies such as this one that challenges our definition of generalizability. That said, progression from a single case to a broader application can be part of the inductive theory building process (Reigeluth & Frick, 1999). To ensure rigor, Reigeluth and Frick recommend the consideration

of situationality, or the applicability of the theory to varying situations. In this study, the analysis is based on a sample chosen via a quota procedure, and all problem situations encountered during the sampling process were readily assigned to a problem category. Thus, within the community being studied the applicability of findings may confidently be extended beyond the sample.

Still, these findings are not generalizable to other communities and settings in the same way that quantitative findings often are. Stake discusses the concept of naturalistic (1978) or petite (1995) generalizations, in which the reader judges for herself the applicability of research findings to her context based on observed similarities. For this reason, rich description of the research setting and findings is important and has been included in this chapter. I would encourage readers of this chapter to determine the suitability of these findings for use in their own virtual community contexts with these considerations in mind.

Further, the classification system was developed using a process of triangulation (described above) that involved looking for evidence of these types of problems in other online communities with which I am familiar. These other communities represent quite different interests than this academic one (animal breeding, wedding planning, motherhood, and crafting), and yet evidence of all of the problem types and related problem-solving discourse was found in each of these communities. I chose to not include these other communities in the analysis and cases presented in this chapter because I have not studied them as closely and systematically as I have the community of academics.

Presentation of Findings

I discuss each classification of problem situation separately, organized by type of problem: unique, shared, or community. Within each of these areas, I discuss the differences that tend to occur when the problem is presented by individual versus multiple blogs. I wish to note, importantly, that the blogs used in this study are not solely focused on problem solving. They are social spaces, and they are journals. While bloggers may post on problems of various magnitudes, they also may write about topics as diverse as hobbies, likes and dislikes, and current events. At other times, they may merely chronicle the events of or tasks completed on a given day. Some posts may be written for the blogger, as a form of personal documentation, whereas others may be written directly to a reader audience. The posts that focus on problem situations thus represent merely a portion of the discourse that takes place on the blogs, much as problem sharing and solving would be merely one of many activities for a face-to-face community.

UNIQUE PROBLEM

Nature of the Problem

A unique problem is one that affects just one blogger and is so specific in its pertinent details that others are unlikely to have a directly related, shared experience. Thus it is the domain of the individual blogger. Typically these problems are interaction or situation specific and not representative of larger systems (e.g. organizations, institutions, or communities). In other words "I don't have enough time to write" is not a unique problem, but "My department chair is continuously favoring another junior faculty member over me" is. Interpersonal relationships tend to be a frequent topic of unique problems, as do particular life events and opportunities. Not all unique problems are simple and potentially solved in a short time frame. Sometimes the problems presented have been plaguing the blogger for a long period of time – years or decades, even.

Problem Presentation

Individual bloggers tend to present their problems in two main ways: reflective and advice seeking. Those who are reflective appear to be writing out their problem situation primarily for themselves. In contrast, those who are advice seeking tend to ask questions of their readers, such as "What would you do?" "Does anyone have any ideas?" or "Am I overreacting?" Additionally, advice-seeking posts are more likely to include a back-story and are more extensively detailed. In other words, they are written with an audience in mind and the awareness that the audience will need sufficient information to provide useful responses.

Many unique problems unfold over the course of multiple posts; this statement is certainly true of the more complex ones. Follow-up posts that engage in problem elaboration tend to either provide updates since the previous post or additional details that were omitted from the previous post. In the latter case, these posts frequently are prompted by comments that either ask for additional information or make clear that the readers do not fully understand the blogger's problem.

About one-half of the time, the blogger will include potential solutions that she has considered so far. This phenomenon occurs regardless of whether the post seems more focused on personal reflection or is clearly seeking advice.

Response and Knowledge Negotiation

When bloggers post about unique problems, they generally do not anticipate that others will be able to provide a precise solution. Advice seekers are most open to comments and a variety of solution ideas. The elements of personal experience and what is called 'I' language in a therapy context (not to be confused with I-language in a linguistics context) are prevalent in the comments. Although commenters occasionally will trend toward the pedantic in their advice, much more often they will share how they would feel and what they might do, and then offer support. Thus, possible solutions are presented as what I would do and what you might or could do, but not as what you should do. The linguistic distinction there is important. Commenters generally do not assume that they know what will work for the blogger, but rather seem to be engaged in a form of brainstorming for the blogger, using their personal experience, problem solving skills and the Internet (they frequently link to resources). Further, commenters frequently will add lines such as "This worked for me" or "YMMV" (your mileage may vary), indicating their awareness that individual differences and preferences may affect the usefulness of any advice. This type of communication is indicative of the underlying sense of community among the participants.

Reflective bloggers tend to seek support and sympathy more than advice or solutions, and when others offer solutions the blogger often is quick to reply with additional details about how the offered solution will not or will only partially fit their specific problem. Even when commenters relate to the blogger's problem through similar experiences, the blogger typically holds onto a notion of his or her problem being unique. In other words, there may be a bit of resistance to accepting that what seemed like a rather unique problem may, actually, be on a fairly universal theme. Commenters and the blogger must negotiate a comfortable space with plenty of room for individual experience in order to successfully discuss the general problem area and perspectives on both the problem and how it might be addressed.

Occasionally, individual bloggers are simply posting about their problems to vent. In these instances they do not seek a response or reaction from others. If not stated clearly in the original post, upon receiving advice in the comments these bloggers may note in the comments that they are not seeking advice.

Regardless of post type advice-seeking or reflective), most commenters are not only commenting on someone else's problem, but also trying to maintain network relationships. This phenomenon is implicit in many of their actions and word choices. They are sensitive to each other's feelings, particularly those who are in need or experiencing a conflict-oriented problem. Occasionally they even make explicit remarks to this effect. In other words, they do not add to the blogger's stress by introducing additional conflict via comments. That said, at times additional conflict arises inadvertently, as noted when the blogger returns with a post or comment to that effect. In such an instance, the commenter typically takes one of two paths: (a) comments no further; or (b) apologizes and tries to smooth over the situation. The third option, which is to argue one's position, is infrequently taken and generally leads to a breakdown in the relationship between the blogger and commenter (supposing one existed). Occasionally a third party will enter the fray, attempting to make peace between the blogger and commenter who are in conflict, validating both of their feelings and actions.

The above holds true when the commenter is a member of the blogger's community. However, trolls (e.g., people looking solely to cause disruption or conflict) and one-time visitors also are possible commenters. Given their lower level of investment in maintaining a relationship with the blogger and, in the case of trolls, a general desire to create conflict, when these visitors comment they are more prone to violate community norms than others. However, comments that come from community outsiders appear to be much easier for insiders to dismiss as irrelevant. The general wisdom of the community is that one should not engage in discussion with the trolls. Community members may, in comments, call them out as trolls and occasionally the blogger will delete the comment altogether.

Resolution

Typically, if a blogger has either posted extensively or received extensive support or feedback about a problem, she will provide an update as to the status of the problem and any resolution that has occurred. In a resolution post, the blogger typically thanks her readers for their support and ideas. She may explain which ones she tried and which she did not, indicating her rationale for their suitability and ultimate effectiveness.

In eight of the ten instances sampled for this analysis the blogger posted about problem resolution within a month of the original post. In some instances, many follow-up posts were provided, describing the ongoing situation in a blow-by-blow manner. The depth of follow-up was related to the depth or severity of the problem in a parabolic manner. In other words the most minor and most severe of problems tended to have the least amount of follow-up, whereas those problems of medium consequence seemed to have the greatest amount of follow-up. This phenomenon seems likely due to the level of investment and emotional charge. Low levels of investment and emotional charge make an extensive update potentially unnecessary or hardly worth the time it takes to type. High levels of investment and emotionally charge, conversely, may make thorough follow-up overly time-consuming and potentially re-open wounds.

Case Example

In the selected case, the individual blogger composed a post about an interpersonal problem she was having with a colleague. Basically, the colleague had taken an action that made the blogger feel as if she were not respected. The initial post had a clearly emotional tone, and was more reflective than advice seeking. While the blogger included lines like "I'm not really sure what to do now," she did not directly ask for advice. However, that did not prevent her commenters from giving advice. Across 18 comments, 16 people addressed the situation. All offered general sympathy and support, whereas nine shared similar stories of conflicts and resolutions and five offered ideas for solving the problem. These comments spanned three days time, but 15 of them were written within the first 24 hours that the post appeared. This immediacy of response, as depicted in Figure 1, is important because it demonstrates how quickly a problem post is noticed and support is offered by regular readers.

On the second day, the blogger wrote another post about the problem. This follow-up post addressed various issues raised in the comments and added details about the problem. The tone was far less emotional with regards to the problem – presumably the previous post had been written in the heat of the moment whereas this second one

Figure 1. Unique problem posting patterns

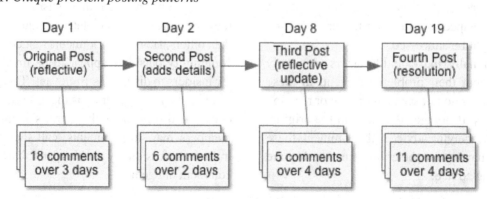

was written after a period of cooling off – although somewhat defensive in response to the comments on the previous post, some of which had gently suggested that the blogger was engaging in behavior that might exacerbate the problem. The first few comments on this post were written concurrently with the comments on the first post from the day before; two readers noticed this second post moments after sharing their thoughts on the first and followed up with an immediate second comment. Each of the six commenters was someone who had posted on the previous post, responding to the additional information posted by the blogger as well as the emotional implications of her tone. Three commenters who perhaps had misinterpreted the first post also made an attempt to smooth over any misunderstandings.

One week after the initial post, the blogger posted an update to the situation. She had both tried a new technique when interacting with her colleague and had spent some time thinking about her initial interpretation of the problem. She had decided that she may have been misreading her colleague's actions. Finally, about 1.5 weeks later, she posted her last post on the problem, providing some resolution. In short, after the third post she decided to open up a dialogue about the conflict with her colleague. It turns out that the colleague was unaware that she had felt there was a conflict between them, and they were able to negotiate an understanding between them. This final post attracted 11 comments, ten of which were written by commenters on the original post.

SHARED PROBLEMS

Nature of the Problem

A shared problem is one that many people have, although the problem may manifest itself in different ways for different people. Within the community of academic bloggers who I have been studying, some shared professional problems revolve around the common work areas of teaching and scholarship as well as issues related to professional development, employment, and sense of self. Some of these problem areas included course design, grading, classroom management, dealing with rejection, job hunting, and imposter syndrome. A variety of personal problems are shared in this community as well, on topics such as personal finance, relationships, and childcare. Note that each of these problems is common to many people, but tends to affect them as individuals. For example, my time management issues and someone else's may be similar but likely are not exactly the same. We know we can anticipate some commonalities as well as some differences in the details. The differences generally are neither so great nor so personal in nature that we feel we cannot relate to each other on the topic.

Problem Presentation

To some degree, the differentiation between unique and shared problem is being made based on the blogger's presentation and perception of the problem. A blogger who is posting about a shared problem will include direct statements about the shared nature of the problem, such as "I know some of you have dealt with this before." The blogger might even indicate or link to specific instances in which others have shared about having the same problem. If the blogger is uncertain as to the degree to which the problem is a shared one, but is inclined to define it as such, she might state something like "Surely I'm not the only person trying to figure this out."

When writing about the problem, the blogger may glide over details with which the intended audience should already be familiar. This phenomenon shows how critical audience is to the problem presentation as well as the definition of the problem as shared. For example, an academic blogger may write about an upcoming dissertation defense without describing committee composition, defense format, or other information that would be known to other academics. That same blogger, when discussing the same problem with non-academic family and friends, would change her problem presentation to either simplify the context, make it analogous to something familiar to the audience, or explain unfamiliar points as asides. The ability to streamline elements of the problem presentation is an appealing one-to-many bloggers, as is knowing that an audience with shared understanding is likely to provide feedback that fits the context. This phenomenon has been noted directly on blogs, with bloggers stating outright why they at times prefer to blog about certain issues rather than discuss them with family and friends as well as why they get frustrated when explaining elements of academic life to outsiders who have ill-formed ideas about what life as a graduate student or professor is really like.

Response and Knowledge Negotiation

Most of the response to shared problems comes from people who share the problem, which is most likely not a blog's overall audience, but rather a subgroup. That said, people with like backgrounds and thus greater shared experience in the problem area tend to gravitate toward each other in the blogosphere. Commenters on shared problems, then, are most likely to fit one of three categories:

1. People who have solved the problem for themselves.
2. People who currently share the problem and commiserate.
3. People who anticipate having the problem.

Each group of commenters will reply in a different way, based on their own experiences. Further, each seems to serve its own important purpose(s) in the problem response process (see Table 1), ranging from proving the problem can be solved to looking up to the person with the problem.

With a shared problem, knowledge negotiation becomes increasingly complex, but also tends to be less emotionally charged. Mutual problem sufferers willingly share their ideas and resources with the expectation that pooled knowledge and alternate perspectives can only help. There tends to be acceptance that everyone will experience the problem differently. Pressure to accept someone else's solution is lessened (even if only in perception), and attention may not be directly focused on the poster at all times. It becomes increasingly appropriate for others to discuss their problems and seek assistance – although briefly – in the comment space.

Resolution

Resolution of shared problems typically comes in increments, if it all. Many of these problems represent ongoing issues that face bloggers, problems that are never fully resolved (e.g. time management issues) or that extend for a pre-determined time period and then go away regardless of resolution (e.g. end of semester or tenure review).

However, when there is a resolution post it can be useful to many readers. Whereas the resolution of a unique problem is merely a point of curiosity ("Ah, so that's what happened!") or satisfaction ("She took my advice and it worked!") to most commenters, a resolution statement about a shared

Table 1. Commenter function in response phase

Category of Respondent	Purpose in Response Phase
Has solved problem	Provide reassurance and support (experienced cheerleader) Share proven solution(s) and resources Set role model (motivational)
Currently shares problem	Remove sense of isolation Share potential solutions and resources Collaborative exploration of solutions
Anticipates having problem	Observe Offer respect and support (inexperienced cheerleader)

problem becomes a post from which others can learn. Given that most blogs are archived and searchable, people who later have that problem (e.g. the initial readers who anticipated facing it at some point) may return to original post when it is more timely for the individual. Further, the resolution can be linked to, and can become part of a knowledge base on the topic.

Individual vs. Multiple Blogs

The key differences between how shared problems are mentioned on individual versus multiple blogs are related to the dispersion of ideas and the indicators of resolution. Simply put, when a problem is discussed across multiple blogs the dialogue becomes more difficult to follow. Individual readers, by choice or by chance, are likely to read parts of the dialogue on a problem topic, but not all of it. As a researcher diligently and systematically following these blog-based conversations I find it difficult to say with 100% certainty that I have found every last related thread of a cross-blog conversation. By contrast, following a single problem issue on an individual blog is rather simple. If you have the first post and the last post, you merely search all posts in-between. Relevant comments will be attached to the posts

It is useful, at this point, to describe how shared problems are manifest across multiple blogs. The genesis of the problem blogging activity can occur in one of two ways. In the first way, an individual blogger posts about a problem that she has. Her audience reads that post, and some people comment. However, her post also inspires some of her readers to compose their own post on the same topic. Why post to a different blog rather than just as a comment? Typically such actions are due to perceptions of ownership and space; specifically, within this community a blog is like a journal and it is considered a space for one person to write about topics of her choosing. There is a term that is used in the blogosphere to describe the act of writing a lengthy, self-centered comment

underneath someone else's blog post: hijacking. Hijacking generally is frowned upon, and people who find themselves inclined to write such comments are encouraged to instead compose a post for publication on their own blog. Additionally, a person may want to attract more attention than would be garnered in the comments or address the specific audience for her own blog, which may differ slightly from the blog where the inspiration was derived.

The rationale of the bloggers is not being intuited here; rather, bloggers outright engage in moments of meta-blogging in which they discuss their intent and actions. For example, one commenter wrote, "Wow. I have a lot to say about this. Rather than hijack your comments, I'm going to post on my own blog." In another case, a blog post begins with an indicator of inspiration and reason for posting on her own blog rather than as a comment elsewhere:

In case you haven't seen (blogger)'s post about (topic), you should. I'm not going to link to her post because I don't want to connect directly to it. I disagree with much of what she says, but I don't want to send people over from here to disagree with her or disrupt the conversation she has going on. I have some very different personal experiences with (topic) and I want to discuss them here, with my peeps.

Alternately, and less common, some bloggers post in comments rather than on their own blogs to achieve a sense of distance between their blogging identity and the comment. As one blogger led in to a comment, "I'm not comfortable saying this on my own blog, and I'm writing anonymously, but you'll know who I am." These forms of meta-blogging are fairly typical among discussions of shared problems and occurred multiple times within the sampled problem cases.

In the second way, individual bloggers post about the same problem or related problems unbeknownst to each other. Eventually one of them sees the other's post or a third party may make the connection and point it out by commenting or composing yet another post that links to the two previous ones and ties them together. In this instance, the conversation has developed in parallel at multiple isolated locations. When the point of connection or synthesis occurs, it is illuminating to see the commonalities that exist (evidence of shared problem). In fact, this process may sway an individual blogger from viewing her problem as unique to viewing it as shared.

Regardless of how and by whom a problem initially is presented within a blogging community, it is typically in other venues and through the actions of other bloggers that synthesis is found. Bloggers use devices such as "round up" posts or "carnivals" that provide links to and sometimes provides synthetic discussion of a group of related and relevant posts. On occasion, a shared problem might result in the formation of a support group, and perhaps even a related group blog. Such was the case when a group of bloggers wanted to share some physical fitness tips and accountability, in one instance, and when a group of graduate students writing their dissertations chose to develop a support group in another.

Finally, in terms of resolution, when a problem is shared across multiple blogs there are multiple locations in which one might look for indicators of resolution. Individual bloggers may or may not follow up with what they did. However, some bloggers may have their initial entry into the fray at the point of resolution. On occasion, a person who did not blog about a problem, but who read about or participated in problem solving comments on other blogs, will write a resolution-style post giving a post-facto account of how she used tips or advice posted elsewhere to improve her own practices.

Case Example

The problem topic of the selected case was about dealing with a disruptive student. Specifically, Blogger 1 (B1) was struggling to maintain control over a class with two outspoken and assertive students who were overrunning the discussion. B1's blog was not heavily trafficked, and the post received two comments during its first two days. However, on the third day two things happened. First, another blogger (B2) also wrote about dealing with a disruptive student and a third blogger (B3) noticed and linked to both of those posts, commenting on how he typically deals with such students. It was after the appearance of B3's post that B1 received an additional 6 comments, two of which explicitly mentioned having come over from B3's blog. Figure 2 depicts this relationship between posts and how they became interlinked by various bloggers.

Although the specific details of B1 and B2's problems were different, the more general issues of instructor persona (how to be someone students will respect) and classroom management techniques (what to do in response) were raised. Most of the commenters on their posts as well as on B3's post shared their own stories of disruptive students and how they have either dealt with them or avoided disruption altogether.

On the fourth day, B1 wrote a very brief post, connecting to B2 and B3's post and noting that apparently he was not alone in his problem. He further commented that it was helpful to have so many ideas and so much support from everyone, and directed his interested readers to check out the other related posts. An additional blogger, B4, entered the discourse on that same day and connected to the earlier posts of all three bloggers. B4 had yet to teach her own course but would be a TA during the next semester. She expressed fear of disruptive students and uncertainty about other classroom management issues. Her post generated 17 comments, most of which were encouraging.

Figure 2. Shared problem case posting patterns

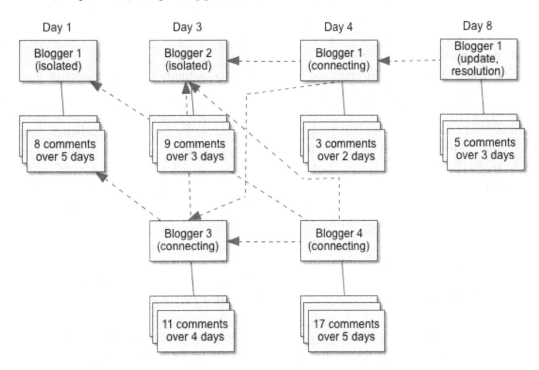

Finally, a week later B1 posted an update to the initial post. He had taught the class with the disruptive students again that morning and had tried two of the recommended techniques. One he felt had been successful and the other he was not sure about, but was planning to try it again. Regardless, he had felt much more in control of the class that morning, which was a positive move. The commenters congratulated him on taking control of the situation.

B1 is arguably the center of this shared problem narrative, but it is important to recall that this perspective is representative of a thorough search for and analysis of all interlinked posts. Any individual blogger or commenter's perspective will differ based on where she entered the narrative and how many of the posts and comments she read.

COMMUNITY PROBLEMS

Nature of the Problem

Community problems are, as the title suggests, problems that face a community as a group. Whereas shared problems are common but occur in each individual's own context, community problems impact people within a shared context. Within the community of academics, two types of community problems have been the topic of posts. First, there are problems that impact the blogging community. These tend to be issues related to blog identity (e.g. privacy and pseudonym use), technology (e.g. a bogging, commenting, or photo sharing platform that many of the bloggers use), or community interactions (e.g. civility, norms, and presence of trolls). Second, there are problems that impact a secondary community in which a number of these bloggers participate. In the instance of academic bloggers, sub-groups belong to the same professional organizations (e.g. Modern Language Association, American Historical Association). Members of the American Historical Association found the location of the organization's annual meeting in 2010 problematic because of the hotel owner's political leanings. Blogs provided a forum for discussion of both the organization's response as well as what individual members might do to both attend the conference and remain true to their own beliefs (see Jaschik, 2005 for additional information). Regardless of whether the problem is focused on the blogging or an off-blog community, the critical point in this definition is that the problem affects people as a byproduct of their membership in the community.

Problem Presentation

The presentation of community problems is not unlike the presentation of shared problems. It may be initiated on one blog or on multiple blogs with points of connected discourse. Individuals will use their own blogs as the platform for presenting their personal opinions about the problem and, at times, their solutions. Since the problems are things that make it challenging to function within the community, they tend to focus on reasons why someone would not want to remain in the community or changes that will be necessary to be a comfortable or productive participant. These problems generally tend to be less personal in nature (unless, of course, the individual interprets a community problem as a personal attack), and more structural or political.

For problems that are related to the blogging community and how it functions, such as trolls, privacy, or technology failures, the first indication of a problem may occur in the comments, not a main post. For example, a conflict may arise when troll-like comments appear or someone posts a counter-productive statement, or a commenter may alert the blogger to a technology malfunction such as a blogroll (a list of links to other blogs that appears on the sidebar) that is not displaying

correctly. Typically these problems then are raised again at the post level.

Posts about external-community problems are likely to link to evidence of the problem (e.g. the other community's web presence or a relevant news story) or quote relevant items within a post. Those presenting the problem demonstrate an awareness of a split in their audience: some who share that external community and some who do not. Some back-story or contextual information might be necessary for that second group of readers and typically is included even though their status typically is that of a distant secondary audience for the post.

Response and Knowledge Negotiation

For within-community problems, there generally is incentive for a strong outpouring of support and free sharing of solutions. Even when one blogger has not been personally afflicted (e.g. a bug has downed comments, but she uses different comment technology, or others have been hit by trolls in the comments but she has not), there is a general sense that any action that weakens the community is worth concern.

A community-level response is evident in many of these cases. The community will respond by pooling its knowledge. Those who have solutions will post them. Early and visible problem solvers will often serve as models for others, and iterations of solutions may appear, each round representing a higher degree of sophistication. The unafflicted will take proactive measures. For example, when one person is affected by a breach of privacy and ratchets up his or her privacy controls and site statistic monitoring, others tend to do so as well in order to avoid the problem. Alternately, a bit of politicking might take place as certain bloggers encourage others to follow their lead (e.g. boycott a particular technology).

It is in response to within-community problems that individual bloggers are most likely to step in and try to directly fix a problem for each other. So, for a technology problem a more tech-savvy blogger may write a piece of code for someone else, call and walk someone else through a fix, or gain access and directly fix the problem themselves.

For a problem like incivility within the comments, which violates a community norm, a commenter from the community may defend the blogger or even attack the negative commenter. Following on to such an event, bloggers may develop or refine their own commenting policies and perhaps even turn on comment moderation. Such actions tend to be noted and discussed when they occur, and often travel through segments of a community like a small wave.

For external-community problems, the within-community response tends to be less great. This phenomenon is not surprising, really; large portions of the community are generally unaffected by the problem and do not anticipate being affected by it, either. Responses from those who share both communities tend to focus on commiseration or presentation of alternate perspectives. Responses from those who do not share the external community tend to focus on support from the outside and examples of how members of other similar communities have addressed like problems. Outsider responses, from people who belong to the external community but not the one where the problem is being discussed, also are likely to appear. One thing is common across all community problem responses: the members flexibly negotiate the allocation of virtual space such that it can at different times be a platform to discuss my individual experience, your individual experience, or our collective experience.

Resolution

Depending on the details of the problem, there may not be a resolution to post. Some of these community problems are ongoing, and some are in areas where there is no individual element of

the problem to solve aside from the decision to continue or discontinue affiliation with the community. Else, resolution patterns and norms mimic those for shared problems (see above).

Multiple vs. Individual Blogs

As with resolution, the patterns and norms for shared problems (see above) tend to apply to community problems. It is far more common to see community problems discussed across multiple blogs. Alternately, when a community-level problem is only mentioned on one blog it might be taken as an indicator that few people perceive or agree with the problem within-community problems, in the case of external community problems, few within-community members share that same external community.

Case Example

In the selected case, the discussion topic was the grant application and awarding process within a national funding agency. Again, one blogger's posts rose to form the central part of the narrative, and it is this one blogger (B1) who posted a total of three times compared to one post each for the other seven bloggers who wrote on the topic. Across a four day period of time, eight different bloggers covered the topic, and only two after the first blogger did so without linking to the other posts (see Figure 3). One of those two cases involved Blogger 6, who offered a contrary view and who stated a desire to stay unconnected to the main discourse.

Among the various posts, also, it was B1 who sought a solution to an individual problem, whereas the other bloggers provided commentary about past problems with the agency and offered their opinions about and sharing experiences with the application process. The others perceived the situation as problematic and recognized how it affects them, but did not state a specific current

Figure 3. Community problem case posting patterns

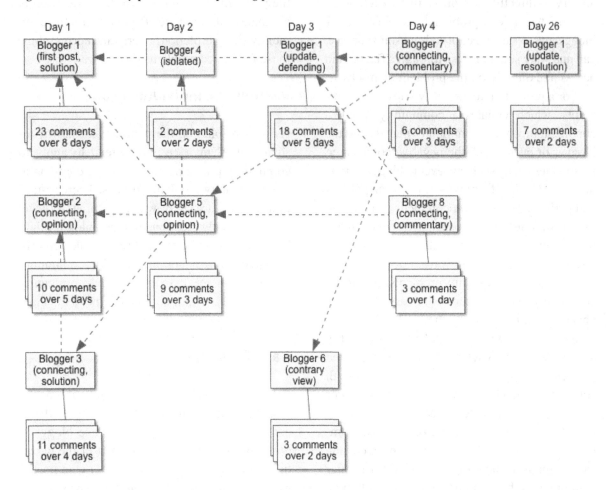

problem of their own from within that context. They called for change writ large, but did not look to take action on their own.

Across the posts, bloggers and commenters alike shared stories and gave advice about dealing with this funding agency. Three commenters represented other disciplines, and commented on how this agency's process differs from the ones with which they are familiar. Collectively, a lot of knowledge was shared across the various posts and comments. In fact, one of the later commenters thanks those who came before him for the wealth of information they have offered which will be useful when he seeks funding in the future.

DISCUSSION

There are some common threads that come out across these categories of problems. First, relationships among the parties involved – specifically the blogger's sense of audience and the commenter's sense of the blogger – permit relatively efficient and effective knowledge exchanges and offerings of support. These relationships are a factor of prolonged engagement within the blogging community, through which co-bloggers typically develop a sense of intersubjectivity (Vygotsky, 1978; Wertsch, 1985). This sense of mutual understanding is dual, in part existing because of community homogeneity (blogger as academic) and in part because of shared experiences within the community.

Second, and highly related to the first point, community discourse norms help structure the manner in which knowledge is negotiated and problems are solved within the virtual community. These norms dictate when it is appropriate to comment and when it is better to write one's own post, how to productively disagree or offer an alternative perspective, and when and how to link to posts written by others. They help focus or expand topics as is contextually appropriate, and keep the knowledge sharing process productive. In order to achieve this goal, they reflect the characteristics that are the underpinnings of community as defined in this study and by others (e.g., Baym, 1998; Hanisch & Churchman, 2008; Josefsson, 2005; Kling & Courtright, 2003), such as trust, reciprocity, identity development and sense of belonging.

Third, problems are recognized as complex narratives to which there may not be a simple solution. Initial problem posts are, in most cases, open-ended stories. This finding supports the usefulness of sharing multiple perspectives and opinions during the knowledge sharing process over focusing on so-called correct solutions. Commenters enter the narrative and weave in their own threads, which may include opinions, resources, emotional support, or somewhat parallel stories, all with a focus on working toward resolution. Still, the solution to the blogged problem is not certain or predetermined and thus must be shared. The natural human desire for story resolution on the behalf of both blogger and audience, as evidenced via meta-blogging activities, frequently leads to at least some form of update or resolution post. Note that this complexity and the desirability of individual perspectives makes community-based problem solving different from activities such as problem solving on a technical forum in which individuals seek and share concrete solutions more than narratives and emotional support

Fourth, the ownership of virtual space impacts who posts what and where they post it. The discourses that take place on these blogs might not work in another forum. Bloggers might not feel as comfortable posting their problems to a general forum as they do to their own space where they write about various topics related to their lives. To some extent, this assertion involves privacy and individual control. Bloggers enjoy the ability to delete entire blogs as well as specific posts or comments from their space as well as to use password protection and prevent search engine indexing (Dennen, 2009a). Additionally, owning the space permits a blogger to be entirely self-centered when discussing a problem, a luxury not always afforded people posting in shared forums. Similarly, commenters may be situationally more comfortable sharing their experiences on someone else's blog as one comment among many than they might be posting at the top level of their own blog. There is a psychological difference between posting a personal story for the sake of posting it and posting it in response to someone else's need or query. There also is a difference between self-disclosure to a known versus unknown audience, which relates to research on self-disclosure and sense of anonymity (Qian & Scott, 2007).

Finally, there are four common roles that emerge from within the discourse. There can be one or more problem owners, who write their own posts about a problem. There are experts, who have previously handled the problem or worked significantly in that area. There are collectors and archivists who bring together ideas, information, and resources from various sites and synthesize them for others to use in the problem solving process. And finally there are the supporters, who may or may not be offering solutions but who more importantly make the problem owners feel less alone.

There are key differences among the problem types, too. In unique problem situations, the focus is maintained on the original blogger. Stories and experiences can be shared as part of generating knowledge and finding a solution for the blogger, but ultimately the interactions do not deviate from that focus on the blogger. Additionally, unique problems are more likely than shared or community ones to demonstrate emotionally charged writing at an interpersonal level. Shared problem posts are more likely than community ones to be ongoing rather than situational, with bloggers often linking to earlier rounds of discourse, knowledge, and solutions on the topic. Finally, the difference between individual and multiple blog problems, whether shared or community oriented, becomes one of scope and interest. Multiple blog problems begin as individual ones. When individual posts inspire others to write on a topic or someone notices and makes explicit the connections between individual posts, the potential for knowledge sharing increases.

Implications

The value of studying these interactions in a naturalistic setting rests in the insights that arise from how people communicate with each other in a problem-solving context. From these blogs, we can see how particular characteristics of a community poise its members to engage in successful ongoing knowledge sharing processes. In other words, the success that these bloggers have in effectively sharing knowledge and supporting each other is in some ways unique to their community context, but is also indicative of the need for community, norms, individually owned space, and role play among participants. A sense of community helps keep participants focused on a bigger picture and encourages collegiality, trust, and reciprocity – all necessary for creating a safe space for sharing problems and anticipating useful solutions. Norms provide shortcuts for behavior and communication and help prevent misunderstandings; when participants intuitively understand each other's actions, they may more readily focus on the ideas being shared. Individually owned space provides a sense of confidence in terms of both platform (no one

else is expected to take center stage) and security or privacy concerns (author controls over editing, deleting, and search engine indexing). Finally, different and dynamic roles allow individuals to contribute to any community conversation in a productive manner regardless of their experience or knowledge (i.e., those who cannot be experts can still be supporters).

These four factors are things that designers of more prescribed online communities might consider in their planning process, and that moderators or facilitators might take into consideration when looking for effective ways of helping people engage in knowledge sharing and problem solving processes in online forums. Although developing a sense of community is heavily reliant on the desire for said community among participants, community builders should be sensitive to the importance of allowing or encouraging that sense of community to form. The other three factors may be more easily supported by deliberate actions. In a highly structured virtual community, desired norms may be set forth as rules, and community facilitators or leaders may readily model both norms and roles. Additionally, tool selection and options may be determined with an eye toward providing users with a sense of both personal space and control over their words once posted.

CONCLUSION

In this chapter I have presented the way in which members of a blogging community present and resolve three classifications of problems – unique, shared, and community – through a knowledge sharing process that is manifest via both blog posts and their resulting comments. Further, I have discussed how problems and knowledge sharing differ based on whether they are centered on one individual blog versus presented across multiple blogs. Regardless of how a problem is classified, this chapter demonstrates that blogs are a useful platform for supporting problem solving and knowledge sharing processes and that both a sense of community and its resulting norms are a key part of making these processes successful.

Classifying problem situations as I have here is important for understanding the relationship between individual bloggers and how support and sharing can occur not despite but rather because of interactions in separate but interlinked virtual spaces. The use of individual blogs as a problem solving and knowledge sharing space is dependent on both blogger and audience, and thus it is the existence of community that helps these virtual spaces thrive. Shared and community knowledge are, in many ways, what bring the community members together. From this sense of belonging, then, and the familiarity and relationships that develop as a result, comes a natural opening for sharing unique problems as well as helping others with unique problems work toward resolution.

The dynamic roles that participants hold in the problem solving process (owner, expert, collector/archivist, and supporter), along with reciprocity (a characteristic of communities) help make these online settings well suited for problem solving and knowledge sharing. Individuals playing the three responsive roles help the problem owner fully develop a problem narrative both implicitly, as intended audience, and explicitly, through their offerings of support and knowledge. The ability to shift in and out of the roles for different problem situations means that one need not be limited to one type of community interaction. Each of these roles is necessary across all three types of problems, and each supports resolution in its own way.

Looking to the future, there is great opportunity for research to be done in this area. The relationship between social elements, individual reflection (e.g. journaling) and knowledge sharing in informal online communities is still not well understood. Many studies – this one included – look at individual communities or populations. More cross-community research is needed to help

increase understanding of what trends and interactions are truly community-specific and which can be generalized to certain types of communities or populations. The challenge to doing such research, however, is the time required to truly come to know an online community. However, an increased focus on research using similar frameworks would, in the long term, facilitate synthesis and triangulation across studies. Similarly, frameworks and classifications themselves might be tested across communities, as was done during the classification development process for this study, to gain confidence in the system's applicability to other people and settings. Finally, related topics such as knowledge management and archiving often are neglected in the context of informal virtual communities, but it seems likely that there is much to be gained from their exploration.

REFERENCES

Baker, J. R., & Moore, S. M. (2008). Blogging as a social tool: A psychosocial examination of the effects of blogging. *Cyberpsychology & Behavior*, *11*(6), 747–749. doi:10.1089/cpb.2008.0053

Baym, N. (1998). The emergence of online community. In Jones, S. G. (Ed.), *Cybersociety 2.0* (pp. 35–68). Thousand Oaks, CA: Sage.

Chayko, M. (2002). *Connecting: How we form social bonds and communities in the Internet age*. Albany, NY: State University of New York Press.

Chayko, M. (2008). *Portable communities: The social dynamics of online and mobile connectedness*. Albany, NY: State University of New York Press.

Cummings, J. N., Butler, B., & Kraut, R. (2002). The quality of online social relationships. *Communications of the ACM*, *45*(7), 103–108. doi:10.1145/514236.514242

Dennen, V. P. (2006). Blogademe: How a group of academics formed and normed an online community of practice. In Mendez-Vilas, A., Martin, A. S., Mesa Gonzalez, J. A., & Mesa Gonzalez, J. (Eds.), *Current developments in technology-assisted education* (*Vol. 1*, pp. 306–310). Badajoz, Spain: Formatex.

Dennen, V. P. (2009a). *Because I know you: Elements of trust among pseudonymous bloggers.* Paper presented at IADIS International Conference on Web-based Communities 2009, Algarve, Portugal.

Dennen, V. P. (2009b). Constructing academic alter-egos: Identity issues in a blog-based community. *Identity in the Information Society, 2*(1), 23–38. doi:10.1007/s12394-009-0020-8

Dennen, V. P., & Pashnyak, T. G. (2008). Finding community in the comments: The role of reader and blogger responses in a weblog community of practice. *International Journal of Web-based Communities, 4*(3), 272–283. doi:10.1504/IJWBC.2008.019189

Dibbell, J. (1998). *My tiny life: Crime and passion in a virtual world.* New York: Henry Holt and Company.

Ellonen, H. K., Kosonen, M., & Henttonen, K. (2007). The development of a sense of virtual community. *International Journal of Web Based Communities, 3*(1), 114–130. doi:10.1504/IJWBC.2007.013778

Fox, S. (2008). The engaged e-patient population. *Pew Internet & American Life Project.* Retrieved January 20, 2009, from http://www.pewinternet.org/Reports/2008/The-Engaged-Epatient-Population.aspx

Greenfield, G., & Campbell, J. (2006). Communicative practices in online communication: A case of agreeing to disagree. *Journal of Organizational Computing and Electronic Commerce, 16*(3-4), 267–277. doi:10.1207/s15327744joce1603&4_6

Hanisch, J., & Churchman, D. (2008). Virtual communities of practice: The communication of knowledge across cultural boundaries. *International Journal of Web-based Communities, 4*(4), 418–433. doi:10.1504/IJWBC.2008.019548

Herring, S. C., Kouper, I., Paolillo, J. C., Scheidt, L. A., Tyworth, M., Welsch, P., et al. (2005). Conversations in the blogosphere: An analysis " from the bottom up". *Proceedings of the 38th Hawaii International Conference on System Sciences.* Los Alamitos: IEEE Press.

Jaschik, S. (2005, January 5). Historians reject proposed boycott. *Inside Higher Ed.* Retrieved from http://www.insidehighered.com/news/2009/01/05/boycott

Josefsson, U. (2005). Coping with illness online: The case of patients' online communities. *The Information Society, 21*, 141–153. doi:10.1080/01972240590925357

Karlsson, L. (2007). Desperately seeking sameness. *Feminist Media Studies, 7*(2), 137–153. doi:10.1080/14680770701287019

Kling, R., & Courtright, C. (2003). Group behavior and learning in electronic forums: A sociotechnical approach. *The Information Society, 19*, 221–235. doi:10.1080/01972240309465

Kosonen, M. (2009). Knowledge sharing in virtual communities: A review of the empirical research. *International Journal of Web Based Communities, 5*(2), 144–163. doi:10.1504/IJWBC.2009.023962

Kumar, R., Novak, J., Raghavan, P., & Tomkins, A. (2004). Structure and evolution of blogspace. *Communications of the ACM, 47*(12), 35–39. doi:10.1145/1035134.1035162

Lee, D. H., Im, S., & Taylor, C. R. (2008). Voluntary self-disclosure of information on the internet: A multimethod study of the motivations and consequences of disclosing information on blogs. *Psychology and Marketing, 25*(7), 692–710. doi:10.1002/mar.20232

Lenhart, A., & Fox, S. (2006). Bloggers: A portrait of the internet's new storytellers. *Pew Internet & American Life Projects*. Retrieved October 20, 2009, from http://www.pewinternet.org/Reports/2006/Bloggers.aspx

Papacharissi, Z. (2006). Audiences as media producers: Content analysis of 260 blogs. In Tremayne, P. (Ed.), *Blogging, citizenship, and the future of media* (pp. 21–38). New York: Routledge.

Qian, H., & Scott, C. R. (2007). Anonymity and self-disclosure on Weblogs. *Journal of Computer-Mediated Communication, 12*(4). Retrieved from http://jcmc.indiana.edu/vol12/issue4/qian.html. doi:10.1111/j.1083-6101.2007.00380.x

Reigeluth, C. M., & Frick, T. W. (1999). Formative research: A methodology for creating and improving design theories. In Reigeluth, C. M. (Ed.), *Instructional design theories and models: A new paradigm of instructional theory* (pp. 633–652). Mahwah, NJ: Lawrence Erlbaum.

Rheingold, H. R. (2000). *The virtual community: Homesteading on the electronic fronteir*. Cambridge, MA: MIT Press.

Ridings, C. M., & Gefen, D. (2004). Virtual community attraction: Why people hang out online. *Journal of Computer-Mediated Communication, 10*(1).

Smith, A., Schlozman, K. L., Verba, S., & Brady, H. (2009). The Internet and civic engagement. *Pew Internet & American Life Project*. Retrieved November 17, 2009, from http://www.pewinternet.org/Reports/2009/15--The-Internet-and-Civic-Engagement.aspx

Stake, R. E. (1978). The case study method in social inquiry. *Educational Researcher, 7*(2), 5–8.

Stake, R. E. (1995). *The art of case study research*. Thousand Oaks, CA: Sage.

Takhteyev, Y., & Hall, J. (2005). *Blogging together: Digital expression in a real-life community*. Paper presented at the Social Software in the Academy Workshop, Los Angeles, CA.

Vygotsky, L. S. (1978). *Mind in society: The development of higher psychological processes*. Cambridge, MA: Harvard University Press.

Wenger, E. (1998). *Communities of practice: Learning, meaning, and identity*. Cambridge, UK: Cambridge University Press.

Wenger, E., White, N., & Smith, J. D. (2009). *Digital habitats: Stewarding technology for communities*. Portland, OR: CPSquare.

Wertsch, J. V. (1985). *Vygotsky and the social formation of mind*. Cambridge, MA: Harvard University Press.

This work was previously published in Technologies for Supporting Reasoning Communities and Collaborative Decision Making: Cooperative Approaches, edited by John Yearwood and Andrew Stranieri, pp. 358-379, copyright 2011 by Information Science Reference (an imprint of IGI Global).

Chapter 11
Social Media (Web 2.0) and Crisis Information:
Case Study Gaza 2008–09

Miranda Dandoulaki
National Centre of Public Administration and Local Government, Greece

Matina Halkia
European Commission Joint Research Centre, Italy

ABSTRACT

Social media technologies such as blogs, social networking sites, microblogs, instant messaging, wikis, widgets, social bookmarking, image/video sharing, virtual worlds, and internet forums, have been identified to have played a role in crises. This chapter examines how social media technologies interact with formal and informal crises communication and information management. We first review the background and history of social media (Web 2.0) in crisis contexts. We then focus on the use of social media in the recent Gaza humanitarian crisis (12.2008-1.2009) in an effort to detect signs of a paradigm shift in crisis information management. Finally, we point to directions in the future development of collaborative intelligence systems for crisis management.

INTRODUCTION: ICTS FOR FORWARD-LOOKING CRISIS MANAGEMENT

Crises in the 21st century are expected to astonish both the experts and the lay people. Lagatec (2005) was succinct in portraying the current situation in his paper titled "Crisis management in the 21st century: 'unthinkable' events in 'inconceivable'

contexts." Therefore, traits such as responsiveness, flexibility, self-organization, improvisation, resilience, agility[1] seem pertinent for forward-looking risk and crisis management.

Crisis information management is no different. Top – down fixed information systems and tools cannot fully capture the spatial - temporal - social dynamics and respond to the uncertainties of future emergencies, disasters and crises.

DOI: 10.4018/978-1-4666-2803-8.ch011

Numerous tools that have recently been developed to support crisis and emergency management[2] have a pre-defined structure and rely primarily on information collected and maintained in normal (i.e. non-emergency) conditions. Although the deficiencies of central, hierarchical architectures have been pointed out and alternative solutions have been proposed (Dandoulaki & Andritsos, 2007), emergency management tools remain habitually centrally administrated and still reflect a top-down approach.

Current information tools, either formal (e.g. local traffic surveillance cameras, global alert and monitoring systems) or informal (mobile phones taking videos on the spot), gradually change the landscape in emergency information (Moss & Townsend, 2006). Social media technologies (social networks, microblogging, blogs, wikis, annotatable maps, image and video sharing, instant messaging, internet forums and other web forms)[3] have also been acknowledged to have played a role in emergencies, disasters and crises (Palen, 2008; Palmer, 2008).

The emergence of such shifts and incursions summon and merit the clarification of the basic concepts in our discussion. The conceptualization of emergency, crisis and disaster[4] is central in the current discussion among scholars (Quarantelli, 1998; Perry & Quarantelli, 2005, Gundel, 2005) and has significant implications in policies and practices. The relationship between crisis and disaster is still to be defined. The two are inextricably linked (Boin, 2005, p.155); however crisis is a general concept encompassing disaster (Quarantelli, 1998 p.235). Boin (2005, p.164) suggests that disaster is a "crisis gone bad," thus including in the disaster category a spectrum of events and processes such as riots, epidemics, acts of terrorism and massacres.

This chapter examines how social media technologies affect crisis information management. We first review the background and history of social media (Web 2.0) in crisis management. We then focus on the use of social media during the recent Gaza humanitarian crisis. Finally, we point to directions in the future development of collaborative intelligence systems for crisis management.

Why focus on Gaza in order to grasp future trends? For one, social media played a major role in redefining the spatial and social locus of the crisis, thus the focus of crisis management, by triggering response all over the world. They also contributed to more resilience of the information system once the formal systems and tools could not or would not perform. Moreover, the Gaza case demonstrates the influence of social media technologies even when the disaster area has huge inadequacies in technological infrastructure. Finally, the vast Gaza mobilization on a worldwide scale exemplifies on one hand, the potential and strengths, as well as the perils and weaknesses of collaborative media, and begs on the other hand the question of a new structured platform for collaborative intelligence systems in decision-making for disaster management. It should be stated, however, that our objective is not to survey the use of social media in this case study; rather, we would like to discuss certain aspects that highlight a paradigm shift in crisis information management.

THEORY AND PRACTICE: SOCIAL MEDIA (WEB 2.0) IN CRISIS CONTEXTS

Crisis Management and the Media in the Risk Society

Crisis management is not separate from understanding the processes of risk and vulnerability that unfold both at micro- and macro- level. Yet, while much effort is put in dealing with the macro-level, especially as regards current information systems, less progress is made in grasping the situation and the dynamics at the micro level.

Being there in a crisis, having experienced the crisis and its multiple realities, having been part of the dynamics of a crisis, is by itself valuable knowledge (Hewitt, 1998, p.87; Barton, 2005 p.136, Buckle, 2005). The knowledge of experience is highly significant and on par with scholar and expert knowledge. Therefore, it is a challenge for crisis management to bring up also the micro-level (i.e. place) and with this, the knowledge and experience of people who "were there."

At the level of working practices, the importance of risk governance is rising. Risk governance encompasses risk assessment and risk management. It concerns how decision-making unfolds when a range of stakeholders, with different expectations, goals, activities and roles, are involved (Renn and Walker 2008). Amongst all of this, communication is at the centre (International Risk Governance Council, 2006).

Risk communication has transitioned to today's internet times (Krimsky, 2007). Risk communicators -a core element of crisis information according to Fischoff (1995)- are not unquestionably central in risk communication any more. There are evident signs of lay people gaining a central role in communication and information through the use of current technological means, among them social media technologies. Social media technologies have already been used for different purposes in crises. Everybody has or can have a voice in the future; this can be significant in times when the challenge is to democratize society's response to risk and disaster (Alexander, 2005, p.35).

Global communications have contributed to the symbolic significance of disaster (Alexander, 2006, p.7). In the present era, disasters are seen as a mindset, their interpretation and symbolic meaning constructed by the mass communication industry (Alexander. 2005, p.38; Saw, 1996). Social media potentially counterbalance mass media.

Today's society is a risk society (Beck, 1992; Adam at al., 2000); risks and society are interwoven and affect one another through a continual negotiation and renegotiation. The risk society is also a reflexive society. Reflexivity entails reflection (knowledge) on risks but also not being aware of risks (Beck, 2009, p.119). In the era of globalization risk reflexivity is shaped by the mass media and the instrumentation of anxiety by a range of political players (Beck, 2009, p.198). Social media offer an opportunity so that everybody can voice a view in the current and future negotiation of risk.

Web 2.0 Technologies and Crisis Information

Although the modern world is technology-dominated, much of the world's population is excluded from the benefits of technology. It is estimated that about a half of the population have never used a telephone (Alexander, 2006, p.3). Then again, peer-to-peer information and communication technologies become increasingly pervasive and among them, social media technologies (Palmer, 2008, p.1; Shankar 2008).

Social media have been identified to have served in crises in many ways. They are attributed with social convergence (Palen, 2008; Hughes et al., 2008), they facilitate people's participation in emergency management (Palen, Hiltz et al., 2007; Vieweg, 2008), they support collective intelligence in a crisis (Palen, Vieweg et al., 2007), and they counterbalance or compliment mass media through citizen journalism (Liu at al., 2008). The emerging role of Web 2.0 technologies in crisis contexts is even a topic of discussion in social media; there are blogs, especially in the USA, where relevant research and experiences are shared.[5]

The topic of if and how social media could change crisis management is a burning one (see for example Tinker & Fouse, 2009); after all, social media have been exploited in one way or another in every recent crisis.

Social Media Use in Crises: Experiences and Trends

There is a mounting number of experiences in the use of social media in crises and a fast-growing quantity of related literature. Depending on the situation, social media has been used for different purposes and in various ways (Sutton et al. 2008). The following selected cases were either studied in-depth and now serve as reference cases in the field, or bring up new dimensions of the issue.

Our reading of these cases is purposeful; the intent is to bring up functionalities served by social media in a crisis. Then we make an effort to identify strong and weak points of social media tools from the point of view of formal crisis management.

9/11 (2001)

Four airplanes were hijacked; two of them hit the WTC towers, one hit the Pentagon and the other crashed in a field. Mass media were transmitting live the situation. E-mail was used by trapped people and by the affected population to communicate with their peers (Palen & Liu 2007, p.729). Harrald et al. (2002) report several uses of web based technologies; web-based technologies were used by corporations to account for and communicate with employees. FEMA, the American Red Cross, The US Army Corps of Engineers, and the Environmental Protection agency all used web communication to inform the public and to provide status reports internally and externally. There is evidence that the terrorists also used the internet to plan the attacks.

Even at this relatively early stage of web use in crisis contexts, the internet proved a significant, although not primary, source of information in the USA with 64% of adults in the USA using the Internet to find information on the situation (Bucher, 2002, p.2).

New Orleans Floods (2005)

Hurricane Katrina (a Category 3 storm) caused damaged in Mississipi and Louisiana. Areas of New Orleans were flooded. The city was evacuated and the population was temporarily relocated throughout USA. Citizen-led online sites were used for aid in the emergency. The site (www.katrina.com) previously used to advertise a small company attracted people seeking information and it was readily converted to link to other useful sites and to serve as a message board to facilitate tracking down missing people. Other sites such as Hurricane Information Maps (www.scipionus.com) were created to collect and share location specific information (Palen & Liu, 2007; Moss & Townsend, 2006; Currion, 2005). The disaster was captured in weblogs that could serve as crisis data sources (James & Rashed, 2006).

Virginia Tech Shootings (2007)

On April 16, a shooter killed 32 people at Virginia Polytechnic Institute and State University. Findings[6] record large-scale social interaction that occurred after this event over multiple sites of interaction; at first peer-to-peer communications and later on-line and off-line IC activities concerning larger sets of data (Palen, Vieweg et al. 2007; Vieweg et al., 2008). Facebook and Wikipedia were the main tools used for the latter. Facebook groups such as "I'm OK at VT" were set for members to post that they were not among the victims; others such as the "We support Chief Flinchum" expressed solidarity to emergency managers (Byrne & Whitmore, 2008, p.8). Online efforts correctly identified deceased victims even before the university released the information (Byrne & Whitmore, 2008). Nonetheless, problems similar to old-fashioned jamming of telephone lines were observed; eventually the bandwidth got overloaded (The Risk Communicator, 2008, p.4).

California Wildfires (2007)

The California wildfires began on October 20, 2007 and burnt over 500.000 acres of land. Sutton et al. (2008) show a wide spectrum of means used for communication and information from mobile phones to contact family and friends, to all available web-based tools. The informal notification process has been identified as one of two significant warning mechanisms in San Diego County, the other being a reverse call-down emergency warning system that San Diego County and the City of San Diego had put in place (Sorensen et al., 2009, p.22).

People were informed through traditional media (television, radio) and new alternative media mainly advertised in the former (websites, information portals). They also posted and shared information and participated in web forums (Palen, 2007, p.78). Websites were created with annotated geo-information of interest to the affected people (burnt areas, evacuation areas, shelters etc.) (Hughes et al., 2008). This website had more than 1.7 million hits (Sutton et al., 2008). Flickr served as a repository of photos both for personal use and group based interaction; As a result, houses were also inventoried for insurance purposes (Liu et al., 2008).

The Greek Riots (2008)

The riots started on 6th December 2008, after a teenager was fatally shot by a policeman at around 9:00 pm. His death triggered large protests and demonstrations, which escalated to widespread rioting. Demonstrations and rioting were propagated to several other cities. Outside Greece, solidarity demonstrations, riots and in some cases clashes with local police also took place in a number of European cities.

Thousands of people were on the streets protesting while established media had not even reported the event (P.Tsimas as referenced in Lam, 2008). SMS messaging and re-broadcasting on the internet played a significant role for the mobilization of people (Gavriilidis, 2009). Twitter was extensively used, with the first twitter message sent around 15 minutes after the event. About three hours later Pathfindernews and Skaigr were the first official media to announce the event.[7] At around 3:00am on December 7th, the hashtag #riots started to be used on Twitter.[8] Thousands of messages were sent over the next days, many of them directly from the streets.[9] Citizen reporting was intense especially in terms of images. Flickr featured more than 1.100 photos (Tziros, 2009). Facebook also played a role in the protests with several protest groups set-up, totalling some 187,000 members. The largest one had around 136,500 members and shared messages about the upcoming protests and commemorating activity (Joyce, 2008).

Wikipedia started reporting on the issue on December 7th at 21:00, and was updated more than 200 times afterwards (http://en.wikipedia. org/wiki/2008_Greek_riots). Collaborative citizen journalism projects, like Global Voices Online, NowPublic, allvoices and CNN's iReport, were used to publish original reports.[10]

L'Aquila Earthquake (2009)

On April 6th 2009, at 3:32am local time, an earthquake (Ml=5.8) struck central Italy in the vicinity of L'Aquila (capital of the Abruzzo region). The earthquake killed 308 people,[11] destroyed more than 10,000 buildings and left more than 24,000 homeless.[12] Approximately 65,000 people had to evacuate their homes and were lodged in tent camps, hotels and other structures located along the Adriatic coast.[13] This quake was the strongest of a sequence that started a few months earlier.

Seismic activity in the area was an issue in social media even before the main event.[14] After the event, information was disseminated immediately by people directly on site that kept feeding the web and the mass media.[15] Wikipedia (Italian edition) had the first entry at 11:52 on the same day, something which is still updated even today[16] and then at 12:03 it appeared also in the English edition. A website was created and offered annotated geo-referenced information on the earthquakes, their effects, emergency management and aid provision activities (information on victims and damages, blocked roads, hospitals, evacuation areas, shelters, coordination centres etc.) (Figure 1).

Bearing in mind that social media have been used in one form or another in every recent crisis, the aforementioned cases are only a few examples.

Formal crisis management cannot but take into consideration emergency management technology advances[17] and the mainstreaming of social media technologies. The response to this challenge differs in different contexts but the signs of change are already there (Collins, 2009).

As early as 2002, FBI established a website after 9/11 to receive information and tips from citizens that might aid the law enforcement investigation

(Harrald et al., 2002). Many USA government agencies use Twitter, among them the Department of Homeland Security (DHS), the Environmental Protection Agency (EPA)[18] and the US Geological Survey, to mention only a few. [19] The Los Angeles Fire Department and Philadelphia Emergency Management are among agencies using social networking to send breaking news and preparedness information. FEMA developed the Integrated Public Alert and Warning System (IPAWS) that expands upon traditional (radio and television) media and adds new media - internet and cell phones[20]. NGOs dealing with emergencies and crises have already put social media in everyday use, among them the American Red Cross.[21]

The Australian Government has announced plans to use social media websites such as Twitter and Facebook, alongside traditional warning mechanisms in order to improve the "quality and

Figure 1. Google maps mashup of the L'Aquila earthquake disaster. (Source: Google.maps-terremoto L'Aquila. Retrieved April 14, 2009, from http://www.google.it/landing/terremoto_abruzzo.html)

timeliness of bushfire warnings" ahead of the upcoming fire season.[22] Toronto's Police Service and Toronto Fire Service are both on Twitter.[23] Initiatives involving social media and government can be currently noticed in many countries all over the world.[24] Recently, a 20-page guide to using twitter was published targeting the UK government agencies.[25] These are initial but important steps that allow testing social media in formal crisis management, although only as peripheral tools for the time being.

Mass media (TV, radio and the press) also use social media to involve people in offering opinions and sharing information; see for example: the CNN user-generated site (http://www.ireport.com/) (Catone 2008), the Guardian blogposts (http://www.guardian.co.uk/tone/blog), The New York Times twitter (http://twitter.com/NyTimes).

Main Functionalities of Social Media in Crises

Studying a number of cases, we can identify three core functions of social media during crises each having strong and weak points as regards formal emergency management. It should be noted however, that these functions do not exclude one another since in many occasions they coexist and work in a complementary way.

Peer-to-Peer Communication

People tend to seek friends and family in a crisis in order to communicate personal information and get or give practical or psychological support; in doing so they also transmit information useful for emergency management. All means of communication are employed from leaving hand-written notes and signs, to mobile phones, e-mail and social networking tools such as Facebook and Twitter. As social media enable everyday networking, they also serve for peer-to-peer communications in a crisis. See Table 1.

Information

Information is vital in a crisis. People seek information. They also offer information demonstrating a socially convergent behavior; in so doing they also participate in emergency management. Social media tools facilitate searching, collecting, posting, sharing information by individuals and groups. Tools such as wikis, blogs, Facebook, Flickr, Twitter, etc., are used. Previous networking of various types makes the activity easier. See Table 2.

Activism

Individuals and groups sharing similar goals in specific circumstances use social media tools for awareness raising and mobilization of others. Ad

Table 1. Strong and weak points of social media use for communication as regards emergency management

Strong Points	Weak Points
• Redundancy in communication • Vast territorial and social coverage • Able to communicate multi-type information • Facilitate social convergence • Potential for fast and broad communication of brief messages (such as alerts and warnings)	• Credibility of information not known • Information, information source and information flows not possible to control • No clear responsibility and no accountability • Risk of malicious spreading of information and easier dissemination of potentially problematic rumours • The bandwidth can get overloaded in the same way as telephone communications suffer in crises.

Table 2. Strong and weak points of social media use for information as regards emergency management

Strong Points	Weak Points
• Multi-source information of various types with vast geographical and social coverage • Timely field information possible • Information can be disseminated easily • Potential for developing new forms of public involvement and attracting volunteers • Social convergence • Potential for raising global awareness and harvesting aid from all over the world with less dependency on mass media	• Credibility of information not known, information can be misleading, confusing or contradictory • Accountability very low, liability is obscure • Information stays on the internet long and might become out-of-date and be misleading • Huge bulk of information from different sources creates an information management nightmare • Authorities and official agencies are not accustomed to make use of possibly non-credible available information • Privacy protection issues. Low security of information • Possible manipulation and malicious use of information • Significant issues of transparency and trust

hoc networks are created to channel messages and disseminate information. Clustering of nodes can usually be detected after some time. Twitter, blogs, wikis, Flickr have been used in different occasions, depending on the context and the situation. See Table 3.

Case Study Gaza 2008-2009

Between December 2008 and January 2009, Israeli military forces carried out a major offensive against the Gaza strip by air, land and sea. Although news of the attack was not readily available through the major media channels – due in part to a lack of foreign journalists in Gaza[26] – news about what was happening on the ground rapidly disseminated worldwide through social media, such as blogs, independent internet news broadcasting and social networking.

Although it might seem premature to assess objectively the impact of social media on the dis-semination of information in the Gaza conflict, it can be said that this was probably the first time social media were used in absence of any other formal crisis management system[27], unlike previ-ous examples discussed. Information about the humanitarian aspects of the disaster were gathered by international organizations and observers on the ground and transmitted to western audiences at large through social media.

"Everyone Has a Voice"

It was immediately apparent to Westerners at-tempting to follow the conflict in its early days through social media, that there was a host of voices reporting, many of them without established credibility. Trust and identity had to be constructed on-the-fly, and very quickly, as news were com-ing in through blogs and websites. Many times during the three weeks of the assault news was conflicting, and the lack of institutional voices on

Table 3. Strong and weak points of social media use for activism as regards emergency management

Strong Points	Weak Points
• Fast notification of large numbers of people. • Independence from official information sources and mass media. • Social and political convergence in crisis • Participation and active involvement easier. • A good platform to democratize society's response to risk and disaster.	• Credibility of information not known • Accountability very low • Information stays on the internet long and might become outdated and misleading • Low security of information • Risk of manipulation of information and of malicious use of social media • Difficult to control

the ground, in this highly controversial conflict, made news evaluation problematic.

The network of nodes in the social network that was reporting on the Gaza crisis was initially built on apparently strong, rather homogeneous ties, already in place before the conflict begun. In the USA, Gazasiege.org and FreeGaza.org and in Italy, Vittorio Arrigoni, an Italian volunteer contributing to the paramedic squads of Al-Quds hospital in Gaza city, were some of the most consistent providers of daily updates of the situation on the ground. Arrigoni's blogs were translated within hours in English and other languages and copied in a host of English-speaking sites including the ones mentioned above. The Gaza social network expanded very rapidly to a worldwide network of now weaker, heterogeneous ties (for definitions and social network analysis see boyd & Ellison, 2007). Facebook groups decrying the Gaza assault gathered rapidly. As these lines are written, the number of groups returning the keyword "Gaza" in Facebook amount to 597, the largest of which counts short of 200,000 members. Of them, three supported or defended the assault. The others mainly in English (but also French, Italian, Spanish, and Norwegian) denounced the humanitarian catastrophe for Gazan children, disapproved the difficulty in access for humanitarian supplies or medical personnel, or praised Al-Jazeera for "providing the truth."[28]

Information spread through daily updates about the number of deaths, even though the actual numbers varied. Personal stories of family strife, of death and despair spread geographically far away from the epicenter of the disaster, and became the theme for household small talk. For example, Facebook chats, wall posts, and comments were discussing Gaza events in the context of family or friend connections. The audience by the end of the three-week assault was international and heterogeneous. The news story on Gaza was constructed collaboratively in a distributed geography, dislocated from the crisis theatre, through opinions voiced in blogs and social network sites.

Slowly and steadily the voices converged and the narrative became common ground. The Gaza disaster network was now forming a community.

Community Narratives, Distributed Collaborative Production of News: Multi-Author Narratives and the Construction of Meaning, Dissemination and Structure of the News Message

The Gaza network prompted the development of another community. Defenders of Israeli politics were quick to respond. While a Facebook group donated their status[29] in counting victims and wounded in the Gaza strip, another community was emphasizing the number of Qassam rockets fired at and dropping on Israel.

Different community narratives dictated different versions of news. The social media consumer had to navigate through these stories with caution. The construction of a coherent view of the crisis became relative to one's personal interest and motivation to follow another kind of story. A story that was multi-threaded, collaboratively constructed, fraught occasionally with conflicting or ambiguous news messages. The passive news consumer was called to active involvement in structuring the different elements of the news message. In practice, this involved navigating from blog to related blog, perusing blog comments, controversies, and criticism, consulting wikis, complimenting information by Google searches, and more often than not revert to "authority voices"[30] to balance, prioritize, and contextualize reported stories. In so doing, the crisis reports created were as many as the reporters; every blogger or blog comment became potentially a report in itself, complimenting the edifice of the crisis from yet another point of view.

Although it is has been said that the Gaza crisis triggered a social media war,[31] this has not been war of winners and losers, as the news story was written by everyone and all: a conglomeration of eyewitness reports, comments, criticisms, background information and historical accounts brought to the international observer mainly through social media.

Establishing Credibility, Constructing Reliability, Accuracy in the Multifaceted Narrative, Discrediting and Slander

With a multitude of voices reporting their different realities of a crisis, trust to news sources is key. What is reported requires careful evaluation in relation to contextual and background information, but often verification is difficult, especially under the time pressures in a crisis situation. This was especially true in the Gaza conflict, where crisis reports were often antagonistic.[32] Credibility was built as more and more information was becoming available to confirm or contrast previous reports. Over a reasonable amount of time, certain blogs or news' sites risked losing credibility as military/state propaganda,[33] and independent eyewitnesses on the ground emerged as dispassionate, objective reporters providing accurate crisis updates.

There was more than multifaceted reporting, which called for attention in the construction of the news message. Often, there was slander and deliberate discrediting of reporting voices through offensive sites. Offence included death threats against bloggers.[34] Although social media resist censorship and this remains one of their perceived advantages, in the Gaza crisis spontaneous informal efforts to control freedom of speech quickly became so serious as to instigate calls for official diplomatic action at state/national level.[35] While state control was exercised in other crises where social media were key,[36] in the Gaza conflict, censorship efforts were targeted towards exerting individual pressure, through offence or slander, in order to obtain self-censorship or at least curtail freedom of speech.

Roles: The Changing Role of the Audience, The Changing Role of the Journalist, The Expert

In social media, collection and dissemination of information transforms the passive crisis onlooker in an active news producer. By commenting on blogs, collecting reports and consulting search engines in order to piece the fragmented news landscape, she actively engages in an editorial effort even this does not materialize in written word. How can it be argued that a Facebook member who donates its status to a cause does not engage in activism for that cause? And how can it be claimed, that in the same act there is not implicit recruitment of supporters among its immediate social network for the said cause? By the same token, funds raised as a result of social media activism can be considered enabling financial instruments for crisis management. Undoubtedly, the case of Gaza resists any crisis management plan, as effective crisis mitigation was not possible. Among other ineffective crisis response mechanisms, humanitarian aid promised by the international community did not reach Gaza due to the practically closed border crossings.[37]

On the other hand, the professional journalist can then be seen as the key factor in filtering and editing the reported information, providing context and background and facilitating interpretation. In the Gaza example, due to access restrictions journalists played a secondary role in reporting from the crisis site. Still, CNN's i-report community provides a moderated social network in which eyewitnesses can provide information about events occurring real-time. Additionally, important texts providing background information and enabling historical interpretation of the crisis climax were published in major western media. An impressive amount of international media coverage resulted, in order to elucidate the hows and whys of the humanitarian crisis. Indeed, academic experts and political analysts became popular reference figures[38] in this crisis, as there was a particular need to untangle complicated dependencies between cause and effect phenomena which would eventually lead and nurture a successful crisis management strategy.

Processes

We examined earlier in the chapter the three main uses of social media in crisis contexts: peer-to-peer communication; information dissemination; mobilization, awareness, and activism. In the Gaza example, these processes were particularly heightened because of the lack of any adequate crisis management system. The population used mobile phones and SMS messages to notify the location of wounded to ambulance services. SMS notifications gave the latest information about whereabouts of Israeli forces, affected quarters and closed streets to family and friends. International awareness was raised through social media because of the access restrictions on international journalists. Activism through fundraising on social media platforms and organization of public support events and gatherings was also observed.

Social Media as a News Centre: The News Consumer as a News Producer and Peace Activist

In stark contrast to cases examined earlier in the chapter, in Gaza there was lack of formal crisis management systems. When the United Nations (UN) headquarters, two UN schools and the Al Quds Hospital[39], were hit by Israeli forces it became apparent that any minimal organized crisis response that might have been in place, was no more. In this sense, the Gaza crisis is very different from other crises where social media played a role. Taken to extremes, this means that social media on Gaza can be likened to a crisis situation room. Because of the lack of international journalists to disseminate information and adequate civil protection mechanisms to respond to emergencies

and support the population, social media provided the missing functions of a centralized emergency response system, as much as that was possible. For this reason, the Gaza example provides poignant lessons for system design of Web 2.0 technologies in crisis contexts: social media with all the weaknesses and defects that we have already examined, was a stand-alone support system in the Gaza disaster. Those participating in the Gaza crisis social network were perceived as emergency workers by the Gaza population.

The Two Faces of an Activist's Blog: A Forum, and a News Service

Vittorio Arrigoni's blog (http://guerrillaradio. iobloggo.com/) became, during the three weeks of the assault, the most visited blog in Italy with peaks of 20.000 visits per day[40]. Translations of his texts were read daily in USA and in South America. Arrigoni, one of the few internationals remaining in Gaza during the assault, and by all accounts the only Italian citizen on the Gaza strip, provided eyewitness accounts from the ambulance service he helped to man in Al Quds hospital. He consistently updated his blog with disaster testimony, and a touching account of the ground experience. His audience expanded rapidly. He started receiving supportive emails from factory workers, university professors, students, housewives and air traffic controllers. He claims that the reason for the heterogeneity of his audience and the international impact of his texts was his independence; he does not belong to any interest group, political faction, or non-governmental organization. He says: "My language is clear, on level of the reader, and respects fundamental human values such as peace, justice, and freedom."[41]

Arrigoni, apart from providing ground reports, built overtime a knowledge of experience that indeed became as important as the scholar or expert knowledge on the Gaza crisis. His blog with its qualities of immediacy, lack of censorship, timeliness, and availability beyond national borders, became a reference node in the cluster of social media activity on the Gaza crisis. The comments following one of his daily blogs extended over several pages. Individual disputes, discussions and stories unfolded between his commentators. Arrigoni claims that, over time, the blog was "infiltrated" by organized "trolls." He says "I believe in freedom of speech and for this reason I allowed trolls transformed into seemingly ordinary visitors to infiltrate and post racist remarks for months on end. When they surpassed any acceptable limits I ceded to my blogs' community pressure and banned the 'disturbing' comments." [42]

Indeed, during the days of the assault, the comments in Arrigoni's blog provided an impressively varied spectrum of different positions on the Gaza crisis; this demonstrated beyond doubt the heterogeneity of his readers. The passionate and occasionally inflammatory style of some comments supplied contextual information as rich as Arrigoni's eyewitness account itself; an indication of how important the blog had become to the Italian public interested in the Gaza crisis.

Summing Up Gaza: Does It Indicate a Paradigm Shift in Crisis Information Management?

In Gaza, social media were used to manage crisis information that was not readily available through state authorities or international journalists. In absence of any formal crisis management system, they replaced emergency response functions on the ground. They also provided a channel of information to the international community about the disaster. Because of the lack of established credibility, information verification was an intensive and elaborate process within the various Gaza crisis social networks. The cluster of network nodes was initially based on strong homogeneous ties, but rapidly grew to become a cluster of weak heterogeneous ties, crossing geographical borders, languages, social and ethno-religious backgrounds. While this hetero-

geneity was sometimes perceived, rightly so, as antagonistic -- so much as to be labeled a social media war -- it reflected an example of a multi-threaded narrative unprecedented so far in crisis and disaster contexts.

Over time, credibility was built as crisis reports converged. Credibility can be problematic in emergency response contexts; however, in Gaza we observed that if given enough time collective social network intelligence can be impressively accurate and coherent even if or because of, lack of any effective emergency response system. Although social networks do not easily facilitate application of censorship, in the Gaza crisis we have seen attempts to limit freedom of speech through self-censorship by exerting significant pressure at individual level.

Traditional roles were redefined. Eye-witnesses were reporting through blogs as international journalists could not. Experts and authority voices became key as they were uniquely positioned to explain conflicting information. Journalists had to provide context to all of this. More importantly though, the news consumer actively sought, produced and disseminated crisis updates, empowered by social media.

In Gaza, because social media functioned as a stand-alone support system, Web 2.0 technologies can be likened to a distributed, collaborative crisis situation centre. Online discussions were as informative about the crisis as were eyewitness reports. Momentous online activity and heated debate were both cause and effect of social media use. Ultimately, this demonstrates the importance of social media in managing and mitigating crises, albeit the effective benefits of the crisis response on the Gaza population are yet to be documented.

For all the above reasons, we sustain that the Gaza crisis may point to a paradigm shift in the use of social media in crisis contexts, the strengths and weaknesses of which we believe to have demonstrated above. We posit that social media use in Gaza can be employed as a poignant precedent for the development of systems, auxiliary to traditional crisis management technologies, which valorize collective distributed intelligence; provide for flexible role/actor definition in crisis management; further intrinsic resistance to censorship by supporting diversity; are complimented by adequate international legal instruments to protect individuals/entities against harmful material and internet crime, and support on-the-fly evaluation of network node connectivity and network cluster development to aid verification and confirmation of crisis information reports.

Towards Future Developments

The use of social media in crisis contexts has already been the focus of experts, academics and professionals. This interest seems compelled by the upsurge of social media use in recent crises: a reaction to unavoidable progress. To be sure, guidelines circulate on the internet on the integration of Web 2.0 technologies in existing risk communication strategies (Linder, 2006; Tinker & Fouse, 2009). We have not seen yet, design aspirations for integrated systems employing respective system strengths and addressing respective system weaknesses.

At the moment, social media technologies develop independently from formal emergency planning and management. What we observe today is facile usage of the former's capacities in the latter's service. For one, social media used for official alerts and warnings seems common sense: they decouple other technologies of communication thus offering redundancy in a crisis; some of them, Twitter for example, are believed to procure resilient means of communication; they offer rapid, timely and wide dissemination of messages, sent both by formal and informal sources.

However, it remains a challenge to see how the two could converge into more than the sum. It is time that the integration of social media in crisis management is considered a system design topic, bringing into focus traditional emergency management tools, together with computer-sup-

ported-collaborative-work (CSCW) know-how, human factors and interaction design experiences (HCI), as well as online communities' expertise.

Some experts call for a networked informed response to disaster, conflict and terrorism (Stephenson & Bonabeau, 2007; Linder, 2006). This cannot but include the successful integration of existing social media applications, to remain responsive to emergency situations where the latter seem to take precedence. Tools and methods developed in adjacent ICT fields are becoming available to make this possible.

Granted, one should acknowledge that social media in crisis contexts pose core issues concerning technology as discussed by Quarantelli (2007). Examples of potential hindrances are: inequalities in access to technology, further diminution of non-verbal communication, security of computer-based systems, social infrastructure and cultures that resist technology in a crisis. In the case of Gaza, these obstacles had a limited effect; on the contrary, issues such as trust and trustworthiness, identity management, reliability and accuracy, cooperation and transparency were key.

To address these concerns, one should think of the mapping of crisis events as a system of events that are interconnected and share similarities or feed into one another. The visualization and system design of such a mapping would provide a bird's eye view of an event when it happens in relation to a context of common or similar elements in space and time.

Given that current command and control systems would be unable to describe the complexity of such networks of events/actors/information, design requirements for such systemic approaches would point to two directions for promoting system flexibility: a. cross-platform, cross-application systems, and super social networks, which visually map crisis information over space and time, and b. network cluster visualization, enabling different types of actors to digest, evaluate, and prioritize information, according to source, reliability, and trustworthiness. Tools and methods developed in adjacent ICT fields are becoming available to make this possible.

As a social network's accessibility and participation increases the more difficult it is to assess credibility. However, Donath (2007) has demonstrated how signals can be used in building trust and demonstrate trustworthiness. She has also demonstrated that the credibility cost can be balanced out by the richness of information obtained through increased connectivity. We argue that visually conveying the social network information flows, network node stability and behaviour, would aid the construction of trustworthiness in crisis contexts. In the case of Gaza the cost of trustworthiness was high because information required time for verification and confirmation. The design goal of an integrated "social network/ crisis management" system would be to lower the costs of credibility assessment against time by increasing transparency of information flows, network connections, and actor definitions.

Kittur (2008) has demonstrated that users' trust of social media is directly proportional to revealing hidden information about elements, element stability and actor behaviour. This finding further supports system transparency to counterbalance credibility costs. Other tools are: visual signaling strategies and *online fashions* (Donath, 2007), which can be used for identity definition of actors. Mapping information types, paths and provenance (clustering coefficients, the degree of bi-directionality in information flows, and types of media sharing) have also been found to play a role in the construction of trust in online communities (Donath, 2007).

Crisis management begs for something more than raw information. "Actionable knowledge" (von Lubitz et al., 2008) is required: usable, useful and immediately relevant information. Social media in crisis contexts have demonstrated the value of collective intelligence that can nourish actionable knowledge; they could effectively fertilize formal crisis information management. Acknowledging the importance of knowledge

experience, integrating expert experience by system design on one hand, while providing metrics and peer-to-peer credibility evaluation, would be important milestones to achieve. Moreover, the convergence of Web 2.0 technologies with crisis management would need concerted effort towards an international legal framework that would provide security and accountability while respecting privacy and civil liberties.

Envisioning a global social network for crisis information, made of many separate interconnected emergency nets where individuals have a role to play, along with state authorities and emergency planners, may not be as far reaching as it seems, especially if the opportunity is seized now. Technological divergence might indeed prove to be a peril if openness is posited against security, and solutions to balance both, are not sought.

CONCLUSION

In the era of risk society the certainties of the past seem to fail us; what remains is a constant negotiation and renegotiation of our understanding of risk. In this, communication is core. Social media offer a platform of communication, sharing information and collaborative construction of meaning that cuts across geographical boundaries, social barriers and political limitations.

Especially in a crisis, social media offer a voice to many. They communicate the knowledge of extreme experience; they bring up the multiple realities of the situation; they counterbalance the dominant role of mass media in the construction of a symbolic meaning of the crisis.

Social media technologies have been used in many ways in different kinds of crisis contexts: disasters triggered by natural hazards, man-made and intended disasters, and conflicts. We identified three main functions of social media in crisis: communication, information and activism. These functions do not exclude one another but work in a complementary manner or in parallel. Each is of interest to formal crisis management, yet has weak and strong points that must be taken into account if more convergence between formal emergency management and social media is to be obtained.

A closer examination of social media use in the Gaza crisis (2008-09) reveals a paradigm shift in the use of social media in crisis contexts. Even in the case of strong technological inadequacies and weak formal crisis management in situ, they served as a substitute to traditional crisis management technologies; moreover they demonstrated the potential for collective distributed intelligence; they backed flexibility on the definition of roles and actors in crisis management; they resisted to censorship and supported expression of multiple realities in a crisis; finally, they enhanced social and geographic convergence in a crisis.

Academics, experts, governments and NGOs currently explore the present and future of Web 2.0 technologies in crisis management; moreover Web 2.0 use in a crisis has already been discussed in social media either occasionally or as a main field of interest. Current developments in the Computer Supported Collaborative Work (CSCW) and Human Computer Interaction (HCI) fields would address many of the technological issues at stake, and help to fertilize further integration of social media in crisis management. A design perspective is needed.

With Web 2.0 becoming more and more pervasive, formal crisis management is compelled to take it into account; but the way the two will be linked to one another is still to be defined. Developments both in crisis management and in social media will shape the future. These developments are not only technological, but are also mainly social and political. Re-examining the command and control structure of crisis management, which has been dominant over the last decade, is fundamental for better agreement between social media and crisis management.

Social media technologies are here to stay and set to change the landscape in crisis information management. They can be a way to democratize

society's responses to risk and crisis, and at the same time boost formal crisis management. The question, however, remains whether the formal crisis communication and information community sees them as an uninvited guest or an intriguing newcomer.

DISCLAIMER

1. The views expressed here represent the individual views of the authors only. They do not necessarily express official positions of the organizations they represent.
2. The authors contributed equally and their names are placed in alphabetical order. Matina Halkia is the corresponding author.

ACKNOWLEDGMENT

We are indebted to the Italian blogger, writer and journalist, Vittorio Arrigoni, who shared with us his eye-witness testimony and social media experience from the Gaza crisis ground-site.

REFERENCES

Adam, B., Beck, U., & van Loon, J. (Eds.). (2000). *The risk society and beyond: Critical issues in social theory*. London: Sage.

Alexander, D. (2005). An interpretation of disaster in terms of changes in culture, society and international relations. In R.W. Perry & E.L. Quarantelli (Eds.), *What is a disaster? New answers to old questions* (pp. 25-38). Philadelphia: XLibris.

Alexander, D. E. (2006). Globalization of disaster: trends, problems and dilemmas. *Journal of International Affairs, 59*(2), 1-22. Retrieved August 3, 2009, from http://www.policyinnovations.org/ideas/policy_library/data/01330/_res/id=sa_File1/alexander_globofdisaster.pdf

Barton, A. H. (2005). Disaster and collective stress. In R.W. Perry & E.L. Quarantelli (Eds.), *What is a disaster? New answers to old questions* (pp.125-152). Philadelphia: XLibris.

Beck, U. (1992). *Risk Society: Towards a new modernity*. London: Sage.

Beck, U. (2009). *World at risk*. Cambridge: Polity.

Boin, A. (2005). From crisis to disaster. In R.W. Perry & E.L. Quarantelli (Eds.), *What is a disaster? New answers to old questions* (pp. 153-172). Philadelphia: XLibris.

Boyd, D. M., & Ellison, N. B. (2007). Social network sites: Definition, history, and scholarship. *Journal of Computer-Mediated Communication, 13*(1), article 11. Retrieved August 3, 2009 from http://jcmc.indiana.edu/vol13/issue1/boyd.ellison.html

Bucher, H. G. (2002). Crisis communication and the internet: Risk and trust in a global media. *First Monday, 7*(4). Retrieved May 15, 2009, from http://outreach.lib.uic.edu/www/issues/issue7_4/bucher/index.html

Buckle, P. (2005). Disaster: Mandated definitions, local knowledge and complexity. In R.W. Perry & E.L. Quarantelli (Eds.), *What is a disaster? New answers to old questions* (pp. 173-200). Philadelphia: XLibris.

Byrne, M., & Whitmore, C. (2008). Crisis informatics. *IAEM Bulletin, February 2008*, 8.

Collins, H. (2009). *Emergency managers and first responders use twitter and facebook to update communities*. Retrieved August, 7, 2009, from http://www.emergencymgmt.com/safety/Emergency-Managers-and-First.html

Currion, P. (2005). *An ill wind? The role of accessible ITC following hurricane Katrina*. Retrieved February, 10, 2009, from http://www.humanitarian.info/itc-and_katrina

Dandoulaki, M., & Andritsos, F. (2007). Autonomous sensors for just in time information supporting search and rescue in the event of a building collapse. *International Journal of Emergency Management, 4*(4), 704–725. doi:10.1504/IJEM.2007.015737

Donath, J. (2007). Signals in social supernets. *Journal of Computer-Mediated Communication, 13*(1), article 12. Retrieved August 3, 2009, from http://jcmc.indiana.edu/vol13/issue1/donath.html

Fischhof, B. (1995). Risk Perception and Communication Unplugged: Twenty Years of Process. *Risk Analysis, 15*(2), 137–145. doi:10.1111/j.1539-6924.1995.tb00308.x

Fischhoff, B. (2006). Bevaviorally realistic risk management. In Daniels, R.J., Kettl, D. F., & Kunreuther, H. (Eds.), *On risk and disaster: Lessons from hurricane Katrina* (pp. 78–88). Philadelphia: University Pennsylvania Press.

Gavriilidis, A. (2009). Greek riots 2008: A mobile Tiananmen. In Economides, S., & Monastiriotis, V. (Eds.), *The return of street politics? Essays on the December riots in Greece* (pp. 15–19). London: LSE Reprographics Department.

Gundel, S. (2005). Towards a new typology of crises. *Journal of Contingencies and Crisis Management, 13*(3), 106–115. doi:10.1111/j.1468-5973.2005.00465.x

Harrald, J. R. (2002). Web enabled disaster and crisis response: What have we learned from the September 11th. In *15th Bled eCommerce Conference Proceedings "eReality: Constructing the eEconomy* (17-19th June 2002). Retrieved October 4, 2008, from http://domino.fov.uni-mb.si/proceedings.nsf/Proceedings/D3A6817C6CC6C4B5C1256E9F003BB2BD/$File/Harrald.pdf

Harrald, J. R. (2009). Achieving agility in disaster management. *International Journal of Information Systems for Crisis Response Management, 1*(1), 1–11.

Hewitt, K. (1998). Excluded perspectives in the social construction of disaster. In Quarantelli, E. L. (Ed.), *What is a disaster? Perspectives on the question* (pp. 75–91). London: Routledge.

Hughes, A. L., Palen, L., Sutton, J., & Vieweg, S. (2008). "Site-Seeing" in Disaster: An examination of on-line social convergence. In F. Fiedrich & B. Van de Walle, (Eds.), *Proceedings of the 5th International ISCRAM Conference.* Washington, DC, USA, May 2008.

International Risk Governance Council. (2005). *Risk governance: Towards and integrative approach.* White paper no.1. Retrieved August 5, 2009, from http://www.irgc.org/IMG/pdf/IRGC_WP_No_1_Risk_Governance__reprinted_version_.pdf

James, A. M., & Rashed, T. (2006). In their own words: Utilizing weblogs in quick response research. In Guibert, G. (Ed.), *Learning from Catastrophe: Quick Response Research in the Wake of Hurricane Katrina* (pp. 57–84). Boulder: Natural Hazards Center Press.

Joyce, M. (2008). *Campaign: Digital tools and the Greek riots.* Retrieved August 1, 2009, from http://www.digiactive.org/2008/12/22/digital-tools-and-the-greek_riots

Kittur, A., Suh, B., & Chi, E. H. (2008). Can you ever trust a wiki?: impacting perceived trustworthiness in wikipedia. In *Proceedings of the ACM 2008 Conference on Computer Supported Cooperative Work (San Diego, CA, USA, November 08 - 12, 2008). CSCW '08.* ACM, New York, NY, (pp.477-480). Retrieved August 7, 2009, from http://doi.acm.org/10.1145/1460563.1460639

Krimsky, S. (2007). Risk communication in the internet age: The rise of disorganized skepticism. *Environmental Hazards, 7,* 157–164. doi:10.1016/j.envhaz.2007.05.006

Lagadec, P. (2005). *Crisis management in the 21st century: "Unthinkable" events in "inconceivable" contexts.* Ecole Polytechnique - Centre National de la Recherche Scientifique, Cahier No 2005-003. Retrieved August 5, 2009, from http://hal.archives-ouvertes.fr/docs/00/24/29/62/PDF/2005-03-14-219.pdf

Lem, A. (2008). *Letter from Athens: Greek riots and the news media in the age of twitter.* Retrieved August 3, 2009, from http://www.alternet.org/media/113389/letter_from_athens:_greek_riots_and_the_news_media_in_the_age_of_twitter

Linder, R. (2006). *Wikis, webs, and networks: Creating connections for conflict-prone settings.* Washington: Council for Strategic and International Studies. Retrieved August 5, 2009, from http://www.csis.org/component/option,com_csis_pubs/task,view/id,3542/type,1/

Liu, S. B., Palen, L., Sutton, J., Hughes, A. L., & Vieweg, S. (2008). In search of the bigger picture: The emergent role of on-line photo sharing in times of disaster. In F. Fiedrich & B. Van de Walle, (Eds.), *Proceedings of the 5th International ISCRAM Conference – Washington,* DC, USA, May 2008.

Moss, M. L., & Townsend, A. M. (2006). Disaster forensics: Leveraging crisis information systems for social science. In Van de Walle, B. and Turoff, M. (Eds.) *Proceedings of the 3rd International ISCRAM Conference,* Newark, NJ (USA), May 2006.

Palen, L. (2008). On line social media in crisis events. *Educase Quartely, 3,* 76–78.

Palen, L., Hiltz, S. R., & Liu, S. B. (2007). On line forums supporting grassroots participation in emergency preparedness and response. *Communications of the ACM, 50*(3), 54–58. doi:10.1145/1226736.1226766

Palen, L., & Liu, S. B. (2007). Citizen communications in crisis: Anticipating the future of ITC-supported public participation. In *CHI 2007 Proceedings* (pp. 727-735), San Jose, CA, USA.

Palen, L., Vieweg, S., Sutton, J., Liu, S. B., & Hughes, A. (2007). *Crisis informatics: Studying crisis in a networked world*. Paper presented at the Third International Conference on e-Social Science (e-SS). Retrieved August 5, 2009, from http://www.cs.colorado.edu/~palen/Papers/iscram08/CollectiveIntelligenceISCRAM08.pdf

Palmer, J. (2008, May 3). Emergency 2.0 is coming to a website near you. *New Scientist*, 24–25. doi:10.1016/S0262-4079(08)61097-0

Perry, R. W., & Quarantelli, E. L. (Eds.). (2005). *What is a disaster? New answers to old questions*. Philadelphia: XLibris.

Quarantelli, E. L. (1998). *What is a disaster? Perspectives on the question*. New York: Routledge.

Quarantelli, E. L. (2007). Problematical aspects of the information/ communication revolution for disaster planning and research: ten non-technical issues and questions. *Disaster Prevention and Management*, *6*(2), 94–106. doi:10.1108/09653569710164053

Renn, O., & Walker, K. (Eds.). (2008). *Global Risk Governance: Concept and Practice Using the IRGC Framework*. Dordrecht, The Netherlands: Springer. doi:10.1007/978-1-4020-6799-0

Rettew, J. (2009). *Crisis communication and social media*. Paper presented at the AIM2009 Conference. Retrieved August 4, 2009, from http://www.slideshare.net/AIM_Conference/crisis-communications-and-social-media-jim-rettew-the-red-cross-2009-aim-conference

Shankar, K. (2008). Wind, Water, and Wi-Fi: New Trends in Community Informatics and Disaster Management. *The Information Society*, *24*(2), 116–120. doi:10.1080/01972240701883963

Shaw, M. (1996). *Civil society and media in global crisis: Representing distant violence*. London: Pinter.

Sorensen, J. H., Sorensen, B. V., Smith, A., & Williams, Z. (2009). *Results of An Investigation of the Effectiveness of Using Reverse Telephone Emergency Warning Systems in the October 2007 San Diego Wildfires*. Report prepared for the U.S. Department of Homeland Security. Retrieved August 4, 2009, from http://galainsolutions.com/resources/San$2520DiegoWildfires$2520Report.pdf

Stephenson, W. D., & Bonabeau, E. (2007). Expecting the unexpected: The need for a networked terrorism and disaster response strategy. *Homeland Security Affairs*, *III*(1). Retrieved August 5, 2009, from http://www.hsaj.org/?article=3.1.3

Sutton, J., Palen, L., & Shklovski, I. (2008). Backchannels on the front lines: Emergent uses of social media in the 2007 Southern California Wildfires. In F. Fiedrich & B. Van de Walle (Eds.), *Proceedings of the 5th International ISCRAM Conference*, Washington, DC, USA, May 2008.

The Risk Communicator. (2008). Social media and your emergency communication efforts. The *Risk Communicator*, *1*, 3-6. Retrieved June 15, 2009, from http://emergency.cdc.gov

Tinker, T., & Fouse, D. (Eds.). (2009). *Expert round table on social media and risk communication during times of crisis: Strategic challenges and opportunities*. Special report. Retrieved August 6, 2009, from http://www.boozallen.com/media/file/Risk_Communications_Times_of_Crisis.pdf

Tziros, T. (2008). *The riots in new media*. Article in Newspaper "Makedonia" on December 15, 2008. [In Greek]. Retrieved August 5, 2009, from http://www.makthes.gr/index.php?name=News&file=article&sid=30225

Vieweg, S. (2008). Social networking sites: Re-interpretation in crisis situations. *Workshop on Social Networking in Organizations CSCW08*, San Diego USA, November 9. Retrieved August 5, 2009, from http://research.ihost.com/cscw08-socialnetworkinginorgs/papers/vieweg_cscw08_workshop.pdf

Vieweg, S., Palen, L., Liu, S. B., Hughes, A. L., & Sutton, J. (2008). Collective intelligence in disaster: Examination of the phenomenon in the aftermath of the 2007 Virginia Tech shooting. In F. Fiedrich & B. Van de Walle (Eds.), *Proceedings of the 5th International ISCRAM Conference*, Washington, DC, USA, May 2008.

von Lubitz, D. K. J. E., Beakley, J. E., & Patricelli, F. (2008). All hazards approach' to disaster management: the role of information and knowledge management, Boyd's OODA Loop, and network-centricity. *Disasters*, *32*(4), 561–585.

von Lubitz, D. K. J. E., Beakley, J. E., & Patricelli, F. (2008). Disaster management: The structure, function, and significance of network-centric operations. *Journal of Homeland Security and Emergency Management, 5*(1). Retrieved August 5, 2009, from http://www.bepress.com/jhsem/vol5/iss1/42/

ENDNOTES

[1] Harrald (2009) advocates for agility in USA emergency management that has emphasized doctrine, process and structure after 9/11.

[2] Such tools serve mainly for alert and early warning, monitoring, decision support etc. To mention only a few, GDACS (Global Disaster Alert and Coordination System), the Indian Ocean Tsunami Warning System, various GIS based Decision Support Systems for disaster management at different scales and levels of administration.

[3] Tinker and Fouse (2009:17) provide a provisional list of social media.

[4] The issue "What is a disaster?" is said to be a "definitional minefield" (Alexander 2005:26).

[5] A google search in the internet (keywords: social media + crisis or disaster or emergency) returns mainly blogs and wikis.

[6] The connectivIT Laboratory of Natural hazards Center, University of Colorado investigated on-site and on-line citizen-site information generation and dissemination activities.

[7] See blog "The riots for the murder of Alexis Grigoropoulos in social media" [in Greek]. Retrieved August, 5, 2009, from http://oneiros.gr/blog/2008/12/07/griotscoverage/

[8] A similar tactic was followed during the Mumbai attacks.

[9] M.Tsimitakis was one of the few journalists who actually used new media to transmit information from the streets. There were claims that twittering, although extensive, contributed little to information as it either reproduced news from the mass media or was used for expressing thoughts and opinions (Tziros 2008).

[10] Blog "Twittering away". Retrieved August 6, 2009, from http://oneiros.gr/blog/2009/01/13/gazasocialmediane/

[11] Local newspaper "Il Centro". Retrieved August 3, 2009, from http://racconta.kataweb.it/terremotoabruzzo/index.php

[12] Announcement of EERI's briefing on the Abruzzo, Italy (L'Aquila) Earthquake. Retrieved August 3, 2009, from http://www.eeri.org/site/meetings/laquila-eq-briefing

[13] http://www.geosynthetica.net/news/article/2009/Interview_Rimoldi_080309.aspx

[14] See for example chat in Italian in the blog http://www.earthquake.it/blog/2009/terremoto-provincia-de-laquila-05-aprile-2009/#more-123 (retrieved 6th April 2009)

15 See for example in English chat in http://
www.italymag.co.uk/forums/general-chat-
about-italy/12088-earthquake.html (retrieved on 6th April 2009) and

16 http://it.wikipedia.org/wiki/Terremoto_
dell'Aquila_del_2009

17 E.Holdman (2009) presents his experiences
on the advances of emergency management
technology in his blog "Typewriters to Twitter: Emergency Management Technology
Through the Decades". Retrieved August
5, 2009, from http://www.emergencymgmt.
com/disaster/From-Typewriters-to-Twitter.
html

18 Governments use twitter for emergency
alerts, traffic notices and more. Retrieved
August 4, 2009, from http://www.govtech.
com/gt/579338?id=579338&full=1&sto
ry_pg=2

19 Alexandra Rampy published a list of US
government Twitter accounts on October
2008. Retrieved August 5, 2009, from http://
fly4change.wordpress.com/2008/10/08/the-
governments-a-twitter-take-2-its-official/

20 Retrieved August 1, 2009, from http://www.
fema.gov/emergency/ipaws

21 http://www.redcross.org/en/connect

22 The Institute of Commercial Management
on 14th July 2009. Retrieved August 5,
2009, from http://news.icm.ac.uk/technol-
ogy/australia-to-use-twitter-for-bush-fire-
alerts/2516/

23 D. Fleet lists Ontario government agencies
trying Twitter. Retrieved August 5, 2009,
from http://davefleet.com/2008/10/twitter-
as-a-hyper-local-emergency-information-
tool/

24 M.Kujawski compiles a list of current initiatives involving social media and government
in several countries. Retrieved August 5,
2009, from http://government20bestprac-
tices.pbworks.com/

25 Retrieved August 5, 2009, from http://www.
guardian.co.uk/uk/2009/jul/27/twitter-
socialnetworking/

26 Al-Jazeera and Palestinian journalists were
reporting; their broadcasts being, overall,
popular in the Arab World; Al-Jazeera,
broadcasting in English, does have potentially an international audience. See also http://
www.haaretz.com/hasen/spages/1054282.
html (retrieved August 4, 2009)

27 Presuming the lack of an effective control
of the territory by the Palestinian Authority.

28 Facebook query on causes (keyword: Gaza)
on 29/7/2009

29 By donating one's status, a Facebook member
becomes a personal advocate of a cause.

30 The different roles and actors in the creation
of the news message will be discussed below.

31 http://www.pbs.org/mediashift/2009/02/
how-social-media-war-was-waged-in-gaza-
israel-conflict044.html (retrieved August 3,
2009).

32 On the number of dead and wounded for
example; or on the use of human shields.

33 In January 2009, the Israeli Ministry of
Immigration announced the international
recruitment of multilingual volunteers to post
comments in blogs and websites in defence
of Israeli policies in Gaza. See also 19th January 2009 "Israel recruits 'army of bloggers'
to combat anti-Zionist Web sites" article on
Haaretz. Retrieved August 6, 2009, from
http://haaretz.com/hasen/spages/1056648.
html

34 Israeli journalist Gideon Levy received death
threats for instigating social media action
over the Gaza crisis, as BBS reports on posts
after Gideon Levy's article on Haaretz (retrieved August 5, 2009, from http://bbsnews.
net/article.php/20090111162755263); so did
Facebook members Joel Leyden and Hamzeh
Abu-Abed (Article "Facebook Users Go to
War" in Time. Retrieved August 5, 2009,

from http://www.time.com/time/world/article/0,8599,1871302,00.html

[35] The presence of an internet site calling for the murder of Italian blogger Vittorio Arrigoni instigated a European Parliament member to formally address the Italian Minister of Foreign Affairs on the protection of Italian citizens. The offensive site was eventually closed down.

[36] As early as 2006, state control was exercised in limiting access to social media. State control of Facebook or Twitter in the 2009 upheavals in Iran and China is an unequivocal indication of the power of social media, perceived as antagonistic to crisis mitigation and management. The US Military recently (August 2009) considered a ban of Web 2.0 technologies in the armed forces.

[37] EU High Representative for Common Foreign and Security Policy on the Gaza Crisis press remarks of 21st January 2009

[38] See for example the MIT Center for International Studies' CIS Starr Forum video (13th January) "Chomsky on Gaza". Retrieved March 10, 2009, from http://web.mit.edu/cis/starr.html

[39] http://www.nytimes.com/2009/01/16/world/middleeast/16mideast.html, http://www.guardian.co.uk/world/2009/jan/07/gaza-israel-obama, http://news.bbc.co.uk/2/hi/middle_east/7833919.stm. All three retrieved August 5, 2009.

[40] http://it.blogbabel.com/metrics/

[41] Arrigoni's personal communication to the authors (4.8.2009 -- original in Italian).

This work was previously published in Advanced ICTs for Disaster Management and Threat Detection: Collaborative and Distributed Frameworks, edited by Eleana Asimakopoulou and Nik Bessis, pp. 143-163, copyright 2010 by Information Science Reference (an imprint of IGI Global).

Chapter 12

A Survey of Trust Use and Modeling in Real Online Systems

Paolo Massa
ITC-IRST, Italy

ABSTRACT

This chapter discusses the concept of trust and how trust is used and modeled in online systems currently available on the Web or on the Internet. It starts by describing the concept of information overload and introducing trust as a possible and powerful way to deal with it. It then provides a classification of the systems that currently use trust and, for each category, presents the most representative examples. In these systems, trust is considered as the judgment expressed by one user about another user, often directly and explicitly, sometimes indirectly through an evaluation of the artifacts produced by that user or his/her activity on the system. We hence use the term "trust" to indicate different types of social relationships between two users, such as friendship, appreciation, and interest. These trust relationships are used by the systems in order to infer some measure of importance about the different users and influence their visibility on the system. We conclude with an overview of the open and interesting challenges for online systems that use and model trust information.

INTRODUCTION

The Internet and the Web are pretty new creations in human history, but they have already produced a lot of changes in the lives of people who use them. One of the most visible effects of these two artifacts is that nowadays everyone with an Internet connection has the possibility to easily create content, put it online, and make it available

to everyone else, possibly forever. If we are to compare this with the situation of some dozens of years ago, the difference is striking. In fact, until recently, only a tiny fraction of the world population had the possibility to "publish" content and distribute it to the public: for instance, few were the authors of books and few the musicians able to publish their music. Conversely, now everyone with an Internet connection can easily publish his/

DOI: 10.4018/978-1-4666-2803-8.ch012

her thoughts on the Web: opening and keeping a blog, for instance, is both very easy and cheap today (actually it is offered for free by many Web sites, for example, blogger.com). Likewise, any band can record its songs in a garage, convert them to MP3 format, and create a Web site for the band to place their song files for the global audience. Moreover, in the future, we can only expect to have these capabilities extended, both on the axis of types of content that can easily be created and shared, and in terms of the range of people that are currently excluded for different reasons, such as location (many countries in the world still have to get the benefit of reliable and cheap Internet connections), age, education level, income.

This phenomenon has been described as the "The Mass Amateurisation of Everything" (Coates, 2003), and we believe this term describes effectively the new situation. However, the easy publishing situation creates a problem, namely "information overload," a term coined in 1970 by Alvin Toffler in his book *Future Shock*. Information overload refers to the state of having too much information to make a decision or keep up to date about a topic. In fact, while it is good to have as many points of view as possible on any topic, it is impossible for a single human being to check them all. So we are faced with the challenge of filtering out the vast majority of the flow of daily created information and experience just the small portion that our limited daily attention and time can manage.

At the present time, it is unreasonable (and luckily almost impossible) to have a centralized quality control authority that decides what is good content, and thus worth our attention, and what instead must be ignored. But of course not all the content has the same degree of worthiness and interestingness for a specific person. What can be done is to infer the quality and value of the content from the "quality" of the content creator. However there is a problem: it is impossible for anyone to have a first-hand opinion about every other single creator of content. Until a few years

ago, before the widespread availability of Internet, it was normal for most of the people to interact just with the people who were living physically close by. Geography was used to shape communities, and a person was able to decide about the neighbors trustworthiness in a lifelong ongoing process based on direct evidence and judgments and opinions shared by trusted people, for example by parents. Physical clues like the dress or the perceived sincerity of the eyes were also used to make decisions about trusting someone or not. Moreover, local authorities had some real power to enforce law in case of unacceptable and illegal behavior.

Instead, nowadays, as an example, it is a realistic possibility for a man in Italy to buy a used guitar from a woman in Taiwan and they will never see each other in the eyes, or even talk. Also, the fact they live in different countries with different law systems makes it very difficult to enter into a legal litigation unless for really huge problems. Thanks to the Internet, we live in the so-called "global village," and in this new and totally different context we need new tools. To date, the most promising solution to this new situation is to a have a decentralized collaborative assessment of the quality of the other unknown people, that is, to share the burden to evaluate them. It is in fact the case that most of the community Web sites nowadays let a user express her opinions about every other user, asking how much she finds her interesting and worth her attention. We call these expressed opinions trust statements. For example, on Epinions (http://epinions.com), a site where users can review products, users can also specify which other users they trust, that is, "reviewers whose reviews and ratings they have consistently found to be valuable" (Epinions.com web of trust FAQ, n.d.) and which ones they do not. Similar patterns can be found in online news communities (for example, on slashdot.org, on which millions of users post news and comments daily), in peer-to-peer networks (where peers can enter corrupted items), in e-marketplace sites (such as eBay.com) and in general, in many open publishing communities (Guha, 2003). Usually judgments entered about other users (trust statements) are used to personalize a specific user's experience on the system, for example by giving more prominence to content created by trusted users. These approaches mimic real-life situations in which it is common habit to rely on opinions of people we trust and value: for instance, it is pretty common to ask like-minded friends their opinions about a new movie while considering if it is worth to go watching it or not. But the Web and the Internet exhibit a huge advantage for information dissemination on a global scale: all the trust statements can be made publicly and permanently visible and fetchable, possibly by everyone or only by some specific users. In this way, it is possible to aggregate in one single graph all the trust statements issued by users. We call such a graph, trust network. Figure 1 shows an example of a simple trust network, containing al the trust statements such as, for instance, the one issued by Alice expressing she trusts Carol as 0.7. In fact, trust statements can be weighted so that it is possible to express different levels of trust on another user. We assume the range of trust weights is [0,1]: small values represent low trust expressed by the issuing user on the target user while large values represent high trust. Precisely, the extremes, 0 and 1, represent respectively total distrust and total trust. Modeling distrust and making explicit its meaning is undoubtedly an open point and will be discussed in section "Open Challenges." Note that the trust network is, by definition, directed, and hence not necessarily symmetric. It is totally normal that a user expresses a trust statement on another user and that this user does not reciprocate: for example, Bob Dylan will hardly reciprocate a trust statement by a fan of his on a music site. In the simple trust network of Figure 1, for example, Bob totally trusts Dave but Dave did not express a trust statement on Bob. We say that Bob is unknown to Dave. Even when users know each other, it can be that their subjective trust statements exhibit different scores, as in the case of Alice and Bob in Figure 1.

On a trust network, it is possible to run a trust metric (Golbeck, Hendler, & Parsia, 2003; Levien, n.d.; Massa & Avesani, 2004; Ziegler & Lausen, 2004). A trust metric is an algorithm that uses the information of the trust network in order to predict the trustworthiness of unknown users. Coming back to Figure 1, since user Alice does not have a direct opinion about user David, a trust metric can be used to predict how much Alice could trust David (represented by the dotted edge in Figure 1). Let us suppose that in Figure 1 the trust statements are expressed by the source user based on perceived reliability of the target user as seller of used products, and let us suppose that Alice wants to buy a used camera. She finds out that David is selling a camera but she does not know David and is not sure about his trustworthiness and reliability. However, Alice knows Bob and Carol, who both know and trust David. In this case, a trust metric can happen to suggest to Alice to trust (or not) David and, as a consequence, to buy (or not) the camera from him. Of course, more complex reasoning involving more users and longer trust paths can happen in more realistic examples. By using trust metrics, even if the users known on a first-hand basis are a small fraction, it is possible to exploit the judgments of other users and figure out how much a certain user (and

Figure 1. Trust network. Nodes are users and edges are trust statements. The dotted edge is one of the undefined and predictable trust statements.

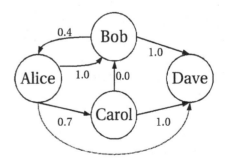

indirectly the content she creates) is interesting for the active user. Trust metrics' common assumption is that trust can be propagated in some way, that is, if user *A* trusts user *B* and user *B* trusts user *C*, something can be said about the level of trust *A* could place in *C*.

The remaining of this chapter is organized as follows: Section "Categories of Online Systems in Which Trust is Modeled and Used" presents a classification of systems in which trust is modeled and used, along with a description of the most representative examples of real systems. For each of them, it describes what are the entities in the system (source and target of trust statement), which social and trust relationships they can express in the system, and how. It also analyzes how trust is used by the system for giving a better experience to the user. In the section "Open Challenges," we discuss the challenges faced by online systems that model and use trust relationships.

CATEGORIES OF ONLINE SYSTEMS IN WHICH TRUST IS MODELED AND USED

This section presents a classification of the online systems in which the concept of trust is modeled and used, and some examples of online systems that fit into the different categories. Even if the listed systems span a large spectrum of purposes and designs, it is possible to recognize some common features, and these drove our classification. In these online systems, visitors are invited to create a user profile so that their online persona is made visible, in general within a "user profile" Web page. Usually this page shows a picture of the user and some information entered by himself/ herself. Often it also shows a summary of the user's activity on the system. This page is very important in these systems since it completely represents the human being behind the online identity, and often users form their opinions about other users only based on this page. These

systems also allow one user to express some level of relationship with the other users, for example, concerning friendship, professional appreciation, commercial satisfaction, or level of acquaintance. We use the term "trust" to represent many slightly different social relationships. Usually the list of expressed relationships with other users is public, but it might also be secret, as in the case of the distrust (or block) list in Epinions.

Trust is a very broad concept that has been investigated for centuries in fields as diverse as sociology, psychology, economics, philosophy, politics, and now computer science (see Mui, 2002 for a detailed summary of contributions from different research fields), and there are no commonly agreed definitions that fit all the purposes and all the investigation lines.

For the purpose of this chapter, we are going to provide an operational definition of trust.

Trust is defined as "the explicit opinion expressed by a user about another user regarding the perceived quality of a certain characteristic of this user." The term "trust statement" will be used as well with the same intended meaning. The user expressing trust (i.e., issuing the trust statement) is the "source user," while the evaluated user is the "target user." We will see in the following how trust is represented and modeled in different ways in the online systems we will explore. For example in some systems, quality refers to the ability to provide reliable and interesting product reviews (as in Epinions). In others systems, it refers to the ability of being a good friend for the user (as in Friendster.com), while in others it is the ability to find interesting new Web sites (as in Del.icio. us). This specificity is the "trust context," and it is the characteristic of the target user evaluated by the user who issues the trust statement. Of course, in different trust contexts, a user can express different trust statements about the same user. For example, the subjective trust expressed by Alice on Bob about his ability of writing an interesting story about computers (the trust context of Slashdot.org) is in general not correlated with the trust expressed by Alice on Bob about his quality of being an honest seller online (the trust context of eBay.com).

In the following we describe different online systems that use and model trust. We have identified few different categories in which the systems can be grouped based on the common features and properties the share. The categories we define are:

- E-marketplaces
- Opinions and activity sharing sites
- Business/job networking sites
- Social/entertainment sites
- News sites
- The Web, the Blogosphere, and the Semantic Web
- Peer-to-Peer (P2P) networks

E-marketplaces are online systems in which a user can sell items owned and can buy offered items. In such a context, typically the buyer does not know the seller, and vice versa. So, in order to decide about a possible commercial exchange that involves the risk of not being paid or of not receiving the products already paid for, it is very important to be able to decide in a quick, reliable, and easy to understand way about the trustworthiness of the possible commercial partner. The success of eBay (http://ebay.com) is largely due to the fact it provides an easy way to do this.

Opinions and activity sharing sites on the other hand are Web sites where users can share with the other users their opinions on items and, in general, make their activities and preferences visible to and usable by other users. The best example of an opinions site is Epinions, in which users can write reviews about products. In activity sharing sites, the user activity is made visible to the other users who can in some way take advantage of it. Two examples of activity sharing sites are *Del.icio.us*, in which users can bookmark URLs they consider interesting, and *Last.fm*, in which users make visible which songs they listen to. Bookmarking a URL and listening to a song can be considered

as the elicitation of a positive opinion about the considered item. However, a user might be more interested in following the reviews and activity of a certain other user, and trust statements can be used exactly for this purpose.

Business/job networking sites are Web sites where users post information about their job skills and ambitions so that other people can find them when they are looking for someone to hire for a specific job. Lately, many systems started to exploit the social (trust) network between users: users can explicitly state their connections, that is, professionals they have already worked with and found reliable and trustworthy. In this way, using the system, a user can enter in contact with the connections of his/her connections and discover potentially interesting new business partners. Linkedin.com and Ryze.com are two examples of such sites.

The idea behind the *social/entertainment sites* is similar to business/job networking sites. However, in this case, the context is more relaxed and informal, and sometimes involves dating and partner search. Here users are, in general, requested to list their friends so that, by browsing the social networks of them, it is possible to discover some friends of friends that might become friends. The first successful example was Friendster.com, soon followed by many other attempts.

News sites are centralized Web sites where users can submit news and stories and comment on them freely. The challenge is to keep the signal-to-noise ratio high. Usually, the users can rate other users' activities (posted news and comments) and these ratings are used to give more visibility to posts and comments the other users appreciate and value. Slashdot.org and Kuro5hin.org are two examples of this category.

The Web, the Blogosphere, and the Semantic Web can be viewed as decentralized news sites. They are systems in which anyone is free to publish whatever content at whatever time in whatever form. Different from the previous examples, in these systems, there is not a single, central point where content is submitted and stored, but the content is published in a decentralized way; for example, it is stored on different Web servers. The challenge in this case is to design a system able to collect all this content and find a suitable algorithm to quickly decide about its importance and value. Google.com was the first company able to achieve this and in fact, a large part of its initial success over search engines of the time is due to the PageRank algorithm (Brin & Page, 1998). PageRank's assumption is to consider a link from page *A* to page *B* as a vote of *A* to *B*, or, in our jargon, a trust statement. The number of received trust statements influences the authority value of every Web page. In the following, we will review also how the concept of trust can be used in the Blogosphere (the collections of all the Web logs) and research efforts for introducing and exploiting trust in the Semantic Web.

The last category of systems that use and model trust is *peer-to-peer (P2P) networks*. P2P networks can be used to share files. P2P networks are, in fact, a controversial technology: large copyright holders claim they are used mainly to violate the copyright of works they own and are fighting to shut down this technology all together. We will not comment on this issue, but present the technological challenges faced by P2P networks. The open, autonomous, and uncontrollable nature of P2P networks in fact opens new challenges: for example, there are peers that enter poisoned content (e.g., corrupted files, songs with annoying noise in the middle) into the network. It has been suggested that a possible way to spot these malicious peers is to let every peer client express their opinions about other peers and share this information using the P2P network in order to isolate them. On a more positive take, a peer can mark as interesting (i.e., trust) another peer when it makes available for download many files that are considered interesting by its human user, or P2P networks can be used to share files only with a controlled and limited community of trusted friends.

Before going on with the discussion, it is worth mentioning that one of the first uses of the concept of trust in computational settings was in PGP (pretty good privacy), a public key encryption program originally written by Phil Zimmermann in 1991. In fact, in order to communicate securely with someone using PGP, the sender needs to be sure that a certain cryptographic key really belongs to the human being he/she wants to communicate with. The sender can verify this by receiving it physically from that person's hands but sometimes this is hard, for example if they live in different continents. The idea of PGP for overcoming this problem was to build a "web of trust": the sender can ask someone whose key he/she already knows to send him/her a certificate confirming that the signed key belongs to that person. In this way, it is possible to validate keys based on the web of trust. The web of trust of course can be longer than two hops in the sense the sender can rely on the certificate received by someone who received it as well as from someone else, and so on. However, in this chapter, we are interested in the concept of trust from a more sociological point of view: trust here represents a social relationship between two entities, usually two users of an online system.

In the following, we present in more details different examples of online systems and, for each of them, what are the entities of the system and which social and trust relationships they can express in the system. We also analyze how trust is used by the system for providing a better experience to the user.

E-Marketplaces

E-marketplaces are Web sites in which users can buy and sell items. The more widely adopted models for arranging a deal are the fixed price (first in first out) and the auction. While the fixed price sell is a fairly straightforward model we are all acquainted with, auctions can take several forms and indulge in a few variants, depending on variables such as visibility of the offer, duration in time, stock availability, start bid, and so on. In an auction, buyers will compete, over the stated period, to put the best bid and win the deal. For the purposes of this chapter, we do not need to go any further in detailing the difference between the types of deals that can be conducted on an e-marketplace, but rather focus on the trust issues between the two roles that users play in this environment: the buyer and the seller. The main complication in conducting a deal in a virtual marketplace is that, in general, the buyer and seller do not know each other, and they only know the information that the Web site is showing about the other user. It is clear that there is a risk involved in a commercial transaction with a total stranger and, in fact, it is not common to give our money to a stranger in the street who promises to send us, days later, a certain product. Akerlof, Nobel Prize in Economy, formalized this idea in his "The market for lemons: Quality uncertainty and the market mechanism" (Akerlof, 1970). He analyzes markets with asymmetry of information, that is, markets in which the seller knows the real quality of the goods for sale but the buyer does not have this information. Using the example of the market of used cars, he argues that people buying used cars do not know whether they are "lemons" (bad cars) or "cherries" (good ones), so they will be willing to pay a price that lies in between the price for lemons and cherries, a willingness based on the probability that a given car is a lemon or a cherry. The seller has incentives to sell bad cars since he/she gets a good price for them, and not good cars since he/she gets a too low price for them. But soon the buyer realizes this situation and that the seller is actually selling only or mainly bad cars. So the price will lower and even less good cars and more bad ones will be put for sale. In the extreme, the sellers of good cars are driven out of the market and only lemons are sold. This effect is the opposite of what a free market should achieve and the only reason for this is asymmetry of information: the buyer has less information about the quality of the goods than the

seller. Here, trust metrics (Golbeck et al., 2003; Levien, n.d.; Massa & Avesani, 2004; Ziegler & Lausen, 2004) and reputation systems (Resnick, Zeckhauser, Friedman, & Kuwabara, 2000) come to provide an escape to this vicious circle by the means of removing or at least reducing asymmetry of information. Giving users the chance to declare their degree of trust in other users makes it possible for future interactions to be influenced by past misbehaviors or good conduct. From this point of view, we can thus say that trust metrics and reputation systems promise to "unsqueeze the bitter lemon" (Resnick et al., 2000) and are a means to even the "risk of prior performance" (Jøsang, 2005).

The prototype of e-marketplace is eBay (http://www.ebay.com), at present the most known and successful example. Let us go through a typical use case of an e-marketplace. Alice found a user whose nickname is CoolJohn12 who accepts bids for his own used guitar. Let us suppose the bid price is fine with Alice. How can Alice be sure that, after she sends the money, CoolJohn12 is going to send her the guitar? How can Alice be sure that the picture on the site is really the picture of the guitar for sale and she is not going to receive another, possibly older, guitar? Unless she finds some evidence reassuring her about these questions, she is probably not going to take the risk and start the commercial exchange. This phenomenon reduces the quantity of commercial exchanges and hence the creation of prosperity (Fukuyama, 1995). But what if the e-marketplace Web site shows Alice that the guitar seller has already sold 187 guitars and banjos, and 100% of the buyers were satisfied by the received product? Or vice versa, what if the site tells Alice that many of the buyers were reporting that seller did not ship the guitar? A simple bit of information shown on the site can make the difference between "It is too risky to buy the guitar" and "I'm going to buy it." This is precisely what eBay does and this is the reason for its worldwide huge success. On eBay, users are allowed to rate other users after every transaction (provide "feedback" in eBay jargon or express trust statements in our jargon). The feedback can be positive, neutral, or negative (1, 0, -1). The main reason for the great success of eBay is due precisely to the idea of assigning a reputation score to every user and showing it. This simple bit of information is shown on the profile page of every eBay user and it is a summary of the judgments of the users who had in past a commercial transaction with that user. It represents what the whole community thinks about the specific user and corresponds to the definition of "reputation" found in Oram (2001). Thanks to this information, everyone can quickly form an opinion about every other user and decide if the risk of conducting a commercial exchange with this user is acceptable. Precisely, on eBay, the reputation score is the sum of positive ratings minus negative ratings. Moreover, the eBay user profile page also shows the total number of positive, negative, and neutral ratings for different time windows: past month, past 6 months, and past 12 months. The purpose is to show the evolution in time of the user's behavior, especially the most recent one.

EBay's feedback ecology is a large and realistic example of a technology-mediated market. The advantage of this is that a large amount of data about users' interactions and behaviors can be recorded in a digital format and can be studied. In fact, there have been many studies on eBay and in particular on how the feedback system influences the market (see for example Resnick & Zeckhauser, 2002). A very interesting observation is related to the distribution of feedback values: "Of feedback provided by buyers, 0.6% of comments were negative, 0.3% were neutral, and 99.1% were positive" (Resnick & Zeckhauser, 2002). This disproportion of positive feedback suggests two considerations: the first is actually a challenge and consists of verifying if these opinions are to be considered realistic or distorted by the interaction with the media and the interface. We will discuss this point later in section "Open Challenges." The second is

about possible weaknesses of the eBay model. The main weakness of this approach is that it considers the feedback of every user with the same weight, and this could be exploited by the malicious user. Since on eBay there are so few negative feedbacks, a user with just a few negative feedbacks is seen as highly suspicious, and it is very likely nobody will risk engaging in a commercial transaction with him/her. Moreover, having an established and reputable identity helps the business activity. A controlled experiment on eBay (Resnick, Zeckhauser, Swanson, & Lockwood, 2003) found that a high reputation identity is able to get a selling price 7.6% higher than a newcomer identity with little reputation. For this reason, there are users who threaten to leave negative feedback, and therefore destroy the other user's reputation, unless they get a discount on their purchase. This activity is called "feedback extortion" on eBay's help pages (EBay help: Feedback extortion, n.d.) and in a November 2004 survey (Steiner, 2004), 38% of the total respondents stated that they had "received retaliatory feedback within the prior 6 months, had been victimized by feedback extortion, or both."

These users are "attacking" the system: as eBay's help page puts it "Feedback is the foundation of trust on eBay. Using eBay feedback to attempt to extort goods or services from another member undermines the integrity of the feedback system" (EBay help: Feedback extortion, n.d.). The system could defend itself by weighting, in different ways, the feedback of different users. For example, if Alice has been directly threatened by CoolJohn12 and thinks the feedback provided by him is not reliable, his feedback about other users should not be taken into account when computing the trust Alice could place in the other users. In fact, a possible way to overcome this problem is to use a local trust metric (Massa & Avesani, 2005; Ziegler & Lausen, 2004), which considers only, or mainly, trust statements given by users trusted by the active user and not all the trust statements with the same, undifferentiated weight. In this way, receiving negative feedback from CoolJohn12 does not influence reputations as seen by the active user if the active user does not trust explicitly CoolJohn12. We will discuss global and local trust metrics, in section "Open Challenges." However, eBay at the moment uses the global trust metric we described before, which is very simple. This simplicity is surely an advantage because it is easy for users to understand it, and the big success of eBay is also due to the fact users easily understand how the system works and hence trust it (note that the meaning of "to trust" here means "to consider reliable and predictable an artifact" and not, as elsewhere in this chapter, "to put some degree of trust in another user"). Nevertheless, in November 2004, a survey on eBay's feedback system (Steiner, 2004) found that only 3% of the respondents found it excellent, 19% felt the system was very good, 39% thought it was adequate, and 39% thought eBay's feedback system was fair or poor. These results are even more interesting when compared with numbers from a January 2003 identical survey. The portion of "excellent" went from 7% to 3%, the "very good" from 29% to 19%, the "adequate" from 35% to 39%, the "fair or poor" from 29% to 39%. Moreover, the portion of total respondents who stated that they had received retaliatory feedback within the prior 6 months passed from 27% of the 2003 survey to 38% of the 2004 survey. These shifts seem to suggest that the time might have come for more sophisticated (and, as a consequence, more complicated to understand) trust metrics. EBay is not the only example of e-marketplace. Following the success of eBay, many other online communities spawned their e-marketplaces, notable examples are Amazon Auctions and Yahoo! Auctions.

Opinions and Activity Sharing Sites

Opinions and activity sharing sites are Web sites where users can share with the other users their opinions on items and, in general, make their activities and preferences visible to and usable by the other users. The best example of an opinion site is Epinions (http://epinions.com). On it, users can write reviews about products (such as books, movies, electronic appliances, and restaurants) and assign them a numeric rating from 1 to 5. The idea behind opinions sites is that every user can check, on the site, what are the opinions of other users about a certain product. In this way, he/she can form an informed opinion about the product in order to decide about buying it or not. However, different reviews have different degrees of interest and reliability for the active user. Reviews are based on subjective tastes and hence, what is judged a good review by a user might be judged as unuseful by another user. So, one goal of Epinions is to differentiate the importance and weight assigned to every single review based on the active user currently served. Epinions reaches this objective by letting users express which other users they trust and which they do not. Epinions' FAQ (Epinions.com web of trust FAQ, n.d.) suggests to place in a user's web of trust "reviewers whose reviews and ratings that user has consistently found to be valuable" and in its block list "authors whose reviews they find consistently offensive, inaccurate, or in general not valuable." Inserting a user in the web of trust is equal to issuing a trust statement in him/her while inserting a user in the block list equals to issuing a distrust statement in him/her. Note that Epinions is one of the few systems that model distrust explicitly. Trust information entered by the users is used to give more visibility and weight to reviews written by trusted users. Reviewers are paid royalties based on how many times their reviews are read. This gives a strong incentive to game the system and this is a serious challenge for Epinions use of trust. Challenges will be analyzed in section "Open Challenges." Epinions' use of trust has been analyzed in Guha (2003). Far from being the only example, other sites implementing metaphors very similar to Epinions are Dooyoo.com and Ciao.com. Note that their business models, heavily based on reviews generated by users, can be threatened by a controversial patent recently acquired by Amazon (Kuchinskas, 2005).

We decided to also place in this category those Web sites where users do not explicitly provide reviews and opinions, but their activity is made visible to the other users who can then take advantage of it. In fact, the activity performed by a user on a system can be seen as an expression of the opinions of that user on what are the most interesting actions to perform on the system, according to his/her personal tastes. Examples of these sites are Del.icio.us (http://del.icio.us), in which users can bookmark URLs they consider interesting, and Last.fm (http://last.fm), in which users make visible which songs they listen to. Bookmarking a URL and listening to a song can be considered as positive opinions about the considered item. On Del.icio.us, the act of trusting another user takes the form of subscribing to the feed of the URLs bookmarked by that user. In this way, it is possible for the active user to follow in a timely manner which URLs the trusted user considers interesting. Flickr (http://flickr.com) is defined by its founders as being part of "massive sharing of what we used to think of as private data" (Koman, 2005). In this scenario, of course, trust is something that really matters. On Flickr, users can upload their photos and comment on those uploaded by other users. Flickr users can then declare their relationship with other users on a role-based taxonomy as friend, family, or contact. Eventually, they can choose to make some photos only visible to or commentable by users of one of these categories. Similarly, Flickr makes use of this information by letting you see the pictures uploaded by your friends in a timely manner. Similar patterns can be seen in the realm of events sharing as well: Web sites such as Upcoming (http://upcoming. org), Rsscalendar (http://rsscalendar.com), and Evdb (http://evdb.org) allow one to submit to the system events the user considers interesting. It is also possible to add other users as friends (i.e., trusted users) in order to see all the public events they have entered. Then, if a user adds another

user as a friend, the second also sees the private events entered by the first. In the domain of music, we already mentioned Last.fm: here the users can declare their friendship to other users by means of a free text sentence connecting user A with B (for example, Alice "goes to concerts with" Bob). Friends are then available to the user who can peek at their recently played tracks, or send and receive recommendations. On the other hand, the tracks played by Last.fm users are recorded in their profiles along with "favorites" (track the users especially likes) and "bans" (tracks the users does not want to listen to anymore). These are used by the system to evaluate a user's musical tastes and to identify his/her "neighbors" (members with interests in similar groups or musical genres). Last. fm members can then exploit their neighborhood by eavesdropping on casual or specific neighbor playlists. Many of these sites, including Del.icio. us, Last.fm, and Flickr and Upcoming, expose very useful application programming interfaces (API) so that the precious data users entered into the system can be used by independent programs and systems (see section "Open Challenges" on problems related to walled gardens). Obviously, by combining two or more dimensions of activity of a specific user, it is possible to aggregate a profile spanning more than one facet about activities of one identity, as the Firefox extension IdentityBurro tries to do (Massa, 2005). The challenge of keeping a single identity under which all users' activities can be tied is briefly addressed in section "Open Challenges." Interestingly enough, Flickr, Del. icio.us, and Upcoming were recently bought by Yahoo!, whose interest in this so-called "social software" seems huge.

Business/Job Networking Sites

On business/job networking sites, users can register and post information about their job skills and ambitions so that other people can find them when they are looking for someone to hire for a specific job. Lately, many systems started to exploit the social (trust) networks between users: users can explicitly state their connections, that is, professionals they have already worked with and found reliable and trustworthy. Notable examples of these sites are LinkedIn (http://linkedin.com) and Ryze (http://ryze.com). On these sites, a user can discover new possible business partners or employees, for example, by entering in contact with the connections of his/her connections. These sites invite users to keep their connections list very realistic and to add as connections only people they really have worked with and deem reliable and recommendable. In order to achieve this purpose, business/job networking sites rely on the fact that user's connections are shown in the profile page and that other users will judge on the basis of the connections (Donath & Boyd, 2004). It is intuitive to say that a user will be better judged as IT consultant if reciprocated connections include Richard Stallman and Steve Jobs than if they contain many random users.

A similar but more playful site is Advogato (http://www.advogato.org). Advogato is a community site of free and open source software developers. The site was designed by Raph Levien, who planned to use it for studying and evaluating his trust metric (Levien, n.d.). On Advogato, users can keep their journal and indicate which free-software projects they are contributing to. A user can also express a judgment on every other user based on their hacking skills by certifying him/her on a three-level basis: Master, Journeyer, and Apprentice. The Advogato trust metric is used to assign to every user a trust level. The trust metric is run once for every level on a trust network consisting only of the certificates not less than that level. Thus, Journeyer certification is computed using Master and Journeyer trust statements (certificates). The computation of the trust metric is based on a network flow operation, also called trust propagation. The trust flow starts from a "seed" of users representing the founders of the site, who are considered reliable ex-ante, and flows across the network with a decay factor

at every hop. The computed certification level of a user (i.e., trust score) is the highest level of certification for which there was a flow who reached him/her; for example, if a user was reached both when propagating trust at level Journeyer and Apprentice, the certification is Journeyer. The trust metric is claimed to be attack-resistant, that is, malicious nodes are not reached by trust propagation (the topic is discussed in section "Open Challenges"). Some other community sites use Advogato's code and hence show similar features. Something notable about Advogato is that it is one of the few sites that let users express a relationship with other users on a weighted base, in this case 3 levels. As a consequence, it is one of the few trust networks with weighted edges. From a research point of view, the availability of the trust network data (at http://www.advogato.org/person/graph.dot) is surely a relevant fact.

On a similar line, Affero (http://www.affero.org) is a peer-based reputation system, combined with a commerce system. It enables individuals to rate other individuals (i.e., express trust statements) and make payments on their behalf. Its goal is to exploit trust elicitation in order to democratically and distributedly decide which projects and foundations are more promising for the community and worth funding. Also, the system does not come bundled with any particular forum or community platform, so any independent community host can integrate the services and individuals can share reputation across various communities. One possible use case is the following: messages written by a user on an independent forum (or via e-mail) are signed with a message such as "Was I helpful? Rate me on Affero." Any individual reading the message and feeling he/she was helped can click on this Affero link and express gratitude by offering ratings, comments, and financial gifts to worthy causes chosen by the helping user on his/her behalf. Affero did not seem to have gotten momentum and is currently used by very few users.

Social/Entertainment Sites

Friendster (http://friendster.com) (Boyd, 2004), founded in 2002, was the first successful site to reach a critical mass of users among the social networking sites. On these sites, every user can create an online identity by filling out a profile form and uploading one or more pictures, and can then express a list of friends. The friends list, along with the user's picture and details, is shown on the user profile page. The idea is that other users can search through the friends lists of their friends and, in this way, discover and be introduced to new people that might be more interesting than a random stranger. We have already called this intuition "trust propagation." In December 2005, Friendster homepage claimed that there were more than 21 million users using the system; however, this is not verifiable. A similar system was Club Nexus (Adamic, Buyukkoten, & Adar, 2003), an online community site introduced at Stanford University in the fall of 2001. Creators were able to use the system to study the real world community structure of the student body. They observed and measured social network phenomena such as the small world effect, clustering, and the strength of weak ties (Adamic et al., 2003). A very interesting and almost unique aspect of Club Nexus was the ability of users to rate other users (express trust statements) on a number of different axis: precisely, based on how "trusty," "nice," "cool," and "sexy" they find their connections (called buddies). Instead, current real online systems in general let users express just a single kind of trust statement and not many facets of it, and we believe this is a strong limitation.

Social sites (and also the previously analyzed business/job networking sites) usually enjoy a rapid growth of their user base due to the viral nature of the invitation process. We have seen that when users register, they can express their trust statements, that is, indicate other users they are connected to. If those users are not on the system they usually receive an e-mail from the system containing an invitation to join the network. This viral invitation strategy is able to rapidly bootstrap the user base. However, one risk is that the social network quickly becomes not representative of the real world because users tend to compete in the game of having more connections than others. Moreover, since everyone can create an identity, fake identities, sometimes called "fakester" (Boyd, 2004), start to emerge and lead the online system even further from a representation of real-world relationships. We will discuss this challenge in section "Open Challenges"; however, let us briefly note how the creator of Club Nexus, Orkut Buyukkokten, later created Orkut (http://www.orkut.com), the social network of Google, and took a different approach. In fact, on Orkut site, it is not possible to create an identity without receiving an invitation from a user who already has an identity in the system. In this way, Orkut staff were able to control Orkut social network's growth, to keep it closer to reality and, as a by-product, to create a desire for users to be inside the system. In fact, the number of social networking sites counts at least in the hundreds, and there are less and less incentives for users to join and reenter their information in YASN (Yet Another Social Network).

Trust statements can be used also for making secure an otherwise risky activity such as hosting unknown people in one's personal house. CouchSurfing (http://couchsurfing.com) and HospitalityClub (http://hospitalityclub.org) are two Web sites in which registered users offer hospitality in their houses to other users, for free, with the goal to make their trips more enjoyable. In order to reduce the risk of unpleasant experiences, there is a trust system in place by which users can express their level of trust in other users (notably, on CouchSurfing the scale is based on 10 different levels ranging from "Only met online and exchanged emails" to "I would trust this person with my life"). The functioning is very similar to the other sites: users can create their profiles, filling in personal details and uploading photos of them. The system shows in the user profile the activity history (who that user hosted, by whom he/she was hosted, how the experiences were in the words and trust statements of the other users) so that, when receiving a request for hospitality from a user, anyone can check his/her history and what other users think about him/her and decide about hosting or denying the request. Additional security mechanisms are possible as well: for example, on CouchSurfing, a user can ask to have his/her physical location address certified by the system by a simple exchange of standard mail with the administrators of the site, and it is also possible to ask administrators to verify personal identity via a small credit card payment. In December 2005, CouchSurfing declared to have almost 44,000 users and HospitalityClub almost 98,000 users.

News Sites

News sites are Web sites where users can write and submit stories and news they want to share with the community. Two notable examples of news sites are Slashdot (http://slashdot.org) and Kuro5hin (http://kuro5hin.org). The most important requirement for such systems is the ability to keep the signal-to-noise ratio high. Slashdot was created in 1997 as a "news for nerds" site. It was a sort of forum in which users could freely post stories and comment on those stories. With popularity and an increased number of users, spam and low-quality stories started to appear and destroy the value of Slashdot. The first countermeasure was to introduce moderation: members of the staff had to approve every single comment before it was displayed. However, the number of users kept increasing and this moderation strategy did not scale. The next phase was the introduction of mass moderation: everyone could act as moderator of other users' posted stories. But in this way there was less control over unfair moderators and hence, metamoderation was introduced.

In December 2005, moderation on Slashdot consists of two levels: M1, moderation, serves for moderating comments, and M2, metamoderation,

serves for moderating M1 moderators. Note that moderation is used only for comments; in fact, it is the staff of Slashdot editors who decide which stories appear on the homepage. Then, once a story is visible, anyone can comment on it. Every comment has an integer comment score from -1 to +5, and Slashdot users can set a personal threshold where no comments with a lesser score are displayed. M1 moderators can increase or decrease the score of a comment depending on the fact they appreciate it or not. Periodically, the system chooses some users among longtime regular logged-in ones and gives them some moderation points. A moderation point can be spent (during the next 3 days) for increasing the score of a comment by 1 point, choosing from a list of positive adjectives (insightful, interesting, informative, funny, underrated), or for decreasing the score of 1 point the score of a comment, choosing from a list of negative adjectives (off-topic, flamebait, troll, redundant, overrated). Moderation points added or subtracted to a comment are also added or subtracted to the reputation of the user who submitted it. User reputation on Slashdot is called Karma and assumes one of the following values: Terrible, Bad, Neutral, Positive, Good, and Excellent. Karma is important on Slashdot since a comment initial score depends on the Karma of its creator. Slashdot editors can moderate comments with no limits at all in order to cope with attacks or malfunctions in a timely fashion. So at M1 level, users rate other users (i.e., express trust statements on them) by rating their comments based on the perceived and subjectively judged ability to provide useful and interesting comments (the trust context of Slashdot M1 level). In fact, the ratings received by a comment directly influence its creator Karma.

Level M2 (called metamoderation) has the purpose to moderate M1 moderators and to help contain abuses by malicious or unreliable moderators. At M2 level, the trust context is related to how good a job a moderator did in moderating a specific comment. Only users whose account is one of the oldest 92.5% of accounts on the system can metamoderate, so that it is ineffective to create a new account just in order to metamoderate and possibly attack the system. Users can volunteer to metamoderate several times per day. They are then taken to a page that shows 10 randomly selected comments on posts along with the rating previously assigned by the M1 moderator. The metamoderator's task is to decide if the moderator's rating was fair, unfair, or neither. M2 is used to remove bad moderators from the M1 eligibility pool and reward good moderators with more moderation points. On Slashdot, there is also the possibility of expressing an explicit trust statement by indicating another user as friend (positive trust statement) or foe (negative trust statement). For every user in the system, it is possible to see friends, and foes (users at distance 1 in the trust network), friends of friends, and foes of friends (distance 2). For every user in the system it is also possible to see which users consider that user a friend (they are called fans) or a foe (called freaks). Every user can specify a comment score for every one of these categories so that, for example, he/she can increase the comment score of friends and be able to place their comments over the threshold, notwithstanding the comment score they received because of moderation.

Kuro5hin is a very similar system, but in December 2005, had a smaller community. However on Kuro5hin, users can directly rate stories and not only comments like on Slashdot. In this way, they influence which stories appear on the homepage, while on Slashdot this is done by editors. Kuro5hin users have the following options for rating a submitted story: "Post it to the Front Page! (+1)," "Post it to the Section Page Only (+1)," "I Don't Care (0)," "Dump It! (-1)." User reputation on Kuro5hin is called Mojo.

The goal of these systems is to keep the signal-to-noise ratio very high, even in the presence of thousands of daily comments, and in order to achieve this they rely on all the users rating other users' contributions and hence, indirectly expressing trust statements on them. The code running Slashdot and Kuro5hin is available as free software (GPL license). We will discuss, in section "Open Challenges," how the fact everyone can analyze and study the code is a positive fact for the overall security of the system and for the ability of the system to evolve continuously and to adapt to new situations and challenges.

The Web, the Blogosphere, and the Semantic Web

Different from the previous examples, in the systems presented in this section, there is not one single central point where content is submitted and stored, but the content is published in a decentralized way; for example, it is stored on different Web servers. The challenge here is to design a system able to collect this vast amount of content, and algorithms able to quickly decide about its relative importance and value. This section is about how the concept of trust can be used and modeled in the Web, the Blogosphere, and the Semantic Web, in order to make them more useful, so that, for example, users can search them and be informed about the quality of the different published information. This might mean either exploiting existing information such as the link structure of the Web or proposing new ways to represent trust information, for example, on the Semantic Web.

The World Wide Web (WWW, or in short simply the Web) is the collection of all the Web pages (or Web resources). It is an artifact created in a decentralized way by millions of different people who decided to publish a Web page written in Hypertext Markup Language (HTML). Web pages are published on billions of different Web servers and are tied together into a giant network by HTML links. Hence, the Web is not controlled in a single point by anybody: the Web can be considered as a giant, decentralized online system. Search engines try to index all the information published on the Web and make it available for searching. Typically, search engines return a list of Web page references that match a user query containing one or more words. Early search engines were using information retrieval (Salton & McGill, 1986) techniques and were considering all the pages published on the Web as equally relevant and worth. However, since search engines are the most used way to locate a page, Webmasters wanted to have their pages on top of the list returned by a search engine for specific keywords. This gave the rise in the mid-1990s to a practice called spamdexing: some Webmasters were inserting into their Web pages chosen keywords in small-point font face the same color as the page background so that they are invisible to humans but not to search engine Web crawlers. In this way, performances of early search engines quickly degraded to very low levels since the returned pages were no more the most relevant ones but just the better manipulated. Note that there is not a single entity with control over the content published on the Web and hence, it was not possible to block this behavior. In 1998, Sergey Brin and Larry Page, at that time students at Stanford University, introduced a new algorithm called PageRank (Brin & Page, 1998) that was very successful for combating spamdexing and producing better search results. They founded a new search engine company, Google (http://google.com), that, thanks to PageRank, was able to quickly become the most used search engine. The simple and genial intuition of PageRank is the following: not all the pages have the same level of authority, and their level of authority can be derived by analyzing the link structure. PageRank assumption is that a link from page *A* to page *B* is a "vote" of *A* on *B* and that authoritative pages either received many incoming links (votes) or even few incoming links but from authoritative pages. As an example, it seems reasonable to assume that a page that received no links is a nonauthoritative page. Based on an iterative algorithm, PageRank is able to assign to every page a score value that represents its predicted authority. This score value

can be used by the search engine in order to give more prominence to more authoritative pages in the results list. PageRank is reported in this chapter because links are essentially what we call "trust statements," and PageRank is performing what we called "trust propagation" over the link network representing the Web. Instead of asking trust statements in order to form the network as the previously introduced online systems did, PageRank's great intuition was to exploit a great amount of information that was already present, the links between Web pages, in a new and effective way.

Other even more explicit trust statements already available on the Web are represented by so-called blogrolls. Web logs (often contracted in blogs) are a very interesting recent phenomenon of the Web. A blog is a sort of online diary, a frequently updated Web page arranged chronologically, that is very easy to create and maintain and does not require knowing HTML or programming. It provides a very low barrier entry for personal Web publishing and so many millions of people in the world maintain their own blog and post on it daily thoughts (Coates, 2003). They pose new challenges and new opportunities for search engines and aggregators due to their continuously changing nature. In general, blogs contain a blogroll: a list of the blogs the blogger usually reads. With the blogroll the blogger is stating: "I trust these other blogs, so, if you like what I write, you will like what they write." What is relevant is that today there are millions of daily updated blogs and that blogs represent, in some sense, a human being identity. So the network of blogrolls really represents an updated, evolving social network of human beings who express their trust relationships via their blogrolls. There is an attempt to add some semantics to blogrolls: XFN (XHTML Friends Network, n.d.) is a microformat that allows representation of human relationships using hyperlinks. XFN enables Web authors to indicate their relationships to the people in their blogrolls simply by adding a rel attribute to their <a> tags.

For example, <a href="http://alice.example.org" rel="met friend"> means that the author of the Web page in which the link is contained has met the person "represented" by http://alice.example.org and considers her a friend. There are also some Semantic Web proposals for expressing, in a semantic format, social relationships. Friend-of-a-friend (FOAF) (Golbeck et al., 2003) is an RDF format that allows anyone to express social relationships and place this file on the Web. There is also a trust extension that allows enrichment of an FOAF file by expressing a trust statement in other people on a 10 level basis (Golbeck et al., 2003). While preliminary research in this field hints its usefulness, the adoption of these semantic formats is slow and not straightforward.

Peer-to-Peer (P2P) Networks

Peer-to-peer (P2P) is "a class of applications that takes advantage of resources (storage, CPU cycles, content, human presence) available at the edges of the Internet" (Shirky, 2000), and has been defined as a disruptive technology (Oram, 2001). Three primary classes of P2P applications have emerged: distributed computing, content sharing, and collaboration. P2P networks are based on decentralized architectures and are composed by a large number of autonomous peers (nobody has control over the overall network) that join and leave the network continuously. In this sense, their open, autonomous, and evolving nature pushes the challenges of the Web to new and harder levels. Just as with Web pages, the reliability of other peers is not uniform. For example, in content-sharing networks, there are peers who insert poisoned content, such as songs with annoying noise in the middle, or files not corresponding to the textual description. And there are peers who share copyrighted content violating the law of some countries. The human controlling the peer, based on his/her subjective judgments, might not want to download files from peers of one of the two categories, that is, she distrusts them. So one possibility is that peers are allowed to express trust statements in other peers in order to communicate their level of desire of interacting in future with those peers. By sharing these trust statements (expressing both trust for appreciated peers and distrust for disliked peers), it is possible to use a trust metric to predict a trust score in unknown peers, before starting to download content from them or upload it to them. Trust metrics can also be used for individuating a close community of friends and share private files just with them.

There are some attempts to build trust-aware systems on top of current P2P networks: on the eDonkey network, it is possible for every peer to mark other peers as friends who are given more priority in downloading files, and a protocol for sharing trust statements has been proposed for the Gnutella P2P network (Cornelli, Damiani, DeCapitani di Vimercati, Paraboschi, & Samarati, 2002). A trust model called Poblano (Chen & Yeager, 2001) was introduced in JXTA, a Java-based P2P framework, and mechanisms based on trust and reputation are present in Free Haven (Oram, 2001) and in BitTorrent (Cohen, 2003).

There is also evidence that P2P networks suffer free riding (Adar & Huberman, 2000), that is, some peers only download files without letting their files available for downloads and in this way they reduce the value of the entire network. The same trust-aware techniques can be used to share information about which peers allow or not to download files and give priority to nonfree riding peers.

Research on reputation and trust in P2P networks is an ongoing effort and many proposals have been made lately. However, due to the autonomous and inherently uncontrollable nature of P2P networks, most of the research papers present results validated with simulations (Kamvar, Schlosser, & Garcia-Molina, 2003; Lee, Sherwood, & Bhattacharjee, 2003), while it is difficult to evaluate the real impact of these strategies on real and running systems.

OPEN CHALLENGES

In this section, we will introduce what are the most interesting challenges related to the use and modeling of trust in online systems. They are divided into three subsections analyzing respectively: (1) differences in how trust relationships are modeled in real and virtual worlds, (2) how trust can be exploited in online systems, and (3) identity, privacy, and attacks in online systems.

Differences in How Trust Relationships are Modeled in Real and Virtual Worlds

It should come as no surprise that social relationships (particularly trust relationships) are different in the "real" world and in the "virtual" world. However, this fact is particularly relevant if the online systems designers want the trust statements expressed in their environment to resemble the real ones. The differences are especially evident with respect to the following issues: how trust relationships can be represented, how they begin and develop over time, and how their representation is perceived by the humans involved. What follows is a list of the most relevant issues.

Explicitness and visibility of trust statements. In a virtual environment, for example, on a community Web site, often trust relationships are explicit. And they are often publicly visible. This means that a user is in general able to check if there is a relationship between two users and, in this case, to see it and refer to it (Donath & Boyd, 2004). This is of course very different from the real world in which almost always interpersonal relationships and their levels are implicit and not publicly stated.

Trust statements realism and social spam. There is also a risk of creating what has been named "social spam" (Shirky, 2004). This happens when a new user in a social network site is allowed to invite, by e-mail, a large number of people into the system, for example, by uploading his/her entire address book. This has happened with at least two social network sites, ZeroDegrees (http://zerodegrees.com) and Multiply (http://multiply.com), and has generated a large vent of protests (Shirky, 2004). New users used this feature and, with a single click, sent an invitation e-mail to all the e-mail addresses in their uploaded contact list, often without realizing this would have resulted in thousands of sent e-mails. Exploiting the viral nature of social networks can be used for passing from zero members to millions, but designers should ask themselves if it is worthwhile to annoy so many users in the hope of retaining a small portion of them, or if this feature is just creating annoying "social spam." Instead, since the beginning, Orkut tried to exploit the same viral nature of sending invitations into the system but in an opposite and more creative way: it was possible to register on the Web site only by being invited by someone already inside the system. By manipulating the number of invitations members could use to invite other people who were still outside the system, Orkut staff was able to create a lot of expectation and a lot of requests for joining the network. In this way, they were also able to control the growth of the network, in order to check if their servers and code were able to handle the load. And another good side effect of this was that, at least at the beginning, the social network was resembling real-world relationships, since every user had a limited number of invitations he/she could use and could not easily engage in the activity of adding as many friends as possible, even if they are not real-world friends. The optimal situation would be the one in which the user remains the owner of his/her social network and trust statements and can export them to every social site instead of having to re-express them every time. We will comment on interoperability at the end of this section.

Disproportion in positive trust. The explicitness and visibility of social relationships represents a huge challenge especially for e-marketplaces. Some reports (see for example Resnick & Zeckhauser, 2002) have found there is high correlation between buyer and seller ratings on eBay, meaning that there is a degree of reciprocation of positive ratings and retaliation of negative ratings. This fact is probably a by-product of the site design and does not closely represent real-world opinions. We have also already commented on how feedbacks on eBay are disproportionately and unrealistically positive (almost 98% of feedback is positive) (Resnick & Zeckhauser, 2002). One explanation of this fact is that, for fear of retaliation, negative feedback is simply not provided in case of a negative experience. Gross and Acquisti (2003) suggest that "no feedback is a proxy for bad feedback," and one solution the authors propose is that the seller gives feedback first and then the buyer is free to give the "real" feedback without fear of retaliation. Anyway, it is easy to argue how often online trust statements do not represent real-world relationships; for example, on many social sites there is a run to have as many friends as possible, and on many e-marketplaces there is an incentive for not providing negative ratings. Psychological and sociological considerations must be taken into account when making available a system that allows one to express relationships online.

Modeling negative trust. Modeling negative relationships (i.e., distrust) is another serious challenge. Few systems attempt to do it: eBay allows users to give negative feedback, but we have seen how this is problematic and seldom used; Epinions allows the active user to place another user in the block list in order to communicate to the system his/her reviews are considered unreliable and should be hidden and not considered. However, Epinions clearly states that "the distrust list is kept private and only you can see your block list. Unlike the web of trust, there is no way you can display your block list to others. This feature was designed to prevent hard feelings or retaliation when you add members to your block list" (Epinions.com web of trust FAQ, n.d.). In a similar way, while in the real world it can happen, for example, that someone expresses, in private, doubts about the

skills of the boss, it is very unlikely that he/she will state a negative trust statement on the boss on a professional site, if this is publicly visible. So surely, the visibility of trust statements changes how users express them and this is something that must be taken into account when designing an online system that models trust. Moreover, on a social site (like Friendster), there are few reasons for entering a negative concept like distrust, since people engaging in a community of friends are there for sharing experiences with people they like and not to punish people they do not appreciate. On the other hand, on P2P systems, trust statements are used both in a positive way in order to keep track of peers whose shared content is reliable and appreciated, but also in a regulative way in order to keep track of peers whose shared content is considered inappropriate and undesirable (for example, depending on the subjective desires, it is poisoned or it is illegally shared). So, in those systems, explicit modeling of negative relationships is necessary since one of the goals is to spot out what are the peers the active peer considers malicious and to warn other peers about them. In short, modeling negative trust statements must be dealt with even more care than positive trust statements, because of the perception humans can have of it and for its great potential of destroying the feeling of community users often look for in online systems. How to exploit negative trust statements, in case they are modeled, will be analyzed in the next section.

Rigidity of language for expressing trust statements. We have also seen in the examples of the previous section that online relationships are represented in a rigid way. For example, it is common to represent friendship as a binary relationship between two users: either it is there or not. Even the richest representations are just a set of numeric values or of predetermined text labels. Anyway, they are rigid. For instance, the evolution in time of a relationship in real life follows a smooth growth or decay and it is often unconscious, or at least not continuously explicitly represented and polled. On the other hand, in virtual environments, the representation is always explicit, and it grows or decays in discrete steps: a possible event on an online community is, for example: "today, I downgrade my friendship to you from level 7 to level 6" and this discreteness hardly models any real-world relationship evolutions. This is surely a challenge for a system that wants to model real-world trust relationships in a reasonable, human-acceptable way. On the other hand, it is possible to keep relationships implicit: for example, the strength of a relationship can be derived on the fly from the number of messages two users have exchanged and hence, this value would closely model changes in the relationship patterns. While this option partially solves the aforementioned issue, in this way the system would become a black box for the user who is not in the condition to know what the system thinks is his/her relationship with the other users and possibly to change it.

Keywords for trust statements conveying undesired meanings. Keywords used in the graphical user interface (GUI) of the system are very important as well. If the system uses the term "friend" for defining a relationship on the system, where a friend is someone who provides timely links to interesting pages (e.g., Del.icio.us), that could be misleading, since the term "friend" in real life means something else. For example, a non-Web savvy but real friend could be unhappy with not seeing himself/herself on the friend list. A reasonable suggestion is to avoid the term "friend" in online systems unless the social relationship really represents friendship, and to use less emotional terms such as "connection" (as LinkedIn and Ryze do) or to use a unique, made-up word with no predefined meaning in the real world (Allen, 2004). We believe this is a key issue for the representativeness of issued trust statements, but we are not aware of research analyzing, with controlled experiments, the impact of different chosen terms in the trust elicitation patterns.

Single-trust context. Moreover, real-world relationships are not embeddable into a single-trust context. A user might appreciate another user for his/her discussions on computer-related topics, but less for his/her always being late or for his/her political ideas. At the moment, it seems very unlikely that an online system that asks users to state trust statements for more than one trust context will be successful; the previously described Club Nexus (Adamic et al., 2003) was an exception in this sense. Even in this case, it is not easy to find the "right" categories for defining a relationship and, as already stated, rigid predefined categories are surely not optimal for representing ongoing real-world situations.

Incentives mechanism for trust elicitation. Another challenge is to find the correct incentives for providing trust statements. The basic assumption of economy, rationality, would suggest that users have the incentive to free ride. In this context, free riding means not providing trust statements and just relying on the trust statements provided by the other users. However, contrary to the basic assumption of economy, many eBay users do provide feedback after a transaction: Resnick and Zeckhauser (2002) found that on average 60.7% of the buyers and 51.7% of the sellers on eBay provided feedback about each other. However, in general, incentives must be envisioned by online-systems designers. On eBay, providing (positive) feedback after a transaction might be seen as an exchange of courtesies (Resnick & Zeckhauser, 2002), or it might be that users perceive the global value of the feedback system and that in order to keep the community healthy, they think they should contribute to it when they can by providing feedback.

On social and activity sharing sites, expressing a trust statement provides a direct benefit to the user since he/she is then able to spend more time on content created by trusted users and less time on not interesting content. In fact, the system in general gives more visibility to trusted users and the content they created. For example, Flickr shows to logged-in users the pictures uploaded by their friends and contacts in a timely manner, and the same happens on Upcoming, which gives visibility to events entered by friends; on Last.fm for songs recently played by friends and on Del.icio.us for URLs recently bookmarked by subscribed users. In a similar way, the Epinions "web of trust" and "block list" give an immediate benefit to the user: when an offensive or unreliable review is found, the user can simply add the reviewer into his/her "block list," telling the system he/she does not want to see his/her reviews again. On the opposite side, users who create interesting, useful reviews can be placed in the "web of trust" so that their reviews are given more prominence. An alternative way for using trust statements would be to exploit already existing information instead of asking it directly of the user. This was the path Google's founders followed when they created PageRank (Brin & Page, 1998). Links between Web pages were already there and PageRank intuition was to consider a link from page *A* to page *B* as a vote of *A* on *B*, or as a trust statement in our jargon.

How to Exploit Trust in Online Systems

In the previous subsection, we discussed challenges in modeling trust in online systems. In this one we assume the trust relationships information is available and concentrate on ways of exploiting it. Based on the subjective trust statements provided by users, we can in fact aggregate the complete trust network (see Figure 1). Trust metrics (Golbeck et al., 2003; Levien, n.d.; Massa & Avesani, 2004; Ziegler & Lausen, 2004) and reputation systems (Resnick et al., 2000) can then be used in order to predict a trust value for every other user based on the opinions of all the users. An important classification of trust metrics (TM) is in local and global (Massa & Avesani, 2004; Ziegler & Lausen, 2004). Global TMs predict the same value of trustworthiness of *A* for every user. PageRank (Brin & Page, 1998) is a global trust metric: the PageRank of the Web page Microsoft.com is, for example, 9/10 for everyone, notwithstanding what the active user querying

Google likes and dislikes. Sometimes this identical value for all the members of the community is called reputation: "reputation is what is generally said or believed about a person's or thing's character or standing" (Oxford Dictionary). On the other hand, local TMs are personalized: they predict the trustworthiness of other users from the point of view of every single different user. A personalized PageRank (Haveliwala, Kamvar, & Jeh, 2003) would predict different trust values for the Web page Microsoft.com for a user who appreciates (trusts) GNU/Linux and a user who appreciates Windows. In fact, a trust statement is personal and subjective: it is absolutely possible for user Bob to be trusted by user Alice and distrusted by Carol, as it is the case in the simple trust network depicted in Figure 1. Actually, it is normal to have, in a real community, controversial users: users trusted by many other users and distrusted by many other ones (Massa & Avesani, 2005). We have seen that reputation is a global, collective measure of trustworthiness (in the sense of reliability) based on trust statements from all the members of the community. Surely, in many situations, it is important to compute an average value. For example, different people can have different opinions on who is the best physicist of the year and nominate different ones. However, only one physicist can get the Nobel Prize and it should be the one that is more appreciated by the community as a whole. The same happens when someone might be elected president of a country: different people would have different preferences but they must be averaged using a global metric in order to identify just one person who will become president. Actually, most of the system we reviewed in section 1 uses a global metric: eBay, Slashdot (on which reputation is called Karma), Kuro5hin (on which it is called Mojo), PageRank, and many others. Reputation is a sort of status that gives additional powers and capabilities in the online system, and it can even be considered a sort of currency. In fact, Cory Doctorow, in his science-fiction novel *Down and Out in the Magic*

Kingdom (Doctorow, 2003) already envisioned a post-scarcity economy in which all the necessities of life are free for the taking, and what people compete for is whuffie, an ephemeral, reputation-based currency. A person's current whuffie is instantly viewable to anyone, as everybody has a brain-implant giving them an interface with the Net. The usual economic incentives have disappeared from the book's world. Whuffie has replaced money, providing a motivation for people to do useful and creative things. A person's whuffie is a general measurement of his or her overall reputation, and whuffie is lost and gained according to a person's favorable or unfavorable actions. Note that Doctorow also acknowledges that a personalized and subjective whuffie can be useful as well, weighting opinions of other people differently depending on one's subjective trust in them. Even if this does not refer to how trust is used and modeled by current online systems, it is an interesting speculation into one of the possible futures and the central role trust would play in it. Coming back to current online systems, Epinions provides personalized results and filtering and hence, exploits a local trust metric, even if precise details about how it is implemented and used are not public (Guha, 2003).

It is worthwhile noting that the largest portion of research papers studying reputation and trust often run simulations on synthesized data representing online systems, and often assume that there are "malicious" peers and "good" peers in the system, and that the goal of the system is just to allow good peers to spot out and isolate malicious peers. In this sense, they also often assume there are "correct" trust statements (a good peer must be trusted) and "wrong" or "unfair" trust statements (if a peer does not trust a good peer, he/she is providing a wrong rating). We would like to point out how these synthesized communities are unrealistic and how reality is more complicated than this; see, for example, a study on controversial users on Epinions (Massa & Avesani, 2005). Assuming that there are globally agreed

good peers and that peers who think differently from the average are malicious encourages herd behavior and penalizes creative thinkers, black sheep, and original, unexpected opinions. This is in essence the "tyranny of the majority" risk, a term coined by Alexis de Tocqueville in his book, *Democracy in America* (1835) (de Tocqueville, 1840). The 19th century philosopher John Stuart Mill, in his philosophical book *On Liberty* (Mill, 1859), analyzes this concept with respect to social conformity. Tyranny of the majority refers to the fact that the opinions of the majority within society contributes to create all the rules valid in that society. On a specific topic, people will express themselves for or against it and the largest subset will overcome, but this does not mean that people in the other subset are wrong. So for one minority, which by definition has opinions that are different from the ones of the majority, there is no way to be protected "against the tyranny of the prevailing opinion and feeling" (Mill, 1859). However, we believe the minority's opinions should be seen as an opportunity and as a point of discussion and not as "wrong" or "unfair" ratings, as often they are modeled in research simulations. However, there is a risk on the opposite extreme as well and it is called "echo chamber" or "daily me" (Sunstein, 1999). Sunstein notes how "technology has greatly increased people's ability to filter what they want to read, see, and hear" (Sunstein, 1999). He warns how in this way, everyone has the ability to just listen and watch what he/she wants to hear and see, to encounter only opinions of like-minded people and never again be confronted with people with different ideas and opinions. In this way, there is a risk of segmentation of society into microgroups who tend to extremize their views, develop their own culture and language, and not be able to communicate with people outside their group anymore. He argues that "people should be exposed to materials that they would not have chosen in advance. Unplanned, unanticipated encounters are central to democracy itself, " and that "many or most citizens should have a range of common experiences. Without shared experiences, [...] people may even find it hard to understand one another" (Sunstein, 1999). Finding the correct balance between these two extremes is surely not an easy task, but something that must be taken into account both for systems designers and researchers.

Creating scalable trust metrics. A challenge for local trust metrics is to be time efficient, that is, to predict trustworthiness of unknown peers in a short time. In fact, in general, local trust metrics must be run one time for every single user propagating trust from his/her point of view, while global ones are just run once for the entire community. In this sense, the load placed on a centralized system (for example, on Google) for predicting the trust scores for every user as seen by every other user seems to be too large to be handled. We believe a much more meaningful situation is the following: every single user predicts the trust scores he/she should place in other users from his/her personal point of view and on his/her behalf. In this way, every user is in charge of aggregating all the trust statements he/she deems relevant (and in this way, he/she can, for example, limit himself/herself to just fetch information expressed by friends of friends) and run the local trust metric on this data just for himself/herself on his/her local device, his/her server or, in the short future, his/her mobile.

Exploiting negative trust statements. We mentioned earlier the challenges in modeling negative trust statements and how few systems attempt to do it. For this reason, research about how to exploit distrust statements is really in its infancy. The lines of early inquiry at the moment are limited to the already cited studies on eBay's feedback system (Resnick & Zeckhauser, 2002), to propagation of distrust (Guha, Kumar, Raghavan, & Tomkins, 2004), and analysis on controversial users (Massa & Avesani, 2005).

Visualization of trust network for explanation. Another open challenge is related to visualization and explanation of how the system used trust information, especially if this affects the user

experience. For example, it is important that the user is aware of the reason a certain review is shown, especially if the system's goal is to let to the user be able to master and guide the process and provide additional information. Visualizing the social network, for example showing to the user a picture similar to Figure 1, might be a powerful option to give awareness to the user of his/her position in the network, and to let him/her navigate it. Surely this kind of interface promises to be useful and enjoyable (see for example, a study on visualization of Friendster network in Heer & Boyd, 2005). However, we note that none of the online systems we introduced earlier use them: the reasons might be that these interfaces are not easily doable with standard HTML, but at the moment require the browser to use external plugins (for example, supporting Java applets, Flash, or SVG), and in this way they also break standard browsing metaphors and linking patterns. Moreover, creating a visualization tool easily understandable and usable is a very difficult task.

Public details of the used algorithms. Another challenge we think should be overcome is related to "security through obscurity" principle. Security through obscurity refers to a situation in which the internal functioning of an artifact is kept secret with the goal to ensure its security. However, if a system's security depends mainly on keeping an exploitable weakness hidden, then, when that weakness is discovered, the security of the system is compromised. In some sense, the system is not secure; it is just temporarily not compromised. This flaw is well acknowledged in cryptography: Kerckoffs' law states that: "a cryptosystem should be secure even if everything about the system, except the key, is public knowledge." Most of the systems we reviewed adopt the security through obscurity principle in the sense that the precise details of how they exploit trust information are kept secret. For example, PageRank is left intentionally obscure. There are early reports about its functioning (Brin & Page, 1998), but Google does not disclose the used algorithm (probably different from the original one) and in particular the parameters used to fine-tune it. Epinions follows the same "security through obscurity" principle: "Income Share is determined by an objective formula that automatically distributes the bonuses. The exact details of the formula must remain vague, however, in order to limit gaming or other attempts to defraud the system" (Epinions. com earnings FAQ, n.d.). Interesting exceptions are Slashdot and Kuro5hin, whose code is free software released under the GNU General Public License and available respectively at SlashCode (http://slashcode.org) and Scoop (http://scoop. kuro5hin.org). In a similar way, Advogato trust metric is described in detail (Levien, n.d.), and the code is available on the Advogato Web site as well. Of course, one problem is related to the fact that commercial companies do not want to disclose their secret algorithms because this would allow any competing company to copy them. Luckily this is not a problem for noncommercial online systems and for systems that do not rely on a central server. However, we believe that a user should be able to know how recommendations are generated (for example, for checking if the system introduces undesired biases) and, in case she desires it, to use trust information as she prefers. We will touch this topic briefly in the following section on walled gardens.

Identities, Privacy, and Attacks

Identity, privacy, and attacks are huge topics by themselves and in this subsection, we are just going to scrape the surface and touch on challenges related to online systems that model and use trust. In general, on these systems, users act under pseudonyms (also called nicknames or usernames). Seldom, the real-world identity of the person using the online system is verified by the system because this would create a huge access barrier, cause great costs, and slow down the process of creating an identity, in a significant

way. Unless there is a great need for the user to enter the system, this will drive him/her away and to the next easier-to-enter online system.

Pseudonymity. As long as a user has some way to decide if another user (as represented by their nickname) is trustworthy, this is often enough. A partial exception in this is represented by eBay. An eBay user can enter credit card details and in this way, eBay can tie the pseudonym with that credit card so that it can be possible to find the person in the real world in case this is needed for some reason, for example, an accusation of fraud or a law suit. Pseudonymity is of course a situation that marks a striking difference between online systems and real world. In real-world interactions, almost always the identity of the other person is known, while this is really the exception in online systems, where it is often possible to interact, communicate, and make business with other users who will never be met in person. In general, users can enter some details that describe themselves, and the system shows this information in their profile page. Note that users can lie in providing this information; for example, a survey found that 24% of interviewed teens that have used instant messaging services and e-mail or been to chat rooms have pretended to be a different person when they were communicating online (Lenhart, Rainie, & Lewis, 2001). The profile page of a user often shows a summary of recent activity in the system and social relationships with other users (Donath & Boyd, 2004), and usually this is the only information available and other users will form an opinion of that user based on this information. The effect of pseudonymity is well captured in the popular cartoon depicting two dogs in front of a computer with one dog saying to the other dog "On the Internet, nobody knows you're a dog." Of course, this situation works until there are no problems, but in case something goes wrong (accusation of fraud, molestation, or any accusation of illegal activity), it is required to identify the real-world identity, and this is not always easy. Moreover, different legal systems make it hard to have justice for crimes perpetuated in the virtual world.

Multiple identities. It is also common for a person to have more than one identity in an online system (Friedman & Resnick, 2001). A recent survey (Aschbacher, 2003) found that users of an informal science learning Web site have more than one identity (60% of girls vs. 41% of boys) on the site. Respondents gave various reasons for the multiple identities including sharing them with school friends, using them to earn more points on the site, and just trying out different identities from day to day. This behavior is quite common in social sites.

Fake identities and attacks. Moreover, sometimes humans create fake identities (also known as fakester) (Boyd, 2004) such as identities representing famous people. However, besides playful reasons, often these multiple identities in control of a single human being are used to game the system. This behavior is often called "pseudospoofing" (a term first coined by L. Detweiler on the Cypherpunks mailing list) or "sybil attack" (Douceur, 2002). Usually these fake identities are used in a concerted way and collaborate with each other in order to achieve a certain result on the system. For example, a person might use them to submit many positive ratings for a friend in order to boost his/her reputation (positive shilling), or negative ratings for the competition in order to nuke his/her reputation (negative shilling) (Oram, 2001).

These multiple identities can also be used, for example, by a book's author for writing many positive reviews of his/her own book. At least an occurrence of this behavior has been revealed publicly because of a computer "glitch" that occurred in February 2004 on the Canadian Amazon site. This mistake revealed for several days the real names of thousands of people who had posted customer reviews of books under pseudonyms (Amazon glitch out, 2004), and it was possible to note that many reviews made about a certain book were in reality created by the author of that book using many different pseudonyms. This possibility seriously mines at the basis the functioning of opinion-sharing sites. Another similar attack occurs on the Web: a link farm is a large group of Web pages that contain hyperlinks to one another or a specific other page. Link farms are normally created by programs, rather than by human beings, and are controlled by a single principal. The purpose of a link farm is to fool search engines into believing that the promoted page is hugely popular, and hence the goal is to maliciously increase its PageRank. A considerable amount of research is devoted to designing methods to spot out these attacks. For example, TrustRank (Gyongyi, Molina, & Pedersen, 2004) is a technique proposed by researchers from Stanford University and Yahoo! to semiautomatically separate reputable, good pages from spam.

Another possible way to deal with link farms is to enrich the language for expressing links, that is HTML (hypertext markup language). In fact, a common practice for increasing the score of a certain page is to use programs that automatically insert links to that page on blogs (in the form of comments) and wikis available on the Web. In order to counter this practice, in early 2005, Google proposed a new solution suggesting that blog and wiki engines should add to every link not directly created by the blog and wiki author a rel="nofollow" attribute. This attribute of the <a> HTML element is a explicit way to tell search engines that the corresponding link should not be considered as a "vote" for the linked page, or a trust statement in our jargon. A related initiative is VoteLinks Microformat (Technorati.com, n.d.), which enriches HTML by proposing a set of three new values for the rev attribute of the <a> HTML element. The values are vote-for, vote-abstain, and vote-against, and represent agreement, abstention or indifference, and disagreement respectively. In fact, as already noted, PageRank's assumption is that a link from page A to page B is a vote of A on B. However, this means that a link created with the purpose of critiquing the linked resource is increasing its PageRank score, and this might induce the author to not link to the criticized page. In short, attention is not necessarily appreciation (Massa & Hayes, 2005). With VoteLinks, it would be possible to tell search engines the reason behind a link so that they could create more useful services. Considering these proposals from a trust point of view, nofollow would express "this is not a trust statement, do not consider it" and VoteLinks would allow authors to express weighted trust statements in a linked page: vote-for is trust, vote-against is distrust, vote-abstain is similar to nofollow. It is interesting to note that Google's nofollow proposal was adopted by most search engines and blog and wiki programs in a few weeks, while VoteLinks proposal seems very little used. This has to do a lot with the authority of the proponent and a little with the proposal itself.

As we already said, local trust metrics can be effective in not letting untrusted nodes influence trust scores (Levien, n.d.; Massa & Avesani, 2004; Ziegler & Lausen, 2004) and in fact, there is research into personalizing PageRank (Haveliwala, Kamvar, & Jeh, 2003) as well. OutFoxed (James, 2005) is exploring ways for a user to use his/her network of trusted friends to determine what is good, bad, and dangerous on the Web. This is done by adding functionality to the Firefox Web browser who is able to predict the trust score of Web pages based on opinions of trusted friends.

Another possible attack is the following. A user could "play" the good behaviored role for a while with an identity and gain a good trust and reputation through a series of perfectly good deals, then try to complete a fraud and eventually drop the identity to start again with a new one. This has been reported at least once in a mid-2000 eBay fraud in which the user "turned evil and cashed out" (Wolverton, 2000). Friedman and Resnick (2001) analyze the phenomenon of multiple pseudonyms and conclude that, in systems in which new pseudonyms can be acquired for free, since new logins could be malicious users who just dropped an identity, the starting reputation

of newcomers should be as low as possible. They prove that "there is no way to achieve substantially more cooperation in equilibrium than that achieved by distrusting all newcomers. Thus, the distrust of newcomers is an inherent social cost of easy identity changes."

Local trust metrics (Massa & Avesani, 2004'Ziegler & Lausen, 2004) can solve the problem introduced by multiple identities. Since with local trust metrics only trusted users (or users trusted by trusted users) are considered, the activity of fake identities not reached by the trust propagation does not influence the active user. In fact, attack resistance of trust metrics and reputation systems is a very important topic that is starting to receive great attention only recently, probably because of the complexity of the problem itself. Some trust metrics are claimed to be resistant to some attacks, for example Advogato (Levien, n.d.): "If a bunch of attackers were to create lots of accounts and mutually certify each other, only a very few would be accepted by the trust metric, assuming there were only a few certificates from legitimate members to the hackers." On the other hand, eBay metric (Resnick et al., 2003) is a very simple one, and we have seen that many attacks can be easily mounted against it. However, it seems to work well in practice, and surely one of the reasons is that, because of its simplicity, every user can understand how it works and get some confidence in the functioning of the system: more complicated metrics would be harder to understand and the user would probably lose confidence in the system altogether. In fact, Resnick and Zeckhauser (2002) consider two explanations related to the success of eBay's feedback system: (1) "The system may still work, even if it is unreliable or unsound, if its participants think it is working. (...) It is the perception of how the system operates, not the facts, that matters" and (2) "Even though the system may not work well in the statistical tabulation sense, it may function successfully if it swiftly turns against undesirable sellers (...), and if it imposes costs for a seller to get established."

They also argue that: "on the other hand, making dissatisfaction more visible might destroy people's overall faith in eBay as a generally safe marketplace." This seems confirmed by a message posted on eBay by its founder in 1996: "Most people are honest. And they mean well. Some people go out of their way to make things right. I've heard great stories about the honesty of people here. But some people are dishonest: or deceptive. This is true here, in the newsgroups, in the classifieds, and right next door. It's a fact of life. But here, those people can't hide. We'll drive them away. Protect others from them. This grand hope depends on your active participation" (Omidyar, 1996). On eBay, whose goal, after all, is to allow a large number of commercial transactions to happen, it seems that positive feelings and perceptions can create a successful and active community more than a sound trust metric and reputation system. This means that the fact that a trust metric or reputation system is proved to be attack resistant does not have an immediate effect on how users perceive it and hence, on how this helps in keeping the community healthy and working.

Another problem with online identities is represented by "identity theft." This refers to the ability of someone to get in control of someone else's identity on an online system. We have seen already how a reputable identity on eBay is valuable by an average 7.6% increase of selling price (Resnick et al,, 2003), and this gives a reason for trying to get into control of them. This phenomenon is also called "account hijacking" and usually happens by phishing or by password guessing. Since online identities have an economic value, they are also sold for real money, often on e-marketplaces.

Privacy. Privacy is another huge issue for online systems, and here we are just going to discuss its main implications. Who can access information users express in an online system undoubtedly modifies which kind of information they will be willing to express and how they express it. As we have already seen, fear of retaliation for a negative trust statement has the consequence of very few negative ratings on eBay and, for this reason for example, Epinions distrust list (block list) is kept secret and not visible. Moreover, trust statements can also be used to model access permission to published information. For example, on Flickr it is possible to make some photo visible only to contacts, friends, or family members. The topic of privacy is very large and has huge psychological implications we cannot address here for reasons of space. Also note that private information a user expresses in an online system can be disclosed by error, as the previously cited example of Amazon Canada showed. The best possible situation for users would be to remain in total control and possession of their information (not only trust statements), and to upload it and show it to who the user wants, when he/she wants.

Portability and interoperability. And in fact, the next challenge we are going to comment about is related to portability of trust and reputation scores across walled gardens. Let us consider the following situation. A person utilizes eBay for some years, provides a lot of trust statements and, even more importantly from his/her point of view, receives a lot of trust statements: he/she built up a good reputation and is recognized by the community. If then, for some reason, he/she would like to change e-marketplaces (for example, eBay could close its operations or the user could prefer a new system that applies smaller fees), he/she has no choice but to start from scratch: there is no way he/she can migrate with his/her activity history (the information he/she entered in the system) and his/her reputation. This is because his/her information is not under his/her control, but under the control of the online system: he/she does not own the information. Clearly, the value of an online system is in the network of users that are using it, and companies prefer to not allow interoperability because competitors would use this information to bootstrap their networks. For example, eBay started a law suit against another e-marketplace who was copying the information about users and their feedback from the eBay Web site or that was, according to eBay, "engaging in the wholesale copying of our content and then using that content without our permission" (Sandoval & Wolverton, 1999). We believe that the content is the users' content and not eBay's content and in fact, users would have all the advantages letting different online systems compete to provide useful and cheap services with the information they expressed. We already discussed about semantic formats for letting users express, on their servers and under their control in a decentralized way, information (for example about the people they know using FOAF or XFN). These attempts have still to gain momentum, and it is surely easier to manage information about millions of users on a centralized server (as eBay, Epinions, Amazon, and almost all the systems we reviewed do at the moment) because there are no problems with formats, retrieval, and update. An attempt to achieve portability of trust and reputation across communities was Affero (see description in section "Categories of Online Systems in Which Trust is Modeled and Used"), but it seems it did not reach a critical mass and has very few users. However, we note how many of the online systems we reviewed in the previous section are starting to expose application programming interfaces (API) so that the precious data users entered into the system can be extracted by them for backups and for migration and, even more interestingly, can be used by independent programs and systems. Flickr, Del.icio.us, Upcoming, and many more systems have already done this or are in the process of doing it and this fact, instead of endangering their existence, has favored a plethora of independent services that are adding value to the original systems.

We believe users are starting to understand that the data they inserted into an online system really belong to them and not to the system, and they will be requiring more and more possibility of directly managing these data and getting them back. When all the systems will export this information, it will be possible to aggregate all the different domains in which a user is acting and get an overall perception of his/her activity in the online world, as the Firefox extension IdentityBurro tries to do (Massa, 2005).

CONCLUSION

In this chapter, we have presented a classification and prototypical examples of online systems that model and use trust information. We have also discussed what are the most important challenges for these systems and some possible solutions. This domain is very active and new initiatives, both commercial startups and research studies, are proposed continuously. New service metaphors and algorithms are invented daily, also based on feedback from users who are becoming more and more aware of their needs. It seems unreasonable to claim that a single approach might fit all the different scenarios we presented in this chapter, and the ones that will emerge in future. Instead, the designers of the online communities will have to continuously rethink basic mechanisms and readapt them to the different needs that emerge. Nevertheless, learning from past experiences, successes, and failures is an important activity, and this is what we tried to do with this chapter. Modeling and exploiting trust in online systems is and will remain an exciting, ongoing challenge.

ACKNOWLEDGMENT

We would like to thank Riccardo Cambiassi for helpful feedback on this chapter and discussions. We would like to thank Jennifer Golbeck for compiling and making public a list of trust-enabled platforms (Golbeck, 2005) that was a good starting point for this chapter. We would also like to thank the creators of the online systems and Web sites we reviewed and commented. Without them, this chapter would have not been, and the Web would have been a less interesting place.

REFERENCES

Adamic, L. A., Buyukkokten, O., & Adar, E. (2003). A social network caught in the web. *First Monday, 8*(6). Retrieved December 28, 2005, from http://www.firstmonday.rog/issues/issues8_6/adamic/

Adar, E., & Huberman, B. (2000). *Free riding on Gnutella*. Technical report, Xerox PARC.

Akerlof, G. A. (1970). The market for lemons: Quality uncertainty and the market mechanism. *The Quarterly Journal of Economics, 84*(3), 488–500. doi:10.2307/1879431

Allen, C. (2004). *My advice to social networking services*. Retrieved December 28, 2005, from http://www.lifewithalacrity.com/2004/02/my_advice_to_so.html

Amazon glitch outs authors reviewing own books. (2004). Retrieved December 28, 2005, from http://www.ctv.ca/servlet/ArticleNews/story/CTVNews/1076990577460_35

Aschbacher, P. R. (2003). Gender differences in the perception and use of an informal science learning Web site. *Journal of the Learning Sciences, 9*.

Boyd, D. (2004). Friendster and publicly articulated social networking. In *CHI '04: CHI '04 extended abstracts on human factors in computing systems* (pp. 1279-1282). New York: ACM Press.

Brin, S., & Page, L. (1998). The anatomy of a largescale hypertextual Web search engine. In *WWW7: Proceedings of the Seventh International Conference on World Wide Web 7*. Elsevier Science Publishers.

Chen, R., & Yeager, W. (2001). *Poblano: A distributed trust model for peer-to-peer networks*. Technical report, Sun Microsystems.

Coates, T. (2003). *(Weblogs and) the mass amateurisation of (nearly) everything*. Retrieved December 28, 2005, from http://www.plasticbag.org/archives/2003/09/weblogs_and_the_mass_amateurisation_of_nearly_everything.shtml

Cohen, B. (2003). Incentives build robustness in BitTorrent. In *Proceedings of the Workshop on Economics of Peer-to-Peer Systems*, Berkeley, CA.

Cornelli, F., Damiani, E., De Capitani di Vimercati, S., Paraboschi, S., & Samarati, P. (2002). Implementing a reputation-aware Gnutella servent. In *Proceedings of the International Workshop on Peer-to-Peer Computing*, Berkeley, CA.

de Tocqueville, A. (1840). *Democracy in America*. (G. Lawrence, Trans.). New York: Doubleday.

Doctorow, C. (2003). *Down and out in the Magic Kingdom*. Tor Books.

Donath, J., & Boyd, D. (2004). Public displays of connection. *BT Technology Journal, 22*(4), 71–82. doi:10.1023/B:BTTJ.0000047585.06264.cc

Douceur, J. (2002). The Sybil attack. In *Proceedings of the 1st International Peer-To-Peer Systems Workshop (IPTPS). eBay help: Feedback extortion*. (n.d.). Retrieved December 28, 2005, from http://pages.ebay.co.uk/help/policies/feedback-extortion.html

Epinions.com earnings FAQ. (n.d.). Retrieved December 28, 2005, from http://www.epinions.com/help/faq/show_faq_earnings

Friedman, E. J., & Resnick, P. (2001). The social cost of cheap pseudonyms. *Journal of Economics & Management Strategy, 10*(2), 173–199. doi:10.1162/105864001300122476

Fukuyama, F. (1995). *Trust: The social virtues and the creation of prosperity*. New York: Free Press Paperbacks.

Golbeck, J. (2005). *Web-based social network survey*. Retrieved December 28, 2005, from http://trust.mindswap.org/cgibin/relationshipTable.cgi

Golbeck, J., Hendler, J., & Parsia, B. (2003). Trust networks on the semantic Web. In *Proceedings of Cooperative Intelligent Agents*.

Gross, B., & Acquisti, A. (2003). *Balances of power on eBay: Peers or unequals? The Berkeley Workshop on Economics of Peer-to-Peer Systems*, Berkeley, CA.

Guha, R. (2003). *Open rating systems*. Technical report, Stanford University, CA.

Guha, R., Kumar, R., Raghavan, P., & Tomkins, A. (2004). Propagation of trust and distrust. In *WWW '04: Proceedings of the 13th International Conference on World Wide Web* (pp. 403-412). ACM Press.

Gyongyi, Z., Molina, H. G., & Pedersen, J. (2004). Combating Web spam with TrustRank. In *Proceedings of the Thirtieth International Conference on Very Large Data Bases (VLDB)* (pp. 576-587). Toronto, Canada: Morgan Kaufmann.

Haveliwala, T., Kamvar, S., & Jeh, G. (2003). An analytical comparison of approaches to personalizing PageRank. In *WWW '02: Proceedings of the 11th International Conference on World Wide Web*. ACM Press.

Heer, J., & Boyd, D. (2005). Vizster: Visualizing online social networks. In *IEEE Symposium on Information Visualization (InfoViz)*.

James, S. (2005). *Outfoxed: Trusted metadata distribution using social networks*. Retrieved December 28, 2005, from http://getoutfoxed.com/about

Jøsang, A., Ismail, R., & Boyd, C. (2005). A survey of trust and reputation systems for online service provision. In *Decision support systems*.

Kamvar, S. D., Schlosser, M. T., & Garcia-Molina, H. (2003). The Eigentrust algorithm for reputation management in P2P Networks. In *WWW'03 Conference*.

Koman, R. (2005). *Stewart Butterfield on Flickr*. Retrieved December 28, 2005, from www.oreillynet.com/pub/a/network/2005/02/04/sb_flckr.html

Kuchinskas, S. (2005). *Amazon gets patents on consumer reviews*. Retrieved December 28, 2005, from http://www.internetnews.com/bus-news/article.php/3563396

Lee, S., Sherwood, R., & Bhattacharjee, B. (2003). Cooperative peer groups in NICE. In *IEEE Infocom*.

Lenhart, A., Rainie, L., & Lewis, O. (2001). *Teenage life online: The rise of the instant-message generation and the Internet's impact on friendships and family relationships*. Retrieved December 28, 2005, from http://www.pewinternet.org/report_display.asp?r=36

Levien, R. (n.d.). *Attack resistant trust metrics*. Retrieved December 28, 2005, from http://www.advogato.org/trust-metric.html

Massa, P. (2005). *Identity burro: Making social sites more social*. Retrieved December 28, 2005, from http://moloko.itc.it/paoloblog/archives/2005/07/17/identity_burro_greasemonkey_extension_for_social_sites.html

Massa, P., & Avesani, P. (2004). Trust-aware collaborativefiltering for recommender systems. In *Proceedings of Federated International Conference on the Move to Meaningful Internet: CoopIS, DOA, ODBASE*.

Massa, P., & Avesani, P. (2005). Controversial users demand local trust metrics: An experimental study on Epinions.com community. In *Proceedings of 25th AAAI Conference*.

Massa, P., & Hayes, C. (2005). Page-rerank: Using trusted links to re-rank authority. In *Proceedings of Web Intelligence Conference.*

Mill, J. S. (1859). *On liberty. History of economic thought books.* McMaster University Archive for the History of Economic Thought.

Mui, L. (2002). *Computational models of trust and reputation: Agents, evolutionary games, and social networks.* PhD thesis, Massachusetts Institute of Technology.

Omidyar, P. (1996). *eBay founders letter to eBay community.* Retrieved December 28, 2005, from http://pages.ebay.com/services/forum/feedback-foundersnote.html

Oram, A. (Ed.). (2001). *Peer-to-peer: Harnessing the power of disruptive technologies.* O'Reilly and Associates.

Resnick, P., & Zeckhauser, R. (2002). Trust among strangers in Internet transactions: Empirical analysis of eBay's reputation system. *The Economics of the Internet and eCommerce. Advances in Applied Microeconomics, 11,* 127–157. doi:10.1016/S0278-0984(02)11030-3

Resnick, P., Zeckhauser, R., Friedman, E., & Kuwabara, K. (2000). Reputation systems. *Communications of the ACM, 43*(12), 45–48. doi:10.1145/355112.355122

Resnick, P., Zeckhauser, R., Swanson, J., & Lockwood, K. (2003). *The value of reputation on eBay: A controlled experiment.*

Salton, G., & McGill, M. J. (1986). *Introduction to modern information retrieval.* New York: McGraw-Hill, Inc.

Sandoval, G., & Wolverton, T. (1999). *eBay files suit against auction site bidder's edge.* Retrieved December 28, 2005, from http://news.com.com/2100-1017-234462.html

Shirky, C. (2000). *What is P2P... and what isn't?* Retrieved December 28, 2005 from http://www.openp2p.com/pub/a/p2p/2000/11/24/shirky1-whatisp2p.html

Shirky, C. (2004). *Multiply and social spam: Time for a boycott.* Retrieved December 28, 2005, from http://many.corante.com/archives/2004/08/20/multiply_and_social_spam_time_for_a_boycott.php

Steiner, D. (2004). *Auctionbytes survey results: Your feedback on eBay's feedback system.* Retrieved December 28, 2005, from http://www.auctionbytes.com/cab/abu/y204/m11/abu0131/s02

Sunstein, C. (1999). *Republic.com.* Princeton, NJ: Princeton University Press.

Technorati.com. (n.d.). *VoteLinks.* Retrieved December 28, 2005, from http://microformats.org/wiki/votelinks

Wolverton, T. (2000). *EBay, authorities probe fraud allegations.* Retrieved December 28, 2005, from http://news.com.com/2100-1017_3-238489.html

XHTML Friends Network. (n.d.). Retrieved December 28, 2005, from http://gmpg.org/xfn/

Ziegler, C., & Lausen, G. (2004). Spreading activation models for trust propagation. In *Proceedings of the IEEE International Conference on E-Technology, E-Commerce, and E-Service (EEE'04).*

This work was previously published in Trust in E-Services: Technologies, Practices and Challenges, edited by Ronggong Song, Larry Korba, and George Yee, pp. 51-83, copyright 2007 by Idea Group Publishing (an imprint of IGI Global).

Chapter 13
The Anonymity of the Internet:
A Problem for E-Commerce and a "Modified" Hobbesian Solution

Eric M. Rovie
Agnes Scott College, USA

ABSTRACT

Commerce performed electronically using the Internet (e-commerce) faces a unique and difficult problem, the anonymity of the Internet. Because the parties are not in physical proximity to one another, there are limited avenues for trust to arise between them, and this leads to the fear of cheating and promise-breaking. To resolve this problem, I explore solutions that are based on Thomas Hobbes's solutions to the problem of the free rider and apply them to e-commerce.

TRUST AND THE INTERNET

A firm handshake and a face-to-face meeting over dinner and drinks used to be the model for business transactions. Buyers and sellers met face-to-face in stores, restaurants, board rooms, and even on front porches to arrange transactions, and it would have seemed bizarre to buy anything 'sight unseen,' much less to buy from a person you couldn't see or hear. But the Internet has changed the face of the world, and commerce is no exception: transactions occur between parties who have never met, will never meet, and do not ever need to speak to each other using anything more than a keyboard. These new, electronic, possibilities for business

and commerce are great, but they can also come at a significant cost: a sense of trust that is (for some) generated in a person-to-person (rather than a machine-to-machine) transaction. In this paper, I will argue that the perceived anonymity of the Internet raises problems for traditional models of commerce, but that the problem can be resolved in a Hobbesian fashion by creating an appropriately authoritative framework under which e-commerce can operate, and by having avenues for recourse should a transaction be unsatisfactory to either (or both) parties. I will attempt to shine a light on a more broadly philosophical problem (the problem of the 'free rider') by looking at the hazards of e-commerce, and I will argue that Hobbes' rec-

DOI: 10.4018/978-1-4666-2803-8.ch013

ognition of this problem, and his solutions to it, are useful for steering us clear of these hazards. My primary concern here is philosophical: I offer advice to participants in e-business using Hobbesian precepts without an in-depth analysis of how the business end of the advice might be cashed out. But I think it is clear that business can learn much from philosophy, and vice versa.

We would be wise to begin by noting some of the essential differences that make the problems facing e-commerce different from those that plague standard versions of commercial interaction. To begin with, I should clarify a crucial point: in this paper, I focus on the issues that plague e-commerce and not with the more broad category of e-business (which includes e-commerce but also includes electronic facets of the internal operation of a business) itself. I include all forms of electronic commerce under the broad heading of e-commerce, including sales from vendor to vendor, vendor to consumer, and consumer to consumer. I also include both fixed price sites and auction sites, and third-party hosted transactions. This means that my argument applies equally when ACME Widgets sells parts to General Industrial Incorporated, or when Annie sells Bill a hand-made scarf or a used copy of *London Calling* on eBay or Amazon's Marketplace. I realize that the scope of the transactions is greatly different, but I do not think the principles that ground them should be.

The thrust of my argument here is that there is a philosophically interesting problem that faces e-commerce, and that this problem is based on the distinct environment of e-commerce: the distance and anonymity of the Internet. The problem is not completely unique to e-commerce, of course, because it would plague any form of mail-order sales and even, to a lesser degree, sales by phone and television. It is amplified in e-commerce, however, because there is rarely any contact with another human being, person-to-person, apart from email. When a customer buys a product from a vendor in a traditional setting, there are certainly

going to be concerns about the transaction, but I argue that concerns are greatly increased in an e-commerce setting for a number of reasons. These reasons include, but are not limited to, the following:

1. The lack of face-to-face connection with another person.
2. The lack of familiarity with the seller and her business practices.
3. The lack of repeat transactions with the same seller.
4. The reputed anonymity of the Internet.
5. The lack of a personal filter on conversations that take place over the Internet.

The lack of face-to-face, interpersonal interaction may impact the feelings of trust that each member of the transaction will have. According to recent psychological research (and echoing common folk psychological attributions) there is something important about having a 'trustworthy face', and visual perception of facial cues works to reinforce or create attributions of traits like 'trustworthy' or 'confident' or 'aggressive' in stranger (Oosterhof & Todorov, 2008, p. 126). The inability to examine a face, according to some views of this work, might impede the ability to trust, or at least cause the trust to develop more slowly than it would under face-to-face conditions (Wilson, Straus, & McEvily, 2006).

Secondly, there is the lack of familiarity with the seller. Being unfamiliar with a seller is not a situation limited to e-commerce: it would, presumably, apply to every first-time pair of transactors. But combining this with the 'lack of face' makes it more problematic: not only do parties not see their partner, but they also know little more than what each party posts on its website. In most cases, without doing much digging, it would be hard to ascertain which online sellers are reliable and which are scam artists.

This leads us to the third problem, where buyers may find themselves moving from seller to seller to get the best price (instead of simply committing to one location as central) and, ultimately, may not be able to develop mutually trusting relationships with any one seller. Again, this problem is not exclusive to e-commerce, but may lead to more 'one-off' transactions between parties, rather than a long-term economic relationship between two parties.

The fourth problem will loom large in my later discussion of trust, but I can point to a general issue here: people will be more likely to act in ways that are not as socially acceptable if they believe they are acting anonymously. Anonymous tip lines encourage people to turn in criminals knowing they won't be subject to retribution, and anonymous chatrooms and message boards allow people to say things they normally might refrain from saying in a face-to-face conversation. According to some social psychological research, hostile remarks may increase by as much as six times if a participant believes she is commenting anonymously (S. Kiesler, Siegel, & McGuire, 1984). Anonymity provides a shield behind which individuals feel free to say what they feel, and often those feelings are amplified by the sense of security they have in their anonymity. This leads one prominent scholar of the psychology of Internet behavior to note "when people believe their actions cannot be attributed to them personally, they tend to become less inhibited by social conventions and restraints" (Wallace, 1999, pp. 124-125).

This point connects to the final problem, that conversations between parties over the Internet might be subject to less restraint and less of a 'filter' than conversations that take place face-to-face or over the phone. It would seem to be much easier to lose one's temper, or say unnecessarily cruel things, to a blinking cursor than to a responsive voice or an emotive face. Despite the fact that rules of Internet etiquette ("netiquette") have been around dating back to at least 1995 (for an early draft of a code of Internet etiquette, see http://tools.ietf.org/html/rfc1855), it is not uncommon for 'flame wars' and malicious comments to be generated from across seemingly anonymous computer screens. The combination of these five factors (in various forms depending on the circumstances) can lead to unique situations that are not as common in traditional face-to-face or phone-to-phone transactions. And this can lead to a breakdown in a central facet in a business relationship: a failure in trust.

It should be noted here, also, that the anonymity of the Internet is not absolute: participants are subject only to as much anonymity as their technology (and Internet Service Providers) will allow (Wallace, 1999). A person with the appropriate level of tech-savvy (or the right court order) will be able to crack through almost any veil of Internet anonymity, and many 'anonymous' Internet participants give away much of their anonymity by using real names, personal details, or personal e-mail addresses in their on-line identities. There seems to be little guarantee that any Internet exchange would remain permanently anonymous, although, presumably, most do remain that way.

TRUST AND E-COMMERCE

The relationship between trust and commerce is deceptively simple: for a transaction to happen, some level of trust must exist between the parties in the exchange. If Ron tells Katie "pay me $30 today, and I'll bring the bike to your house tomorrow," Katie needs to have some level of trust that Ron will hold up his end of the deal, or Katie may have just given away $30 for nothing. In business, that trust may come in various forms, from warranties, work orders, and receipts, to interpersonal relationships (Katie is far more likely to trust Ron's cash-for-bike-tomorrow exchange if she knows and trusts Ron as a person) or even through legal institutions and business organizations (tort law, the Universal Commercial Code, etc) Regardless of the form, trust is crucial for such exchanges to be able to occur. As Carson puts it in a discussion of the ethics of advertising and sales, "Deceptive advertising is also harmful in that it lowers the general level of trust and truthfulness essential to a flourishing society and economy. The law alone cannot secure the level of honesty and trust in business necessary for people to be sufficiently willing to enter into mutually beneficial market transactions" (Carson, 2002, p. 41).

The 'deceptively' simple aspect of the trust relationship in commerce, however, stems from the desires of both parties to succeed at the expense of the other. The parties are, of course, competing with one another in the exchange. Ideally, both parties will walk away from the transaction in an improved position. If only one party walks away in an improved position, but the other party is not made any worse (known in economics as a Pareto improvement) and no further improvements can be made, the situation is economically (Pareto) optimal and efficient. But, in a practical sense, most participants in economic exchanges may not care about (or know about) Pareto improvements, and might be inclined to view the transaction as a "me versus them" battle, particularly if there isn't a trust-informed relationship already present between the two parties. If I desire a basket of fresh peaches, and I venture to a local fruit stand, while I may not have any strong desire to see the fruit vendor put out of business (in fact, I may want him to stay open because of the quality and convenience his stand provides me), I might not object if he grossly undercharged me.[1] The trust we need to give (and get) in commercial transaction is, at the very least, a little puzzling if not fully paradoxical, but it seems fully necessary to have something upon which we can rest our hopes for mutual fulfillment of the goals of the transaction. This need for trust is made even more problematic if we consider the possibility, raised by Eric Uslaner (2002), that trust is drastically reduced when parties are on unequal social ground, and that many participants in commerce have lost much of the trust they might have had with successful trade partners. In other words, the bigger and more successful you are in your

business endeavors, the more likely it might end up being that people fail to trust you.

If we take, as a central premise in the argument for commerce, that trust (of some sort) is necessary for transactions to occur, and add to it the problem of the anonymity of the Internet for trust, we face an argument that looks something like this:

1. Trust is one essential component to successful commercial interactions between parties.
2. Trust is best obtained in situations where parties have some familiarity with one another, either through personal (face-to-face) interaction or through repeated business dealing.
3. Therefore, face-to-face interaction makes commerce between parties more likely to occur and/or succeed.
    But since most e-commerce transactions do not occur through personal (face-to-face) interactions, this leads us to what I call the Problem of E-Commerce, or the Anonymity Problem:
4. E-Commerce, in general, does not utilize face-to-face interactions and is, generally, 'anonymous' in most relevant senses, and is, therefore, lacking in the materials that develop trust.
5. Therefore, e-commerce, is less likely to occur and/or succeed than the traditional face-to-face transaction.

Patricia Wallace has noted that all e-commerce transactions, but particularly 'grassroots' e-commerce transactions (in situations where individuals transact with one another rather than with or through a corporation) are open to abuse, fraud, and even violence (Wallace, 1999, pp. 244-245), and this certainly hampers to overall level of trust in the e-marketplace. Successful e-commerce would have to be mostly free of fraud, dishonesty, and bad faith, but anonymity and trust gaps also seem to encourage such shady dealings. So, how can we salvage e-commerce in the wake of this problem? Or, should we simply write off all successful e-commerce as aberrant? I think the answer to this question can be drawn from the work of Thomas Hobbes.

HOBBES'S ARGUMENT

Thomas Hobbes's impact on social, political, and moral philosophy cannot be understated. His answers to the problems of social and political co-ordination, his explanations of moral motivation, and his views on moral authority should be central parts of any discussion of these topics. Hobbes's arguments are most often used as the basis for political theories, but they need not be used exclusively so. His insights on human behavior and moral psychology, for instance, impact the theory of games, and his moral philosophy was greatly influential to several traditions of moral theory, notably contractarianism. Despite the fact that his arguments were developed in response to the dire strife of the English Civil War, his insights are still appropriate to contemporary issues. Social norms, in general, may rest heavily on Hobbesian arguments, providing a philosophical answer to the question "Why observe non-legal rules and norms?" Of course, Hobbes did not explicitly address the problem of e-commerce, because he lived (1588-1679) roughly five hundred years before such a thing even existed, but his thoughts on other topics can guide us in the direction of a useful Hobbesian solution to the Anonymity Problem, as he has done for other problems of the social sciences (Hollis, 2002). In what follows, I will provide an extremely brief sketch of several key elements of Hobbes's argument as developed in *Leviathan* (Hobbes, 1996).

For Hobbes, the crucial reason to enter into civil society, to consent to give up absolute freedom in favor of being governed, is because a life outside of civil society is a brutal and terrifying

constant struggle. The 'state of nature', as Hobbes famously puts it, is a state where the life of all man is "solitary, poor, nasty, brutish, and short" (p. 89). One of Hobbes's central theses is that the state of nature is, for all practical purposes, a state of constant war and strife. Even when there is not active combat going on, all parties in the state of nature must be ready to go to battle with the rest of the world, and this is an exhausting and terrifying way to live. The real reason why self-interested agents would give up the freedoms of the state of nature for the structure and discipline of government is to get a reprieve from the constant struggles of life in the state of nature.

How might these struggles be abated, given the apparent human drive to be in constant conflict? For Hobbes, this problem is resolved by having all parties give up their rights, provided that all other parties do so as well, in the form of large-scale social contract. This contract will be protected by a powerful and people-authorized political leader (or assembly of leaders), an 'artificial man' who keeps the peace and protects the interests of the citizens, called the sovereign. For Hobbes, this means the collective will of the people is "united in one Person" and this serves as the "Generation of the great Leviathan, or rather (to speak more reverently) of that Mortal God, to which we owe under the Immortal God, our peace and defense" (p. 120). The Leviathan will be a feared and powerful political leader whose power and strength is so awe-inspiring and fear-inducing that his rules, whatever they may be, will surely be followed. The Leviathan will provide both law and order, and will do so with an iron fist, if necessary. Hence, the chaos and uncertainty of the state of nature is replaced by order and structure under the rule of the sovereign. Put even more simply, the sovereign replaces the free-for-all that exists prior to civil society with an order that is (even if brutal and cruel) better and more predictable than the disorder of the state of nature. We consent to be governed because even bad government is better than the brutality and uncertainty of the state of nature. This allows us to have agreements, mutually beneficial arrangements and, most importantly, contracts.

Chapter XIV of *Leviathan* sets up a classic philosophical defense of the contract and, by extension, commerce. Hobbes argues that for mutual performance of the two ends of a contract, there needs to be some sort of coercive force to ensure that each participant follows through on their end. It would be foolish, for instance, for one party to provide their service to another before receiving the payment because "he which performeth first, does but betray himself to his enemy" (p. 96). But, short of a simultaneous exchange, this would seem to leave commerce and other sorts of contracts to be difficult to achieve. Hobbes has the solution to this problem at the ready: the sovereign, and his awe-inspiring power. Since both parties to a contract are aiming to see their desires fulfilled without paying any costs, both would be willing to take what they want and leave their end of the contract unfulfilled. According to Hobbes, the third Law of Nature is "that men perform their Covenants made" and this will be achieved, when men are unwilling to do so on their own, by the implementation of "some coercive power, to compel men equally to the performance of their Covenants, by the terror of some punishment, greater than the benefit they expect by the breach of their Covenant" (pp. 100-101). In other words, there must be some force (the sovereign) in place to force parties to abide by contracts made. Without such a force, it would be foolish to expect the other party to your contract to abide by his end, and the whole structure and institution of contracts would collapse on itself in a self-interested muddle. Without contracts, we cannot be assured of any social cooperation at all, but without a sovereign to enforce the contracts made, we cannot consider contracts to be of any practical use at all: they would just be meaningless words, said for show, but broken at will when it would be in one's interest to do so. [2] And, if contracts are meaningless, commerce and trade are meaningless as well. He describes the state of nature as being a world where "there is no place for Industry, for the fruit thereof is uncertain" (p. 89). Hobbes's entire political theory is based on the structure of a social contract between the sovereign and his citizens, and this becomes one version of a moral and political theory known as contractarianism.

For contracts and agreements to work, however, all parties must be expected to hold to them under all circumstances. If one party provided payment, but the other failed to deliver the paid-for service, the contract would be broken. Clever contractors might regularly fail to follow through on their contracts, which could put the whole institution of contracting at risk. This is an issue that Hobbes takes quite seriously, and is often referred to as the Free Rider Problem. Hobbes never directly refers to 'free riders' but speaks of a character called "The Foole", who has "said in his heart, there is no such thing as Justice" (p. 101). The Foole, Hobbes argues, would pretend to be playing by the rules of society, and would appear to be following the laws, but would, in secret, be violating the rules at every beneficial opportunity. The Foole wants to receive the benefits of living in a civil society (namely, being out of the brutal state of nature) without truly paying the costs (by following the rules, which may be inconvenient to his desires). Of course, he can only do this secretly, for if he were caught violating the rules, he would be subject to severe punishment, so the Foole appears to be a law-abiding citizen.

So what, if anything, is there to deter individuals acting as Hobbesian Fooles? Quite simply, the answer is the fear of being caught.[3] All Fooles will face the worry that their violations of the laws of civil society will be found out by their fellow citizens, or by those in society who enforce the laws, and this worry should be enough to keep them in line for at least two main reasons: the sovereign's fearful wrath, or the punishment of being cast out of civil society to perish back in the state of nature (p. 102). Whenever a person is tempted to act like a Foole, they should be reminded of the dangers of being found out, and the hazards of being excommunicated from society. This, says Hobbes, should be enough to discourage their violation of the social order, and give us reason to trust in those who make contracts with us. Although it might appear that it would benefit someone, in the short term, to be a Foole, the long-term consequences might include the collapse of the very social networks and systems that support the Foole's ability to free ride, and this should be enough (from a Hobbesian view) to provide a disincentive for free riding.

These, then, are the core elements of the Hobbesian argument for social order, which tell us a story about the genesis of contracts and the requirements for commerce and trade to occur. In the next section, I move to connect the Hobbesian argument to the Anonymity Problem.

HOBBESIAN SOLUTIONS TO THE ANONYMITY PROBLEM

The Anonymity Problem, as we saw earlier, is essentially a problem of trust, and is exacerbated by the presumed anonymity of the Internet. In some ways, the free-floating world of the Internet is akin to Hobbes's state of nature, where there are not universal protections for all participants, particularly for e-commerce partners. Even from a legal and social perspective, the Internet has really become a sort of 'Wild West' frontier where boundaries, social norms, and legal institutions are tested and shattered on a regular basis, and this makes trust difficult to develop online. Hobbes clearly had the problem of trust in mind when he was developing a theory that would protect contracts and preserve social order in the face of predominantly self-interested agents. He provides us answers to the questions "Whom should I trust?" and "Why should I trust them?" by invoking the twin fears of the sovereign and the state of nature, and provides an argument for even the most selfish and egoistic agents to follow the rules imposed by

them by social forces. Hobbes has provided what some social theorists (Hollis, 1982) have called a 'bottom-up' argument for social order, by arguing that social order is generated by individual agents, acting in particular (and predictable) ways, and Hobbes is claiming that because humans are predictably self-interested, we can control their behavior by providing the right incentives (or disincentives, as the case may be) for their actions. If we can take Hobbes's argument and apply it to the Problem, and particularly premise 4 from our initial argument ("E-Commerce, in general, does not utilize face-to-face interactions and is, generally, 'anonymous' in most relevant senses, and is, therefore, lacking in the materials that develop trust."), we might be able to avoid the unpleasant conclusion of the argument, namely, that e-commerce is more likely to fail than other forms of commerce.

The anonymity of e-commerce, and of the Internet in general, seems to support some Hobbesian claims about human nature. When one perceives oneself to be anonymous, or unable to be identified, it would seem one would be more inclined to act in socially unacceptable ways than if one were well-known to the people affected by the socially unacceptable action. Additionally, the absence of facial cues and the increased physical distance can lead us to feel "safer and more immune to a counterattack" (Wallace, 1999, p. 126) than we would feel if we were having a person-to-person interaction. So, for instance, it is much easier to launch an expletive-laden outburst at a passing motorist, or give them the middle finger, than it would be to do the same thing to a nearby pedestrian or the passenger in the bus seat next to you. And it would even easier to launch an expletive-laden outburst at someone we assume is hundreds of miles away from us, staring at a computer screen. Keisler and Sproull, studying small groups trying to reach consensus on difficult tasks, found a higher frequency of insults, personal attacks, hostile comments, and name-calling in the groups who were communicating by computer only. They found that computer mediated discussions seemed to generate a considerably larger amount of hostility than face-to-face discussions, so much so that on at least one occasion, the anonymous computer discussants had to be escorted from the research facility for fear of violence (S. Kiesler & Sproull, 1992). And, as Hobbes would be careful to inform us at this point, it will be much easier for us to play the role of the Foole if we are anonymous agents in society. So, while our anonymity is helpful in some respects, it might also encourage us to act in socially unacceptable ways (including cheating parties in e-commerce transactions) while online.

I think we can address some of these worries (at least from the perspective of e-commerce) about anonymity by considering some of the Hobbesian safeguards against socially problematic behavior. This is particularly important given the size and distance of the Internet community. If the Internet had, for example, a Leviathan, or a reasonable set of community norms that, if broken, would lead to expulsion from the community, we might be spared these worries (for a proposal of such "Netiquette" see Johnson, 2009), although it might be difficult to enforce an Internet ban. It is true that there are some of these safeguards already in place, in some cases, to protect e-commerce from Fooles and Free Riders. These safeguards can be roughly analogous to the Hobbesian safeguards of the fear of the sovereign and fear of being rejection by fellow citizens. The first type we could call roughly "institutional" remedies to violations of e-commerce norms, and would roughly be equivalent to the punishments that would arise were the sovereign aware of the transgression, although that need not specifically be remedies with the law itself, provided they were remedies offered by some large, powerful body or institution. The second type we could call "social" remedies to violations of e-commerce norms, and they do not feature a formal punishment from an authority figure, but instead are centered on various forms of stigmatization from non-authoritative fellow

participants in commerce. I will briefly examine a few examples of each type of remedy.

The 'institutional' type of remedy to e-commerce problems is quite common, and most individual buyers and sellers are probably likely to utilize something like it if they run into problems with a transaction. If a party to a transaction can appeal to an authoritative body to help resolve the problem, whether that body is a legal system, a business institution, a financial network, or some other body with the power and authority to offer a resolution to the situation, they are making an institutional appeal. This kind of appeal may include governmental intervention (local, state, Federal, international law) for violations of laws and codes dealing with commerce (The Universal Commercial Code, The United Nations Convention on Contracts for the International Sale of Goods), private intervention on behalf of consumers (the Better Business Bureau, local news stations and consumer advocates), financial institution remedies (credit card and bank card guarantees and charge backs), or business guarantees (eBay's Trust and Safety Team, Amazon Marketplace's Guarantee program). A party who did not receive goods, services, or payment appropriate to their electronic transaction may have several institutional outlets to choose from for the appropriate correction. To take just one example, an item purchased from a seller on Amazon's Marketplace (new and used goods being sold by third parties, not by Amazon.com itself, with Amazon collecting fees from sellers) is subject to the "A-to-Z Guarantee" which protects purchasers if the goods are not delivered on time, are damaged or defective, or if the goods are returned but the appropriate refund has not been given. But, failing that, a consumer may also have recourse from her credit card or bank, and may be able to stop payment or receive a refund if an item is defective or undelivered. Clearly, these 'institutional' remedies are only loosely connected to the Hobbesian version, where the 'institution' in question is an all-powerful Leviathan, but they fit the profile of an authoritative set of powers to keep individual agents in line.

The 'social' remedies that are available often use informal social networks, with no specific power or authority over the parties involved, but can provide useful information to other potential buyers and sellers about the habits of their prospective partners. Many e-commerce websites (notably Ebay and Amazon Marketplace) ask for buyers to leave 'feedback' ratings on sellers, and allow them to leave short comments about their transactions, and often have mechanism built in to prevent excessive or unwarranted negative feedback. When a buyer examines a seller's item on one of these sites, they have some information about the seller, based on their feedback rating, the number of sales they have made on the site, and the comments that have been made by buyers about the seller. A potential buyer might avoid buying from a seller who is relatively new to the site, or who has a less than satisfactory feedback rating, or who has had a recent history of unsatisfied customers. An additional advantage is that these systems are built into the framework of the communities themselves, and the information is handily available for all users of that community. Outside of the e-commerce sites themselves, there are other possible sources for 'social' remedies. These include social networks, (Facebook, Twitter and MySpace), internet discussion groups, chatrooms and forums, and private companies (like the Better Business Bureau or Consumer Reports) that track and post relevant data about customer complaints. Some websites allow consumers to enter their own information about problematic transactions and create, essentially, their own webpage about the problems they have encountered, while others simply allow the date and information about the claim to be noted. Professional consumer advocates in the mass media sometimes bridge the gap between institutional and social remedies by allowing both the sharing of information and, sometimes, providing legal and media pressure to help customers who have been wronged.

This pair of solutions, "institutional" and "social," roughly paralleling the twin Hobbesian solutions to the problem of the Foole (the sovereign's wrath and the fear of social ostracism), provide us with a framework to protect trust, override fears of Fooles and free riders, and allow us to be as comfortable with e-commerce as we would normally be making a purchase directly. Combining the institutional solutions with social solutions gives consumers and sellers resources to protect their economic well-being if a transaction goes bad, and also provides them with research possibilities before making transaction decisions, allowing them to avoid potentially troublesome business partners. Clearly, the most Hobbesian solution for The Problem of Anonymity would be an all-encompassing global authority that would enforce these e-commerce contracts, but something of that sort seems to be quite unlikely to occur in the near future, so this version of a "modified Hobbesian solution" serves as an attempted solution to the Anonymity Problem. It is 'modified' in the sense that it takes the ideas behind Hobbes's own argument and recasts them in light of current social and political realities.

One final point might be noted here, to highlight the Hobbesian strain of thought as it relates to problems like this. One could take Hobbes's argument against Fooles farther than just one's e-commerce partners, and apply it broadly, to one's role as participant in the Internet. The Internet is a very large global community, with a fluidity of membership that makes it appear to change at each moment, but like many loose collectivities, there is constancy. Violations of social norms on the Internet that are not specifically related to e-commerce (spamming, trolling, e-bullying) are just as subject to the Hobbesian argument as violations of e-commerce norms are, although there are fewer 'institutional' solutions to non-commerce violations available. If we take Hobbes's social contract argument seriously, as I suggest we do, and apply it to our participation in the Internet, I think Hobbes provides us with an argument not only for following the basic rules of e-commerce (pay on time, don't falsely advertise, etc) but also an argument for following the basic norms of civility and decency as part of the larger community of Internet participants. We are, as participants in a geographically unbounded world of information and ideas, truly 'citizens of the world,' as Diogenes of Sinope was said to describe himself (Book VI, Laertius, 1985). The Internet enables us to be cosmopolitan citizens of a global world, and if we accept that we have 'contracted' to treat our fellow citizens (even those we only meet on the Internet) with respect, we can apply our Hobbesian social contract argument to our behavior on the Internet in general, and not merely to our purchases and sales. We could call this a contractarian view of Internet participation, and I think it would serve to curtail problematic behavior on the Internet, provided that reasonable standards of 'acceptable' Internet behavior were to exist. I think more needs to be said to make this brief argument complete, but I will not do so here.

FUTURE RESEARCH DIRECTIONS

Philosophers, social scientists, and business scholars should watch the development of e-commerce closely. Despite the fact that the Internet is now a crucial and well-integrated part of our lives, e-commerce is still in its relative infancy, and much more study needs to be done to determine whether it will follow the same paths as traditional commerce. Sociologists, psychologists, and economists working on issues in the methodology of trust should be particularly interested in the patterns of 'institutional' and 'social' remedies. In the brief final section that follows, I offer a few other suggestions and make some concluding notes.

CONCLUSION

If the argument I have developed works as I think it does, we have been able to avoid the challenge of premise 4 in the initial argument by offering a supplementary premise (4*) to salvage e-commerce. That supplementary premise is something like this:

(4*): In situations where trust is not readily available for partners in commerce, 'institutional' and/or 'social' remedies (which can include such things as laws, business guarantees, consumer advocacy groups, etc.) can be used to supply the missing elements of trust for both parties in the transaction.

This supplementary premise allows us to replace the conclusion of the argument (5) with:

(5*): E-Commerce is no less likely to occur and succeed than traditional business, provided the appropriate institutional and social remedies are applied to bolster the missing trust in the relationship.

Hence, the Anonymity Problem is rendered a non-problem, by involving a strain of Hobbesian thought to a very 21st Century problem. This solution is made more philosophically rich if we take the contractarian view of Internet participation, under which we have agreed, as cosmopolitan citizens of a global world order, to follow certain rules to keep our online 'society' running well. But even without this deeper view of Internet participation, we can still make strong claims about what should be done to solve the Anonymity Problem. We would be wise to help undercut future problems by, for example, encouraging better and more comprehensive legislation (an 'institutional' solution), locally and internationally, to protect buyers and sellers from fraud. Rooting out Internet scammers from countries with little or no legislation against their actions will engender more trust across the global Internet community. The development of better, open access online resources and databases ('social' solutions) that will allow consumers to share information about their e-commerce partners will create a more informed set of buyers and sellers. All of this will, hopefully, render the Problem solved, and make a better, and more efficient and effective, e-marketplace.

REFERENCES

Carson, T. L. (2002). Ethical Issues in Selling and Advertising. In Bowie, N. E. (Ed.), *The Blackwell Guide to Business Ethics* (pp. 186–205). Malden, MA: Blackwell Publishing.

Hobbes, T. (1996). *Leviathan* (Rev. student ed.). Cambridge, UK: Cambridge University Press.

Hollis, M. (1982). Dirty Hands. *British Journal of Political Science*, *12*, 385–398. doi:10.1017/S0007123400003033

Hollis, M. (2002). *The Philosophy of Social Science: An Introduction (Revised and Updated)*. Cambridge, UK: Cambridge University Press.

Johnson, D. G. (2009). *Computer Ethics* (4th ed.). Upper Saddle River, NJ: Prentice Hall.

Kiesler, S., Siegel, J., & McGuire, T. W. (1984). Social psychological aspects of computer-mediated communication. *The American Psychologist*, *39*(10), 1123–1134. doi:10.1037/0003-066X.39.10.1123

Kiesler, S., & Sproull, L. (1992). Group Decision Making and Communication Technology. *Organizational Behavior and Human Decision Processes*, *52*, 96–123. doi:10.1016/0749-5978(92)90047-B

Laertius, D. (1985). *The Lives and Opinions of Eminent Philosophers*. In C. D. Yonge (Eds.), Retrieved from http://classicpersuasion.org/pw/diogenes/

Oosterhof, N. N., & Todorov, A. (2008). The Functional Basis of Face Evaluation. *Proceedings of the National Academy of Sciences of the United States of America*, *105*, 11087–11092. doi:10.1073/pnas.0805664105

Wallace, P. M. (1999). *The psychology of the Internet*. Cambridge, UK: Cambridge University Press.

Wilson, J. M., Straus, & McEvily, B. (2006). All In due time: the development of trust in computer-mediated and face-to-face teams. *Organizational Behavior and Human Decision Processes*, *99*, 16–33. doi:10.1016/j.obhdp.2005.08.001

ENDNOTES

[1] In a very recent study of honesty in England and Wales, for example, more than two-thirds of participants admitted to keeping quiet after being undercharged in a shop. Full results of the recent "Honesty Lab" study performed by researchers at Brunel University were not available at the time of this writing, but preliminary comments can be found at http://www.guardian.co.uk/science/2009/sep/07/survey-lawyers-honesty-public-attitudes.

[2] Hobbes does offer at least one other argument (other than fear of the sovereign) as to why parties to a social contract should keep their promise to maintain order, although it is not often invoked by Hobbes's defenders. He argues (XIV, p. 93) that it would be 'absurd' to contradict what one maintained in the beginning by promising, and that some

version of the law of non-self-contradiction would apply to social contractors. This would imbue Hobbes's perceived egoism with some semblance of 'morality' that is sometimes overlooked because of the primary importance of the fear of punishment by the sovereign. I find this argument to be often overlooked and compelling, but it need not be accepted here for the purposes of my argument.

3   I leave out, once again, the possibility that aversion to self-contradiction would also be a reason to avoid such behavior, although I am sympathetic to this reading of Hobbes, as noted in footnote 6.

Section 4
Possible Solutions for Dealing with Internet Trolling

Chapter 14

Online Behavior Modeling:
An Effective and Affordable Software Training Method

Charlie Chen
Appalachian State University, USA

Terry Ryan
Claremont Graduate University, USA

Lorne Olfman
Claremont Graduate University, USA

ABSTRACT

Organizations need effective and affordable software training. In face-to-face settings, behavior modeling is an effective, but expensive, training method. Can behavior modeling be employed effectively, and more affordably, for software training in the online environment? An experiment was conducted to compare the effectiveness of online behavior modeling with that of face-to-face behavior modeling for software training. Results indicate that online behavior modeling and face-to-face behavior modeling provide essentially the same outcomes in terms of knowledge near transfer, immediate knowledge for transfer, delayed knowledge for transfer, perceived ease of use, perceived usefulness, and satisfaction. Observed differences were not significant, nor were their patterns consistent, despite sufficient power in the experimental design to detect meaningful differences, if any were present. These results suggest that organizations should consider online behavior modeling as a primary method of software training.

INTRODUCTION

Investment in software training can improve productivity, boost employee morale (Bell, 2004), and reduce employee turnover rate (Heller, 2003). End users who have not received proper software training often feel insecure about their jobs, and this insecurity can contribute to turnover costs and productivity losses (Aytes & Connolly, 2004). The departure of a newly hired IT employee within 180 days of hiring can cost a company as much as $100,000 (Brown, 2000). The departure of em-

DOI: 10.4018/978-1-4666-2803-8.ch014

ployees who leave their companies due to a lack of proper training can have a variety of negative consequences (McEvoy & Cascio, 1987).

In contrast, properly trained end users often feel confident and secure, with positive implications for productivity. Increases in individual performance can add up to substantial improvements for businesses. The American Society for Training and Development (ASTD) conducted a study of 575 U.S.-based, publicly traded firms between 1996 and 1998 to examine the relationship between organizational training investments and the total shareholder return. This study found an 86% higher return on such investments for the top half of firms (in terms of training investment) than for the bottom half of firms (Bassi, Ludwig, McMurrer, & Van Buren, 2000).

Software training requires a significant financial outlay. The most effective software training at present involves face-to-face behavior modeling, but such training is expensive to deliver. One possible way to reduce delivery costs is by offering similar software training, but through less expensive online delivery.

Allen and Seaman (2003) forecast that online learning would grow at a rate approaching 20% per year. The world corporate online learning market has been predicted to grow to nearly $24 billion by 2006, from $6.6 billion in 2002, an annual increase of 35.6% (International Data Corporation, 2002). The continuous growth of the online training market has prompted discussion about the effectiveness of Web-based virtual learning environments (Piccoli, Ahmad, & Ives, 2001).

While it is commonly agreed that online software training is less expensive and more flexible, it may also be less effective. Online software training continues to be of great interest to organizations, but significant challenges remain in implementing online solutions. These challenges include: (1) the cost of acquiring online learning systems, (2) the

time for developing online learning materials, and (3) the need to be convinced of online learning's effectiveness compared to other training models (Bloom, 2004).

Three general training methods have been compared experimentally in face-to-face settings: instruction based, exploration based, and behavior modeling. Instruction-based training occurs when trainers tell trainees about software, but do not model the use of it. Exploration-based training teaches trainees through practice by trainees on relevant examples, also without trainer modeling of software use. Behavior modeling training teaches trainees via demonstrations, in which trainers model the use of software for trainees. Evidence exists that behavior modeling is the most effective method for face-to-face software training (Compeau & Higgins, 1995; Simon, Grover, Teng, & Whitcomb, 1996).

This research compares experimentally the relative effectiveness of face-to-face behavior modeling and online behavior modeling. Since prior research has indicated that the behavior modeling method dominates the instruction-based and the exploration-based methods in face-to-face settings, this study does not include the latter two methods. Online asynchronous methods of software training, because they allow more favorable ratios of trainers to trainees and do not require training participants to meet, have the potential to achieve significant cost savings over face-to-face approaches. On the other hand, given that live trainers are not present in online asynchronous software training, there can be no direct interaction between trainers and trainees. This difference in direct interaction could mean that face-to-face training might be more effective than online training. Knowledge about the relative effectiveness of these methods will be valuable to people who must make decisions about how to provide software training.

THEORETICAL BACKGROUND

Software Training Method

As mentioned above, three methods are common in face-to-face software training: exploration-based training, instruction-based training, and behavior modeling training (Simon et al., 1996). In exploration-based training, the assumption is made that learning is "a matter of rearranging or transforming evidence in such a way that one is enabled to go beyond the evidence so reassembled to additional new insights" (Burner, 1966, p. 22). Exploration-based training involves an inductive process through which individuals learn general concepts by trying to solve specific tasks (Taba, 1963). In instruction-based training, "the entire content of what is to be learned is presented to the learner in the final form" (Ausubel, 1963, p. 16). Instruction-based training is deductive and programmed, with low trainee control and a focus on software features (Davis & Davis, 1990). The behavior modeling method is in some ways a hybrid of exploration-based training and instruction-based training, and is centered on having trainees treat the behavior of their trainer as a model for their own (Simon et al., 1996).

When assessing the applicability of training methods for online asynchronous software training, researchers must bear in mind that some key elements of face-to-face software training may be lost in a movement to the online asynchronous setting. The opportunities for direct interactions between trainers and trainees are necessarily fewer, if they exist at all, in online asynchronous software training. Thus, the beneficial effects of trainer-trainee interactions typical of face-to-face software training may be missing. Videotapes, transcriptions, simulations, or virtual reality are unlikely to serve as complete substitutes for live interactions between trainer and trainee. Along these lines, the features that distinguish behavior modeling training from exploration-based training and instruction-based training may perhaps be less evident in the online asynchronous software training situation.

Online asynchronous software training does not provide close monitoring of trainees by the trainer, as is common in face-to-face software training. This lack of monitoring can be expected to lead to increased levels of distraction among trainees. Trainees in online asynchronous software training might be inclined to attend to matters other than their training, such as 'surfing the Web', to a much greater extent than they would in face-to-face software training. Furthermore, online asynchronous software training may deliver to trainees content that is less tailored to their interests than content provided in face-to-face software training. As a result, trainees may experience a higher degree of boredom in online asynchronous settings than in face-to-face settings, leading to poorer performance and more negative reactions to training experiences.

Online Training Modes

There are two temporal modes of online training: synchronous and asynchronous. Either mode can be used for software training. Text messaging, audio conferencing, and videoconferencing are examples of online applications that can be used for training purposes in a synchronous mode. Web pages, files to be downloaded, e-mail, newsgroups, and discussion forums are examples of applications that can support training asynchronously.

Horton (2000) suggests that synchronous training and asynchronous training must be designed differently. Synchronous training demands the control of schedule, time, people, class size, video and audio equipment, and place. These factors limit the possibilities for reaching trainees in a cost-effective manner. This study focuses on online asynchronous software training, in part to avoid the influence of these factors and in part to concentrate on the methods most likely to keep costs low. This choice reflects the reality that most online training delivered across continents is provided in the asynchronous mode and that use of the online asynchronous mode is destined to grow.

Behavior Modeling in Face-to-Face and Online Asynchronous Software Training

Social cognitive theory (Bandura, 1986) serves as a theoretical basis for behavior modeling. According to this theory, "most human behavior is learned by observation through modeling" (p. 47). Observational learning allows one to form rules to guide future behavior by watching what others do and noting what consequences such behavior has for them. Further, observational learning can make use of symbolic models, allowing people to consider words and images in coming to appreciate what happens to others when they behave in particular ways, thereby extending what can be learned beyond the immediate environment. Learning by modeling involves four kinds of processes: (1) attentional, (2) retention, (3) production, and (4) motivational (Bandura, 1986, p. 52). These processes are influenced both by characteristics of the events that are observed and characteristics of the observer.

Learning by modeling or observing people's behaviors may be more effective than learning by trial-and-error because the former approach can avoid unnecessary mistakes and harm. Modeling a trainer's behavior empowers trainees to: (1) learn new behavior from the trainer, (2) self-evaluate their behavior against the trainer's, and (3) reinforce their current adequate behaviors. The behavior modeling approach is different from learning by adaptation. The former approach teaches via demonstration, while the latter approach influences the behaviors of learners by reward and punishment (Skinner, 1938).

An early application of behavior modeling training was in the area of interpersonal communication and management skills (Decker & Nathan, 1985). In the realm of training for software and computer usage, behavior modeling training has been shown consistently to be more effective than instruction-based training or exploration-based training (Compeau & Higgins, 1995; Gist, Schwoerer, & Rosen, 1989; Simon et al., 1996). Yi and Davis (2001) found that the effects of behavior modeling could be enhanced by the provision of training features to support retention enhancement and practice.

Behavior modeling is readily employed in face-to-face training, but may be difficult to apply in online settings, which may be less suited to demonstrations of behavior. Limitations of the media typically used in online synchronous instruction in terms of their richness constitute one possible constraint. Another is a reduced level of reinforcement possible in online settings, compared with face-to-face instruction. For example, in a live training class, the trainer is able to demonstrate a software process and immediately ask the trainees to repeat the activity under the trainer's close supervision. However, in an online asynchronous situation, in which there is no live trainer, demonstrations lose the benefit of immediate feedback. In an online synchronous situation, bandwidth constraints and compromised reciprocity may undermine the effectiveness of demonstrations. In online environments, effectiveness can be further compromised with the absence of learning by doing, another key element of face-to-face behavior modeling training (McGehee & Tullar, 1978). Table 1 summarizes the strengths and weaknesses of face-to-face behavior modeling and online behavior modeling.

Trying to use behavior modeling for online software training would involve the issues mentioned above. Therefore, there is a strong possibility that the behavior modeling approach cannot be fully replicated in the online asynchronous setting and therefore will not be as effective in the online environment as in the traditional environment. On the other hand, if online behavior modeling were to prove as effective as face-to-face behavior modeling, organizations would prefer it,

Table 1. Face-to-face behavior modeling vs. online behavior modeling: strengths and weaknesses

	Behavior Modeling Approaches	
	Face-to-Face Behavior Modeling	**Online Behavior Modeling**
Strengths	• Regardless of learning styles, trainees are motivated to use the training approach. • The approach is less influenced by learning style and is the most effective training approach. • The teaching quality is contingent upon both the quality of entire course materials packaged in final form, and supplementary materials (e.g., hand-outs) • Trainer's teaching abilities and trainees' participation are major determinants for the training outcomes. • Real-time two-way feedbacks. • Both trainers and trainees have equal power to control the learning pace.	• Regardless of learning styles, trainees may be motivated to use the training approach. • The approach may be less influenced by learning style and should be an effective training approach. • The teaching quality may be contingent upon both the quality of entire course materials packaged in final form, and supplementary materials on the Internet • Trainer's teaching abilities and trainees' participation may be major determinants for the training outcomes. • Web-based peer-to-peer interaction may improve commitment and participation. • Both trainers and trainees may have equal power to control the learning pace.
Weaknesses	• Needs longer time to exercise the training approach. • Constrained by the length of time, a trade-off between instruction and exploration learning is needed. • A trainer's experience will influence the decision quality of the trade-off, thereby affects trainees' motivation and their learning outcomes.	• Trainees may have trouble modeling the behavior of the trainer without direct interaction. • Demonstrative lecturing may need to rely on videotaped or scripted course materials. • Asynchronous, one-way communication may lower motivation, thereby degrade learning outcomes. • Real-time reiterative learning process to confirm understanding may be lost.

because it would cost less and promise higher returns on investment. The increasing use of online asynchronous software training by businesses and schools raises questions, including: What training methods should be used under what circumstances in the online asynchronous environment?

RESEARCH MODEL AND HYPOTHESES

This study tested the relative efficacy of behavior modeling training—done in two different environments—experimentally, in a field setting,. Experimentation can allow testing of causal relationships among variables. A field setting can provide greater confidence in the meaning of experimental tasks than a laboratory setting does. A field experiment methodology has the merits of being able to test theory and being able to answer practical questions (Kerlinger & Lee, 2000).

The independent variable for the experiment was *training method,* set at two levels of behavior modeling—face-to-face and online. Training materials were designed to operationalize each level by integrating key elements of behavior modeling training, as illustrated in Figure 1.

The online behavior modeling treatment was designed to be video-demonstration oriented. Trainees watched a recording of the trainer using the software to perform tasks. All communication between the trainer and trainees was one-way. Following the video, trainees completed tasks to demonstrate their level of mastery of the training materials. Online reference sources were available to them as they completed the tasks. Control over what occurred was shared between the system and trainees. Online behavior modeling, as conceived here, shares features with both exploration and instruction approaches, including both inductive and deductive aspects.

The face-to-face behavior modeling treatment was designed to be direct-demonstration oriented.

Figure 1. Key elements of behavior modeling training

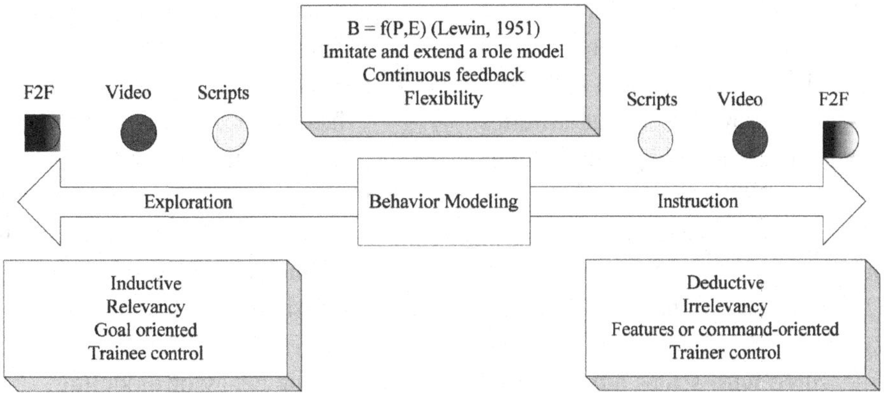

Trainees watched the trainer, in person, performing tasks. Two-way communication between the trainer and trainees was possible. Following the demonstration, trainees performed a task assigned by the trainer. No online reference sources were available. Control over what occurred was shared between the trainer and trainees. The face-to-face condition included a trainer, other than one of the researchers, to reduce chances of awareness of the hypotheses being tested. The trainer followed the same script as the one designed for the online condition.

The length of time allotted to training was the same for both conditions. Both conditions included the same pre- and post-training tests. All training sessions were conducted in the same computer classroom.

Dependent Variables

Regardless of teaching environment, most training is intended to instill a competency of some kind. Software competency depends on the kind of knowledge acquired in training. Learning effectiveness can be evaluated through trainees' reactions and knowledge transfer (Kirpatrick, 1967). Knowledge levels can be categorized as near transfer or far transfer (Simon et al., 1996). Near transfer of knowledge is necessary for understanding basic software commands and procedures. Far transfer of knowledge allows the solving of problems different from those worked out in training. The measurement of near-transfer knowledge involves direct assessment of what was learned about the specific objects (such as software features and commands) covered in training. The measurement of far-transfer knowledge has to do with evaluating the extent to which what was learned is available to trainees in their solution of problems similar to those included in training. For far-transfer knowledge, it is interesting to assess learning both immediately after training and after some delay, because software competency is intended to be a long-term effect of training.

In this study, software competency was measured in terms of near-transfer knowledge, far-transfer knowledge (assessed immediately), and far-transfer knowledge (assessed with delay). The measure of *near transfer (NT)* consisted of 10 multiple-choice questions concerning the details of the software covered in training. *Immediate far transfer (IFT)* was measured with a problem to be solved with the software during the experimental session. *Delayed far transfer (DFT)* was measured with a problem administered later in the academic term as part of the final exam.

In addition to how well it instills software competency, software training should also be judged by the reactions that trainees have to it. It is common to use satisfaction as a surrogate for the effectiveness of information systems (Ives, Olson, & Baroudi, 1983) and it has been adopted as an indicator of success in software training (Simon et al., 1996). Perceived usefulness and perceived ease of use have been shown to predict attitudes and behaviors with information systems (Davis, Bagozzi, & Warshaw, 1989).

In this study, the reactions of trainees were captured through three measurement scales: *satisfaction (SAT), perceived usefulness (PU),* and *perceived ease of use (PEOU)*. SAT, PU, and PEOU were measured with scales administered during the experimental session, as described below.

Hypotheses

As discussed above, online the behavior modeling approach replaces the live instructor with the scripted demonstration, and some key elements of the face-to-face behavior modeling approach may be lost. Characteristics of the online asynchronous environment—particularly limitations on trainer-trainee interactions—suggest that behavior modeling training done in a face-to-face manner should be superior to behavior modeling training done in the online mode. Behavior modeling training in the face-to-face mode may be more effective at improving the learning outcome for a trainee than the behavior modeling approach in the online asynchronous mode. Proving that the face-to-face behavior modeling is more or less effective than online behavior modeling could justify the validity of replicating same pattern in the online asynchronous environment. Hence, it is only hypothesized that the face-to-face behavior modeling approach is more effective than online behavior modeling to improve learning outcomes for trainees of all learning styles. The hypotheses listed below correspond to this expectation with respect to NT, IFT, DFT, SAT, PU, and PEOU.

Hypothesis 1: NT scores will be greater with face-to-face behavior modeling than with online behavior modeling.

Hypothesis 2: IFT scores will be greater with face-to-face behavior modeling than with online behavior modeling.

Hypothesis 3: DFT scores will be greater with face-to-face behavior modeling than with online behavior modeling.

Hypothesis 4: SAT scores will be greater with face-to-face behavior modeling than with online behavior modeling.

Hypothesis 5: PU scores will be greater with face-to-face behavior modeling than with online behavior modeling.

Hypothesis 6: PEOU scores will be greater with face-to-face behavior modeling than with online behavior modeling.

Trainees and Setting

The setting for the experiment was an introductory computer course requiring trainees to learn spreadsheet software (Excel 2000™). Trainees in this course were freshmen, majoring in MIS or Accounting. Participation in the study was voluntary. The faculty for the course agreed to run the experiment near the end of the students' academic term. As a result of this timing, none of the trainees in the experiment were complete novices with respect to computers; all had some literacy and experience.

It was not desirable in this study for experimental trainees to have prior knowledge of spreadsheets. Therefore, as part of the experimental procedure, a pre-test of 10 multiple-choice questions about Excel 2000 was given to each trainee; trainees who scored higher than 50% were excluded.

Training Procedure

The trainer conducted the traditional face-to-face behavior modeling by following the same procedures as delineated in the online behavior modeling training package (Figure 2). The first stage of the package provides examples of how to manage a database using Microsoft Excel 2000. The second stage lists online asynchronous training tools that can be used to assist the trainee. These tools are: (1) examples that are relevant to trainees' backgrounds, (2) a self-practice worksheet for each function of database management, and (3) online reference sources. The third stage allows trainees to choose relevant examples to use for practice throughout the training session.

Two examples were prepared for each major (Accounting and MIS). Trainees were encouraged to practice with examples relevant to their majors. The training covered five database management features of Excel. These features allow one to: (1) create a database without using a data form, (2) create a database with a data form, (3) sort data in a data list, (4) filter data in a data list, and (5) add subtotals to a data list. Step-by-step instructions were adopted to illustrate each function.

Additionally, trainees could choose to watch or not to watch a demonstration for all database functions. On the same page as the demonstration, a self-practice worksheet was presented. Practice results could also be carried over to the next prac-

tice. Trainees of the online behavior modeling group took the same quizzes as the face-to-face group and then concluded the study. Figure 3 and Figure 4 are two screenshots of course materials used for the online behavior modeling approach. The demonstration of a live instructor was substituted for fix-time (12 seconds) transition of page presentations. Each scripted demonstration focused on one particular function of the database management topic. At each learning point of a particular function, trainees were encouraged to experiment by practicing the related exercises.

RESULTS

A total of 114 trainees completed the study by submitting valid questionnaires. This amounted to approximately 46% of the trainees who had been registered for the course. Although a higher rate of participation in the experiment was expected, the achieved rate may have been due to the experiment being a voluntary activity at the end of an academic term. Of the trainees who did participate, 31 had too much prior spreadsheet knowledge to qualify as trainees for the experiment, leaving 83 trainees whose data were analyzed.

Table 2 shows the number of trainees for each experimental treatment. Note that the counts of trainees in each condition, although not exactly the same, are similar enough not to cause analytic

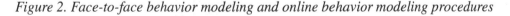

Figure 2. Face-to-face behavior modeling and online behavior modeling procedures

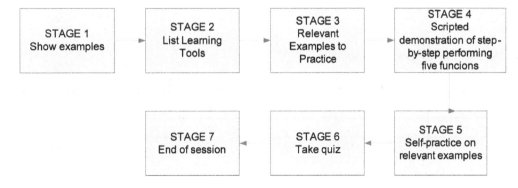

Figure 3. Scripted demonstration of step-by-step performing five functions: The fourth step of OBM method (enabled by Microsoft Excel's macro functions)

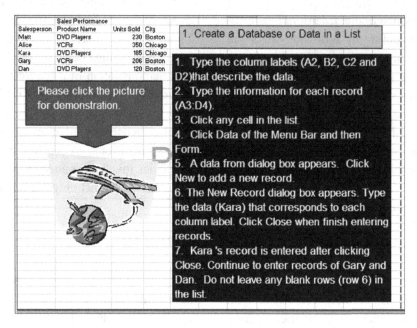

Figure 4. The first step of scripted demonstration for the first subject: Sort data in a list

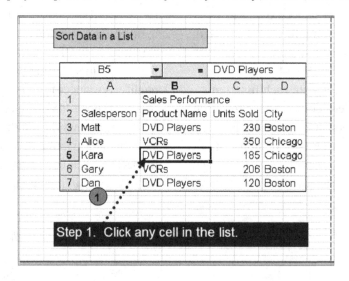

Table 2. Number of trainees by treatments

Training Approaches	Total Completing Study	Excess Prior Knowledge	Number for Data Analysis
Online	44	9	35
Face-to-Face	70	22	48
Total	114	31	83

difficulties. The sample size, although not as large as might be desired, also is not a source of analytic problems.

Multivariate analysis of covariance (MAN-COVA) has advantages over an analysis of variance (ANOVA) in removing some systematic errors and uncontrolled individual differences. Although the researchers planned to adopt MANCOVA, a cursory investigation showed no interaction effect between two dependent variables—user satisfaction and learning performance. This indicates that learning performance and satisfaction effects are separated. Interaction among them is not the issue in this study. The possibility of inflating type I error due to the analyses of multiple univariate ANOVAs is minimum. Furthermore, due to the small sample size of some cells and the unequal cell sizes, complying with the assumptions of two-way ANOVA can improve the reliability of data analysis. Each dependent variable was treated and analyzed independently with ANOVA as a result. Data analysis using a Histogram graph shows that each dependent variable complies with univariate normality assumptions. Additionally, Levene's test at $\alpha = 0.10$ shows that the null hypothesis, "the error variance of the dependent variable is equal across groups," is not violated. The tests indicate that the data is normally distributed. An ANOVA was conducted for each of the dependent variables. ANOVA is robust for situations having a limited number of data points (Moore & McCabe, 1989). Table 3 provides descriptive statistics for dependent variables by experimental treatment. Table 4 summarizes ANOVA results for training methods in terms of learning outcomes. Direction and significance of differences between treatments are indicated.

From a hypothesis-testing standpoint, four out of six hypotheses, all concerning the superiority of face-to-face behavior modeling over online behavior modeling (H1, H3, H4, and H5), were in the direction hypothesized. Despite this, these hypotheses were not supported in a statistical sense.

Table 3. Descriptive statistics by treatments

D.V.	Mean/ S.D.	Online Behavior Modeling	Face-to-Face Behavior Modeling
NT	Mean	45.33	49.38
	S.D.	19.43	16.94
IFT	Mean	39.33	25.00
	S.D.	24.90	18.22
DFT	Mean	36.00	44.63
	S.D.	39.23	36.41
SAT	Mean	2.86	3.05
	S.D.	0.61	0.63
PU	Mean	2.72	2.97
	S.D.	0.69	0.62
PEOU	Mean	3.02	2.93
	S.D.	0.32	0.34

DISCUSSION

Although none of the hypothesized relationships are fully supported, the results obtained are interesting. The most intriguing result is that—contrary to expectations—there are no statistical reasons for preferring face-to-face behavior modeling to online behavior modeling for software training of this kind. The pattern of results indicates that while face-to-face behavior modeling results in better outcomes than online behavior modeling for four

Table 4. Results for training methods

Variable	Hypothesis	Result in Correct Direction?	Significant p-value?
NT	Face-to-face > online	T	n.s.
IFT	Face-to-face > online	F	n.s.
DFT	Face-to-face > online	T	n.s.
SAT	Face-to-face > online	T	n.s.
PU	Face-to-face > online	T	n.s.
PEOU	Face-to-face > online	F	n.s.

of the six dependent variables, it never does so at a statistically significant level. One interpretation of this is that online behavior modeling training is no worse than face-to-face behavior modeling training across all dependent variables. (Trainees in the online behavior modeling condition actually score higher than face-to-face behavior modeling trainees in IFT.) The pattern of results for face-to-face behavior modeling suggests that trainers might choose online behavior modeling—which ought to be a less costly alternative to face-to-face behavior modeling—without making any significant sacrifice in either learning or trainee reaction outcomes. The complex picture of the implications of these four treatments must be more clearly illustrated.

Two methods can illustrate this complexity. The first is the "insufficient difference" finding between online behavior modeling and face-to-face behavior modeling. The second is the beginning of a strategy for online asynchronous software training. As a first result, the conclusion of "insufficient difference" between online behavior modeling and face-to-face behavior modeling depends on being able to say there is not enough difference between their effects to justify the difference in their costs. A practical difference between face-to-face behavior modeling and online behavior modeling—one that matters in cost/benefit terms—must have some minimum size. Specifying a practical difference involves knowing the costs of face-to-face behavior modeling and online behavior modeling, as well as how effect size maps to benefits.

Ability to detect effect sizes is nothing more than statistical power (Cohen, 1977). In IS research, "studies are unlikely to display large effects and that, typically, small to medium effect sizes should be anticipated" (Baroudi & Orlikowski, 1989, p. 90). Because this study exercised due care with experimental procedure and made use of reliable instruments, there is justification in addressing statistically insignificant results. Before

executing the experiment, efforts were made to maximize the difference between face-to-face behavior modeling and online behavior modeling conditions; a Delphi study was conducted regarding the design of course materials to reflect the different training approaches. Despite this careful control over operationalization, there was not enough difference between face-to-face behavior modeling and online behavior modeling effects to justify the difference in their costs.

Due to the undeveloped nature of research in this area, it may be inappropriate to establish an index for effect size based on prior research on software-training strategy in the traditional environment (Mazen et al, 1987). To explain the phenomenon carefully, we employ Cohen's (1977) approach to estimate proxy effect-size levels based on the standardized difference d of two populations taking different training approaches (see Table 5).

The estimated effect size is 0.5 for all dependent variables except KNT and PEOU. Since PEOU is related to the design of an e-learning system rather than the treatment of a training approach, smaller effect size across different groups is understandable. However, the study cannot detect differences of effect size for KNT. This indicates that it makes no difference whether face-to-face behavior modeling or online behavior modeling is employed to improve KNT. Online trainers can choose either face-to-face behavior

Table 5. Effect size estimation

Dependent Variables	Standardized difference between face-to-face behavior modeling and online behavior modeling	* Estimated Effect Size
KNT	25.47%	0.20
KFT	-46.45%	0.50
OS	58.82%	0.50
PEOU	-21.05%	0.20
PU	57.97%	0.50

* Calculated based on Cohen's (1977) effect size conventions

modeling or online behavior modeling to improve end users' KNT if either approach has relatively similar costs.

Contrary to the expectation of hypotheses, a larger effect size was detected for KFT in the short and long term. This practical difference indicates that the benefits of online behavior modeling outweigh face-to-face behavior modeling for KFT in the short and long term. Larger effect size was also detected for the measures of end user satisfaction: OS and PU. This practical difference supports that face-to-face behavior modeling is a better approach than online behavior modeling to improve end user satisfaction. The difference between knowledge absorption capability and end user satisfaction poses many interesting questions.

As a second result, this study can offer concrete suggestions about the beginning of a strategy for online asynchronous software training. One result of interest is that face-to-face behavior modeling might be better than online behavior modeling. Of the six hypotheses concerning relationships between these two methods, four are in the expected direction, none significantly so.

These findings indicate that use of online behavior modeling may be the best software training strategy for the online asynchronous setting. To confidently offer such suggestions, the study needs to discuss the design decisions that trainers face in the online asynchronous environment. The study provides support for using online behavior modeling over exploration-based training and instruction-based training, given that the prior contribution makes the point of favoring online behavior modeling over face-to-face behavior modeling. Since our suggestions are a start on an online asynchronous software training strategy, we will present the outline of the strategy that includes "to-do" and "not-to-do" lists. This online asynchronous software training strategy will allow trainers and vendors to capitalize on these opportunities and avoid costly mistakes.

The largest implication for practice is that online behavior modeling may provide a cost-effective substitute for face-to-face behavior modeling without significant reductions in training outcomes. Compared to face-to-face behavior modeling, online behavior modeling allows trainees to have more control over their learning. Cognitive learning theory indicates that the learning process can be improved via active learning (Shuell, 1986) and problem-solving learning (Alavi, 1994). In the virtual learning environment (VLE), trainees have higher control of learning and can choose to use the exploration-based training or instruction-based training approach depending on tasks and individual needs. For instance, trainees with more experience and knowledge related to a particular trainee may resort to meaningful learning and use relevant examples to practice. Trainees with little knowledge about another trainee may resort to rote learning and use generic examples to practice. The VLE allows trainees to switch freely between meaningful and rote learning, to their advantages.

Since trainees have the control flexibility, online behavior modeling can be viewed as more effective than the face-to-face behavior modeling in helping trainees perform well on near-transfer and far-transfer tasks. In the VLE, the individualized and contextual learning provides anchoring for trainees to transform their mental models. While more must be learned about this relationship, it is encouraging to see evidence that there may be a desirable leverage from online asynchronous software training.

Another thing trainers need to bear in mind when designing an online asynchronous software training strategy is that the effectiveness of online asynchronous software training methods does not necessarily go hand-in-hand with overall satisfaction, perceived ease of use, and perceived usefulness. In particular, it may still be the case that learning effectiveness is neutral to learning style. Improving satisfaction by customizing learning approaches may be the right decision to make, but performance might not be the deciding factor.

Online behavior modeling and face-to-face behavior modeling allow trainees to have some control of the learning process and information acquisition regarding its content, accuracy, format, ease of use, and timeliness (Doll, Xia, & Torkzadeh, 1994), which leads to somewhat higher satisfaction levels. In itself, higher levels of satisfaction may be justification for online behavior modeling and face-to-face behavior modeling use, but much remains to be learned about the effects of these methods for training performance.

Assimilation Theory suggests that being receptive to new procedural knowledge on how to operate a new target system is the prerequisite to meaningful learning (Mayer, 1981) or far knowledge transfer. With the time constraints, a more focused learning approach can be useful at assimilating new knowledge. Hence, the online behavior modeling approach is a logical solution for meaningful learning because the approach allows trainees not only to acquire new knowledge, but also give trainees flexibility to search their long-term memory for "appropriate anchoring ideas or concepts" (p. 64) and to use the ideas to interact with new knowledge (Davis & Bostrom, 1993).

Limitations

While it seems unlikely, given the care taken in the design of the study, there is always the possibility that the independent variable training method could inadequately represent its intended construct. With any complex treatment, such as the establishment of training method here, there is a chance that operationalization can be less than what is needed for effects to occur. Additional research is required to refine and perfect the training method treatments as much as possible. There is no simple manipulation check for verifying the efficacy of this kind of treatment, but continued investigation should reveal the extent to which the manipulation is a successful one.

Future research can attempt to improve the reliability of the findings by controlling the experimental environment more tightly (e.g., equal cell size, larger cell size, and longer training sessions) or by improving the strategy's generalizability through the examination of other variables (e.g., trainees vs. professional workers, number of training duration sessions, type of training media, self-efficacy, experiences of using the online learning system, and software types).

Implications for Research

The findings here raise additional questions for research. Some of the ones that might be addressed in the immediate future include:

- To replicate the experimental equivalence of face-to-face behavior modeling and online behavior modeling methods of software training with different software and trainees. With this, to demonstrate a practical (i.e., cost-based) advantage of online behavior modeling over face-to-face behavior modeling for software training in practical settings.

- To study the impact of training duration on performance and trainee reactions. Trainees should be exposed to the same training methods for different durations.

- To improve the reliability of the study by manipulating some useful blocking variables. A series of comparative studies can be conducted to assess the impact of individualism as a cultural characteristic, computer self-efficacy, task complexity, professional backgrounds, and the ratio of the training duration to the quantity of information to be processed among others.

- To investigate the impacts of social presence and information richness (SPIR) (Fulk, 1993) features of online asynchronous software training media on training outcomes. Future studies might vary the

SPIR features of training media (e.g., face-to-face vs. online asynchronous scripted or Web cam modes).

- To conduct longitudinal studies of the influence of learning style on learning performance and trainee reaction.

- To continue to study the relationship between learning style, training methods, and training outcomes. Learning style is somewhat associated with the cultural backgrounds of online trainees. Trainees with varying cultural backgrounds may prefer to adopt training media with different SPIR features. Cultural differences, such as relative degree of individualism, may affect preference for SPIR characteristics. Some combination of training methods, learning style, and SPIR attributes may jointly determine learning outcomes.

Implications for Practice

The largest implication for practice is that online behavior modeling may provide a cost-effective substitute for face-to-face behavior modeling without significant reductions in training outcomes. While more must be learned about this relationship, it is encouraging to see evidence that there may be a desirable leverage from online asynchronous software training. Also, when designing an online asynchronous software training strategy, trainers need to bear in mind that both face-to-face behavior modeling and online behavior modeling are equally effective to improve learning outcomes (including satisfaction), and performance might not be the decision factor if these two approaches need to be chosen from. Other decision factors, such as trainer's preference, equipment availability, budget, and scheduling, could be more important than the efficacy issue. Online behavior modeling and face-to-face behavior modeling allow trainees to have some control of the learning process, leading to somewhat higher satisfaction levels. This

advantage in itself may be justification for their use, but much remains to be learned about the effects of these methods for training performance.

CONCLUSION

The success of an online asynchronous software training strategy depends on its effectiveness in improving learning outcomes. This study builds on a well-accepted framework for training research (Bostrom, Olfman, & Sein, 1990; Simon et al., 1996), examines the relative effectiveness of four training methods, and begins to derive a strategy for online asynchronous software training. Testing the following hypotheses provides an empirical basis for the development of an online asynchronous software training strategy: (1) face-to-face behavior modeling is more effective than online behavior modeling for learning performance and trainee reactions, and (2) online behavior modeling is more cost effective than face-to-face behavior modeling.

While these hypotheses are not fully supported statistically, and while many of the observed results are difficult to interpret, the study discovers important potential implications for practitioners and researchers. The formulated online asynchronous software training strategy suggests that trainers customize their training methods based on desired learning outcomes.

What is learned from this study can be summarized as follows: When conducting software training, it may be as effective to use an online behavior modeling method as it is to use a more costly face-to-face behavior modeling method. Although somewhat better results are sometimes evident for face-to-face behavior modeling, observed differences are not significant, nor are their patterns consistent.

The study has accomplished its major goal—it provides evidence as to the relative effectiveness of various methods, particularly those of online asynchronous nature, for software training. Within its limits, this research takes a first step in developing a strategy for online asynchronous software training.

REFERENCES

Allen, I. E., & Seaman, J. (2003). *Sizing the opportunity: The quality and extent of online education in the United States, 2002-2003.* Alfred P. Sloan Foundation.

Ausubel, D. P. (1963). *The psychology of meaningful verbal learning.* New York: Grune and Stratton.

Aytes, K., & Connolly, T. (2004). Computer security and risky computing practices: A rational choice perspective. *Journal of Organizational and End User Computing, 16*(3), 22–40.

Bandura, A. (1986). *Social foundations of thought & action.* Englewood Cliffs, NJ: Prentice Hall.

Baroudi, J. J., & Orlikowski, W. J. (1988). A short form measure of user information satisfaction: A psychometric evaluation and notes. *Journal of Management Information Systems, 4*, 45–59.

Bassi, L.J., Ludwig, J., McMurrer, D.P., & Van Buren, M. (2000, September). *Profiting from learning: Do firms' investments in education and training pay off?* American Society for Training and Development (ASTD) and Saba.

Bell, J. (2004). Why software training is a priority? *Booktech the Magazine, 7*, 8.

Bielefield, A., & Cheeseman, L. (1997). *Technology and copyright law.* New York: Neal-Schuman.

Bloom, M. (2003, April 2). *E-learning in Canada, findings from 2003 e-survey.*

Brown, J. (2000). Employee turnover costs billions annually. *Computing Canada, 26*, 25.

Bruner, J. (1966). *Toward a theory of instruction.* New York: Norton.

Cohen, J. (1977). *Statistical power analysis for the behavioral sciences.* New York: Academic Press.

Compeau, D. R., & Higgins, C. A. (1995). Computer self-efficacy: Development of a measure and initial test. *MIS Quarterly, 19*, 189–211. doi:10.2307/249688

Compeau, D. R., Higgins, C. A., & Huff, S. (1999). Social cognitive theory and individual reactions to computing technology: A longitudinal study. *MIS Quarterly, 23*(2), 145–158. doi:10.2307/249749

Davis, D. L., & Davis, D. F. (1990). The effect of training techniques and personal characteristics on training end users of information systems. *Journal of Management Information Systems, 7*(2), 93–110.

Davis, F. D. (1989). Perceived usefulness, perceived ease of use, and user acceptance of information technology. *MIS Quarterly*, (September): 319–339. doi:10.2307/249008

Decker, P. J., & Nathan, B. R. (1985). *Behavior modeling training.* New York: Praeger.

Fulk, J. (1993). Social construction of communication technology. *Academy of Management Journal, 36*, 921–950. doi:10.2307/256641

Gist, M. E., Schwoerer, C., & Rosen, B. (1989). Effects of alternative training methods on self-efficacy and performance in computer software training. *The Journal of Applied Psychology, 74*(6), 884–891. doi:10.1037/0021-9010.74.6.884

Heller, M. (2003). Six ways to boost morale. *CIO Magazine*, (November 15), 1.

Horton, W. (2000). *Designing Web-based training.* New York: John Wiley & Sons.

International Data Corporation. (2002, September 30). *While corporate training markets will not live up to earlier forecasts, IDC suggests reasons for optimism, particularly e-learning.* Author.

Ives, B., Olson, M., & Baroudi, S. (1983). The measurement of user information satisfaction. *Communications of the ACM, 26,* 785–793. doi:10.1145/358413.358430

Kerlinger, F.N., & Lee. H.B. (2000). *Foundations of behavioral research.* New York: Harcourt Brace.

Kirpatrick, D. L. (Ed.). (1967). *Evaluation of training. Training and development handbook.* New York: McGraw-Hill.

Leidner, D. E., & Jarvenpaa, S. L. (1995). The use of information technology to enhance management school education: A theoretical view. *MIS Quarterly, 19,* 265–291. doi:10.2307/249596

McEvoy, G. M., & Cascio, W. F. (1987). Do good or poor performers leave? A meta-analysis of the relationship between performance and turnover. *Academy of Management Journal, 30*(4), 744–762. doi:10.2307/256158

McGehee, W., & Tullar, W. (1978). A note on evaluating behavior modification and behavior modeling as industrial training techniques. *Personnel Psychology, 31,* 477–484. doi:10.1111/j.1744-6570.1978.tb00457.x

Piccoli, G., Ahmad, R., & Ives, B. (2001). Web-based virtual learning environments: A research framework and a preliminary assessment of effectiveness in basic IT skills training. *MIS Quarterly, 25*(4). doi:10.2307/3250989

Simon, S. J., Grover, V., Teng, J. T. C., & Whitcomb, K. (1996). The relationship of information system training methods and cognitive ability to end-user satisfaction, comprehension, and skill transfer: A longitudinal field study. *Information Systems Research, 7*(4), 466–490. doi:10.1287/isre.7.4.466

Skinner, B. F. (1938). *The behavior of organisms: An experimental analysis.* B. F. Skinner Foundation.

Taba, H. (1963). Learning by discovery: Psychological and educational rationale. *Elementary School,* 308-316.

Wexley, K. N., & Baldwin, T. T. (1986). Post-training strategies for facilitating positive transfer: An empirical exploration. *Academy of Management Journal, 29,* 503–520. doi:10.2307/256221

Yi, M. Y., & Davis, F. D. (2001). Improving computer training effectiveness for decision technologies: Behavior modeling and retention enhancement. *Decision Sciences, 32*(3), 521–544. doi:10.1111/j.1540-5915.2001.tb00970.x

Yi, Y. (Ed.). (1990). *A critical review of consumer satisfaction. Review of marketing.* Chicago: American Marketing Association.

This work was previously published in Web-Based Education and Pedagogical Technologies: Solutions for Learning Applications, edited by Liliane Esnault, pp. 137-157, copyright 2008 by IGI Publishing (an imprint of IGI Global).

Chapter 15
A Proposed Framework for Designing Sustainable Communities for Knowledge Management Systems

Lakshmi Goel
University of Houston, USA

Elham Mousavidin
University of Houston, USA

ABSTRACT

Despite considerable academic and practitioner interest in knowledge management, success of knowledge management systems is elusive. This chapter provides a framework which suggests that KM success can be achieved by designing sustainable communities of practice. Communities of practice have proven to have significant economic and practical implications on organizational practices. A growing body of literature in KM recognizes the importance of communities that foster collaborative learning in organizations and almost all KMS have a 'network' component that facilitates connecting people in communities of practice. Evidence has shown that communities have been a key element in KMS of many companies including Xerox PARC, British Petroleum Co., Shell Oil Company, Halliburton, IBM, Proctor and Gamble, and Hewlett Packard.

INTRODUCTION

Despite considerable academic and practitioner interest in knowledge management (KM), success of knowledge management systems (KMS) is elusive (Akhavan et al., 2005; Hammer et al., 2004). There is a considerable body of literature

that has studied factors for KMS success. Jennex and Olfman (2005) provide a review of KMS success literature and propose a comprehensive framework for evaluation of KMS success. In this chapter, our goal is to contribute to this line of research by identifying how these success factors may be achieved. Specifically, we restrict our

DOI: 10.4018/978-1-4666-2803-8.ch015

scope of inquiry to a certain type of knowledge management systems; those that are designed to support communities of practice (CoP).

Prior literature that has sought to identify important factors in KM success has adopted either the individual level of analysis (e.g., Bock et al., 2005; Kankanhalli et al., 2005), the organizational level of analysis (e.g., Brown & Duguid, 2000), or the technological level of analysis (e.g., Markus et al., 2002). We propose an approach that incorporates research on individuals, organizations, and the technology pertaining to knowledge management to suggest a set of design principles for sustainable communities of practice. Communities of practice have proven to have significant economic and practical implications on organizational practice (Brown & Duguid, 1999, 2000). A growing body of literature in knowledge management recognizes the importance of communities that foster collaborative learning in organizations and almost all knowledge management systems have a 'network' component that facilitates connecting people in communities of practice. Evidence has shown that community has been a key element in knowledge management systems of many companies including Xerox PARC, British Petroleum Co., Shell Oil Company, Halliburton, IBM, Proctor and Gamble, and Hewlett Packard (Brown & Gray, 1995; Cohen, 2006; Cross et al., 2006; McDermott, 1999a,1999b).

Attributes of communities of practice, which we believe determine the success or failure of KM initiatives, have been thus far under-researched. KM can benefit from literature in virtual communities that looks at what properties of a community make it sustainable. These properties can then be viewed as a blueprint of what a community needs to have to achieve its function of fostering collaboration and hence, generating knowledge. In sum, this research is intended to help practitioners arrive at how best to design communities in KMS in order to achieve KM success.

KMS success models provide a strategic level process approach to achieving success. KMS suc-cess factors provide a means for evaluation of KMS success. Our goal is to suggest how these success factors could be achieved at an operational level. We draw on Jennex and Olfman's (2005b, 2006) work to arrive at a list of eight success factors that are applicable to our conceptualization of a KMS that supports CoPs. Table 1 below provides a list of these factors.

This chapter is structured as follows. In the next section we present a review of literature in knowledge management, KM success, and communities of practice. The literature helps provide the theoretical basis for our research. Our research methodology section follows the literature review. We elaborate on the process and method for arriving at our design recommendations, and discuss each recommendation in detail. We next provide a discussion, and conclude with our suggestions for future research.

LITERATURE REVIEW

The primary goal of this research is to contribute to literature in KM success. We provide a brief review of literature in knowledge management and knowledge management success to summarize the state of current research. We then focus on the

Table 1. KMS success factors adopted from Jennex and Olfman (2005b, 2006)

Success Factor	Description
SF1	Identification of users, sources, knowledge, and links
SF2	Clear articulation of knowledge structure
SF3	Motivation and commitment of users
SF4	Senior management support
SF5	Measures for assessment of appropriate use
SF6	Clear goal and purpose
SF7	Support for easy knowledge use
SF8	Designing work processes to incorporate knowledge capture and use

literature in communities of practice, which we use to augment research on KM success.

Knowledge Management and Knowledge Management Systems

Managing knowledge is a focal task for organizations today. Appreciating the importance of knowledge as a core capability or resource (Alavi & Leidner, 2001; Grant, 1996) has underscored the need for managing it strategically. Though the effort to manage what a company 'knows' is not a recent phenomenon, new technology and greater awareness fueled by a competitive business landscape has resulted in substantive attention paid to KM (Prusak, 2001).

Knowledge can be conceptualized in different ways. It can be seen as embedded in practices (Orlikowski, 2002) or processes (Epple & Argote, 1996), or as a separate entity or object (Schultze & Stabell, 2004). Another view of knowledge is that which is embedded in people's heads and is a "fluid mix of framed experience, values, contextual information and expert insight that provide a framework for evaluation and incorporating new experiences and information" (Davenport & Prusak, 1997, p. 5). This type of knowledge is referred to as tacit. While explicit knowledge is easily codified, stored, and transferred, by the use of technology (such as knowledge repositories, document control systems, or databases) tacit knowledge is 'stickier' (Hippel, 1994). Tacit knowledge, as conceptualized by Polanyi (1958) refers to knowing-how or embodied knowledge[1], and is the characteristic of an 'expert'[2] who can perform a task without deliberation of the principles or rules involved (Ryle, 1949/1984). This goes beyond a mere technical or physical know-how (Dretske, 1991) as it is highly contextual. Employees of a certain culture may have tacit knowledge about practices that employees of other cultures do not. Being able to deliberately leverage such tacit knowledge is hypothesized to generate value and be a key differentiator for an organiza-

tion (Alavi & Leidner, 1999, 2001; Grant, 1996). It is tacit knowledge that resides in employees' heads, which is vital for problem solving and organizational learning (Davenport & Prusak, 1997). Due to the uneven distribution of expertise, the task of managing tacit knowledge is especially essential in today's dynamic and global business landscape. In the context of large, geographically distributed, multi-cultural firms where employees do not have the opportunity to interact face-to-face, communication and transfer of knowledge becomes even more challenging (Lapre & Van Wassenhove, 2003). Therefore, designing systems that facilitate tacit knowledge management is important. The focus of many design articles is on information or content management, which deals with explicit knowledge. In this paper, we focus on managing tacit knowledge.

We conceptualize knowledge management systems (KMS) as systems designed specifically with the intent to manage organizational knowledge, in line with Jennex and Olfman's (2005b) infrastructure/generic approach to KM, by connecting people. In this approach, KMS are primarily designed to support communities of practice. We elaborate on this further in the subsequent sections.

Knowledge Management Success

KMS success has been defined in many ways. Jennex and Olfman (2004, 2005b) provide an integrated framework of KM success factors identified by previous authors. We adopt a definition of KMS success as proposed by Jennex (2005) as being able to reuse knowledge to improve organizational effectiveness by providing the required knowledge to those that need it when it is needed.

The attempt to manage knowledge is not always successful. IDC[3] estimated an expenditure of $12.7 billion on KM in 2005. However, approximately 70 percent of KM initiatives are deemed unsuccessful (Akhavan et al., 2005; Hammer et al., 2004). Stories such as the struggle of General Motors

(GM) and NUMMI[4] in the initial stages of their initiatives for learning and knowledge transfer highlight the challenges (Inkpen, 2005). Though significant research in MIS is directed toward how to successfully implement information systems in general, KM presents unique challenges that are more complex than what models such as technology acceptance (TAM) (Davis, 1989) and task technology fit (TTF) (Goodhue & Thompson, 1995) can explain issues such as employees' reluctance to share knowledge, coordination of knowledge management efforts, and adoption of the right model for knowledge creation, management, and transfer, present unique difficulties for KM system success (Jennex & Olfman, 2005a; Jennex et al., 2007).

Most research on factors of KM success can be seen to fall within one of three categories. First, using the individual as the unit of analysis, employee reluctance to share knowledge has been studied using frameworks such as identity theory (e.g., Constant et al., 1994), social-cognitive theory (e.g., Constant et al., 1994), social exchange theory (e.g., Kankanhalli et al., 2005), and the theory of reasoned behavior (e.g., Bock et al., 2005). Power has also been discussed as an important factor (Gordon & Grant, 2005; Porra & Goel, 2006). Findings indicate mixed results where employees want to share for reasons of 'showing-off' (Constant et al., 1994), altruism (Constant et al., 1994), or feeling of social obligation while employees are reluctant to share knowledge for reasons of fear of loss of hegemony (Shin, 2004), and costs (such as time and effort) (Butler, 2001) involved. Motivation (Ardichvili, 2003), and extrinsic reward structures (Shin, 2004; Yeu Wah et al., 2007) as predictors of knowledge sharing have also been studied.

Second, employing an organizational unit of analysis, management of KM activities has been examined. In particular, research in knowledge transfer (Argote & Ingram, 2000), organizational culture and norms (Constant et al., 1994; Faraj & Wasko, Forthcoming; Hart & Warne, 2006; Usoro

& Kuofie, 2006), and senior and middle management support (Brown & Duguid, 2000) have been studied. Here, KM success is related to favorable organizational factors for knowledge sharing.

Third, using the system as the unit of analysis, technical characteristics of KMS such as repository structure, directory capabilities, and collaboration tools have been investigated (King, 2006; Markus, 2001; Markus et al., 2002). Hence, KM success is explained by theories such as task-technology fit (TTF) (Goodhue & Thompson, 1995), which focus on choosing the right technology.

This paper investigates KMS from a perspective that incorporates the individual, organizational, and system units of analysis by using the lens of communities of practice (Brown & Duguid, 1999; Lave & Wenger, 1991; Wenger, 1998).

Communities of Practice

Communities of practice have been used by many to study KM practices (for example, see Cheuk, 2006; Koeglreiter et al., 2006); however, this perspective has not been applied to designing KMS per se (an exception is Stein, 2006 who proposes a descriptive model of the functions and structure of a successful CoP). Knowledge management systems are not particularly complex or technically different from other information systems (Alavi & Leidner, 1999). The difference between knowledge management systems and other systems such as group decision support systems (GDSS), electronic meeting systems (EMS), and expert systems lies not primarily in the technology, but in the purpose for their use. GDSS focus on connecting a particular group of employees for the goal of solving particular problems, or arriving at a decision. EMS focuses on facilitating meetings and collaborative work among a certain group of people. Expert systems are typically rule-based, where the knowledge of an expert/s (where experts are identified by the organization) is captured in the system's knowledge base, and then queried by users. The goal of KMS, as we conceptualize them, is to connect all employees in an organization at all times. Unlike in expert systems, the roles of knowledge producers (experts) and consumers (users) are flexible. Groups are not dictated by organizational structures, but emerge ad-hoc as communities of employees with common interests and problems. Interaction within these communities may yield solutions to specific problems, but it is the interaction for the purpose of tacit knowledge exchange that is the goal of the system.

Since the central problem of KM is the creation and transfer of tacit knowledge, it is necessary to look at what facilitates these processes. Experiences and contextual insights have been traditionally transferred through methods such as story-telling (Brown & Duguid, 2000), sense-making (Brown & Duguid, 1999), or through conversations in informal social networks. Communities of practice are informal networks of like-minded individuals, where the process of learning and transfer of tacit knowledge is essentially social, involving a deepening process of participation (Lave & Wenger, 1991). Research shows that in the absence of decisive first-hand knowledge, an individual looks at successful decisions made by other like-minded, similarly-situated people (Nidumolu & Subramani, 2001) as filters or guides to identify potentially good choices (Hill et al., 1995). Prior case studies have shown that even for individuals armed with extensive know-what (explicit knowledge), collective know-how (tacit knowledge) can be highly significant (Brown & Duguid, 1999; Orr, 1989). KM practitioners and researchers recognize the importance of communities that foster collaborative learning in organizations (Pan & Leidner, 2003) and almost all knowledge management systems have a 'network' component that facilitates connecting people in communities of practice (Faraj & Wasko, Forthcoming). The community perspective of knowledge management, which acknowledges the importance of informal networks and emphasizes collaboration, started in the late 1990s (Cross et al., 2006). Evidence has shown that communities have been a key

element in knowledge management systems of many companies including Xerox PARC, British Petroleum Co., Shell Oil Company, Halliburton, IBM, Proctor and Gamble, and Hewlett Packard (Brown & Gray, 1995; Cohen, 2006; Cross et al., 2006; McDermott, 1999a, 1999b). Most of the companies that used IBM's first Web-based knowledge management system organized their activities around communities, an element that IBM had not deliberately implemented in the system initially (McDermott, 1999b).

While studying design characteristics of communities, defining what is meant by a community and its sustainability is important. In attempting to do so, we refer to prior IS literature on virtual communities, since we consider communities in knowledge management systems as virtual. The term 'virtual' has been used here to distinguish these communities from real-life communities with face-to-face interaction. Borrowing from biology, virtual community sustainability has been regarded as the 'intrinsic longevity' of the membership (e.g., Butler, 2001; Porra & Parks, 2005). Hence research has been devoted to studying how members can be encouraged to stay in a community. The concept of a 'community' has received much attention and there are different ideas as to what brings about a 'sense of community' (Blanchard & Markus, 2004). Reasons such as support (Blanchard & Markus, 2004), recognition (Blanchard & Markus, 2004), intimacy (Blanchard & Markus, 2004) and obligation (Wasko & Faraj, 2000), have all been studied as motivators of staying in a community. It has been acknowledged that in organizations tacit knowledge is shared in ad-hoc, informal settings (Brown & Duguid, 2000) that may not be replicable (such as brainstorming sessions, or when new hires approach veterans for specific advice). While the composition of members might change in a community, the community still serves the purpose as a platform for knowledge management. The existing literature does not reflect this functional aspect of communities. Therefore, we define a community as a platform for knowledge creation and exchange, and sustainability as how successful a community is in achieving its function of facilitating knowledge generation and exchange. This research attempts to draw on relevant literature and apply it to an organizational KM context in order to suggest how sustainable communities can be designed. Hence, the output of this chapter is design recommendations or guidelines for communities of practice that are the central part of KMSs.

One view of organizing CoPs is that the design structure should be emergent rather than imposed (Barab et al., 2003). However, it has been recognized that too little structure can also yield negative benefits, and a 'minimalist design' or a tentative platform for a community is needed (Wenger, 1998). Hence, while it is not recommended to have tightly controlled formal designs in place, informal structures and basic guidelines that allow flexibility, diversity, autonomy, creativity, and scalability are necessary (Wenger, 1998).

RESEARCH METHODOLOGY AND ANALYSIS

In this section, we first elaborate on our method for arriving at design guidelines for sustainable communities in KMS. We next present our design recommendations each preceded by supporting literature. These are followed by examples in prior research, which evidence the recommendations.

Method

We conducted an extensive search in literature using engines such as Google Scholar and Social Sciences Index using key words such as knowledge, knowledge management, knowledge management system, knowledge management system success, communities, virtual communities, and communities of practice. Keeping in line with our research objective, we narrowed the results to papers that

were relevant to the design of communities in the context of knowledge management. We support each proposed design guideline with results from prior studies, both qualitative and quantitative. The qualitative data includes quotes from case studies in knowledge management and virtual communities. Data from quantitative research mainly includes their supported hypotheses. The data immediately follows each design recommendation. The goal of using data from previous research was solely for clarification and better understanding, as well as demonstrating that the importance of these issues has been acknowledged explicitly by other researchers. Though the evidence was implicitly or explicitly observed in different papers, the related design guidelines were not the primary focus of these studies. One of our key contributions is thus the synthesis of the findings across the different studies and bringing to the forefront the importance of the design guidelines.

Design Recommendations

A community of practice is defined by the commonality of an interest shared by its members. Members of the community form communal bonds, construct collective identities through communicative action (Blanchard & Markus, 2004; Donath, 1999; Postmes et al., 2000), and reach common understanding (Habermas, 1984), thus generating collaborative knowledge. The concept of a community boundary is hence important and delineates what a community is about, and what it is not. In organizations, it is especially important to make sure a community is 'specialized' or focused on a particular topic in order to maximize the signal[5] to noise ratio and minimize the costs of obtaining relevant information[6]. Thus:

Each community in a KMS should have a central focus which should be maintained throughout the life of the community by ensuring that all posts are relevant to the community's focus.

As will be discussed in the ensuing recommendations, a community manager, or a segmentation strategy, could play a key role in ensuring a topical focus. Jones and Rafaeli (2000) and Jones et al. (2004) conclude that in order for a virtual community to be sustainable, it has to have a topical focus because otherwise members might experience information overload, which is not tolerable by their cognition capacity. They add that:

It logically follows that beyond a particular communication processing-load, the behavioural stress zones encountered will make group communication unsustainable (Jones & Rafaeli 2000, p. 219).

Overall, the empirical findings support the assertion that individual information-overload coping strategies have an observable impact on mass interaction discourse dynamics. Evidence was found [that] users are more likely to end active participation as the overloading of mass interaction increases (Jones et al. 2004, p. 206).

Two complementary concepts discussed in literature are those of information overload (Rogers & Agarwala-Rogers, 1975) and critical mass (Licklider & Taylor, 1968; Markus, 1987). Critical mass refers to the required group size threshold for a sustainable community (Bieber et al., 2002). Information overload results from a higher number of messages, or messages that are not sufficiently organized in a sensible, linked structure, which makes it difficult for individuals to process. While critical mass indicates that a community that is too small will fail, information overload suggests that one that is too big will also fail. Hence, there is a maximum limit to the size of a community beyond which a low signal to noise ratio (Malhotra et al., 1997) and an upper bound on an individual's cognitive processing limits (Jones, 1997) will render it unsuccessful. This topic was the focus of studies conducted specifically on membership limits of communi-

ties (e.g., Jones & Rafaeli, 2000). While the limit is context dependent and varies with the nature of the community, a manager could determine an approximate threshold for when information overload or topical deviation occurs. When a community grows too large, a segmentation strategy can be employed to create an interrelated space (Jones & Rafaeli, 2000). Hence:

There should be a maximum limit set for the membership in the community in a KMS, beyond which the community should be split into interrelated sub-communities each with a central focus or topic.

Jones and Rafaeli (2000) provide examples of communities such as Amazon or Excite, in which a segmentation strategy has been used to keep the communities focused and prevent information overload.

Segmentation strategy' refers here to any systematic method used to split discourse spaces with the aim of creating a system of interrelated virtual publics. As studies of usage show (e.g., Butler, 1999), virtual publics are not 'scalable.' Therefore, a 'mega virtual public' cannot be sustained. Rather, virtual metropolises emerge from the creation of a series of related virtual publics, via the appropriate segmentation of discourse in different related cyberspaces. In turn, the resulting system of interconnected virtual publics encourages the expansion of user populations, while reducing the likelihood of overloaded virtual public discourse (p. 221).

Moderation has been studied in virtual community literature (Markus, 2001). Human intervention is considered necessary for tasks such as maintaining community focus, preventing 'trolling' or 'flaming' (Malhotra et al., 1997), encouraging participation (Preece et al., 2003), and sanitizing data (Markus, 2001). Even though trolling and flaming are unlikely in organizational settings, moderation is required for the other reasons. Since moderators (often referred to as community or knowledge managers) are usually employees with regular duties, it is important that their work for the KM effort be recognized and that they are allowed to devote time to the community as part of their job (Davenport et al., 1998; Silva et al., 2006). Hence:

Each community should have at least one community manager with the authority and resources to manage and act for the benefit of the community.

The following remarks demonstrate the importance of the role of a community manager and the resources (e.g., people and time) that are available to him or her.

Knowledge managers have content leads that evaluate external knowledge resources and establish pointers to the new resources for specific domains. The searches are synthesized to allow higher level of reuse. Knowledge researchers strongly support sharing and reuse of knowledge probes: 'The first thing I do is go check that database to make sure that no one hasn't already pulled that information before I go and start a whole new search' (Sherif & Xing 2006, p. 538).

One person per project is in charge of posting project deliverables onto the repository ... On a weekly basis the knowledge manager goes in and makes sure it has been categorized correctly and the document has been zipped. He just makes sure it has been submitted according to all the standards and guidelines (Sherif & Xing 2006, p. 536).

As seen in the following, lack of sufficient resources such as time can adversely affect the quality of community manager's job.

With the amount of time available to produce high quality and sanitized knowledge for dissemina-

tion ... It's not even just writing it ... I'm on the review committee and that's where a lot of time is as I've got to review every document ... The delay in implementing this second knowledge dissemination plan was due to ... the lack of resources to provide high quality, sanitized knowledge for consumption by customers (Markus, 2001, p. 80).

Research on information overload posits that information should be organized hierarchically in order to place new information in the correct context quickly. Also, costs of obtaining new knowledge are lower when messages are simple and contextually relevant (Jones et al., 2004). Blogs, wikis, and net forums use chronological and topical organizing successfully to map information. Site maps on websites provide a graphical view of how information is arranged, reducing the time and effort required to access it while also providing an overall picture of the Web site. More recent work on the semantic Web enables tagging knowledge objects with keywords pertaining to the context and relevance of the topic (Daconta et al., 2003). Using descriptive tags enables searching and archiving information in a way that is most applicable to the topic. Semantic Web technologies make a static hierarchical representation extraneous. However, to make searches efficient, there needs to be standardization in the tags used to describe the same types of objects. The standards, while imposing a common structure, should also allow flexibility to reflect unique contextual information in the knowledge object (Geroimenko & Chen, 2003). Site maps can be used to provide community-specific keywords that can be used to tag and search for community related topics. The importance of site maps in knowledge management systems has been suggested in prior research (e.g., Jennex & Olfman, 2000, 2005). This leads to the following design recommendations:

Each community in a KMS should have a site map showing how knowledge is arranged and accessible in the community.

All posts in the community should be categorizable according to the site map.

Here is a description of the importance of a well-designed site map (or an equivalent).

We've got to put tags on that content in a way that it's retrievable, in a way that makes sense as to how we do business. When you've done a particular type of project, creating [a] particular set of deliverables to a particular type of client, then you want to be able to hand this next person a path or a navigational metaphor or structure so that they can find what they are looking for and [if it]..is meaningful to them they can reuse it with a slight change (Sherif & Xing 2006, p. 536).

You cannot just willy-nilly grab content and throw it into a big pot and then hope that it can be reused. Obviously, you've got to put tags on that content in a way that it is retrievable. You have to be able to provide the knowledge contributor with the analog on the front end, so that they can categorize and catalog their knowledge in a way that makes it meaningful for the person who is now coming and going to repeat it all (Sherif & Xing 2006, p. 536).

Costs in terms of effort and time can also be reduced by maintaining a community home page which provides a quick overview of things such as the latest news related to the community topic, latest posts and replies. An effective and efficient search function is important in order for members to be able to locate relevant information quickly, which also could reduce the amount of effort made (Markus, 2001). Hence:

The home page for the community should be fresh and dynamic, presenting all new relevant information for the community in a concise and easy-to-read manner.

The KMS should support an efficient and effective search function.

The following statements elaborate on these guidelines.

You go to this site that has the method and you click on the phases and read about the phases, you read about stages, and it shows what work products and what deliverables come out of those activities and stages. You can click on that and it will bring up the best examples of those deliverables and work products (Sherif & Xing 2006, p. 537).

Say you have someone on the ExxonMobil team, and they need to keep up with what's going on with ExxonMobil, so we set up a search term for them and we tie it into our Dow Jones service that we use and everyday, Dow Jones uses that search term that we put in there and dumps these articles about ExxonMobil into a profiling container so the team member can use it. They don't have to go to Dow Jones, they don't have to look through 50 different articles, they go right here and they see right here, here's the 10 articles that came in today about ExxonMobil (Sherif & Xing 2006, p. 537).

A core capability of the tool the team was provided was the ability to reference-link entries and apply multiple keywords, and then use powerful search capabilities to identify similar entries (Markus 2001, p. 80).

Version control system for documents can be used to keep track of dates of multiple submissions of the same document. In addition to automatic checking, members can be encouraged by community managers to periodically review their submissions to the KMS and make sure that they are current and delete the obsolete, inconsistent entries (Damodaran & Olphert, 2000; Kohlhase & Anghelache, 2004).

Accuracy and currency of the information should be maintained by using tools such as version control systems, as well as by encouraging members to review their submissions.

The importance of the ability to revisit and revise posts and to keep the information accurate and up-to-date is seen below.

...the database author might know that the answer could be made more general in order to answer more questions. This might involve abstracting both the question and the answer. Occasionally, the author would feel it necessary to correct incorrect, incomplete, or incoherent answers (Markus 2001, p. 76).

The following shows how version control functionality benefits users of a web-based virtual learning environment for students.

Although this approach for providing group awareness is very simple, feedback from users of the BSCW system indicates that information such as 'A uploaded a new version of document X', or 'B has read document Y' is often very useful for group members in coordinating their work and gaining an overview of what has happened since they last logged in (Appelt & Mambrey 1999, p. 1710).

Social network theory (Granovetter, 1973) and social identity theory (Tajfel & Turner, 1986) have been used to explain knowledge exchange and interactions in communities. According to social identity theory and theories of collective action, gaining self esteem and recognition for their expertise, and enhancing their reputation (Wasko & Faraj, 2005) motivate members to participate in communities (Donath, 1999; Douglas & McGarty, 2001; Postmes et al., 2000). 'Old-timers' in communities are frequently those whose advice is valued due to their track record in the community. Social network theory looks

at relationships between members (Chae et al., 2005; Garton & Haythornthwaite, 1997; Wasko & Faraj, 2004; Wellman, 1996; Wellman & Guila, 1997; Wellman et al., 1996) and the normative structure of content and form in a group (Postmes et al., 2000; Wellman, 1996). While weak ties provide more opportunities for new knowledge (Hansen, 1999; Wellman, 1996), on-line strong ties provide more confidence in the information and reinforce real-life relationships (Wellman, 1996). With social network analysis (Wellman, 1996), a well connected member with a high social capital can be identified as a good information resource (Wasko & Faraj, 2005).

Technology should support multiple ways to connect people (Garton an&d Haythornthwaite, 1997) who need the information to those who have it. Social technologies such as online networking sites (e.g., Facebook, MySpace) and virtual worlds (e.g., Second Life) can be used to support collaboration (Goel & Mousavidin, 2007). Expert directories are often an integral part of KMS and serve as transactive memory systems to identify 'who knows what' in the organization (Argote & Ingram, 2000). If an employee knows exactly who to ask a particular question from, he/she should be able to contact the expert directly, and in the absence of the knowledge, the member can rely on the community to help arrive at an answer. Since members in a KMS can belong to multiple communities, these features apply to the entire system. This discussion leads to the following design recommendations:

The KMS should have the facility to maintain member profiles that indicate the number of posts, other member links, replies, and usage history of a member.

Members should be able to connect to other members directly through tools such as instant messaging as well as indirectly through forums and directories.

The user profile contains compulsory and optional information that a member provides upon registration. If a member of the community publishes a contribution or asks a question, the contributor's name is shown as a hyperlink. By clicking on this hyperlink, one obtains the user profile of the corresponding member. The extent of information other members see on the user profile depends on the level of anonymity the member has chosen (Leimeister et al. 2005, p. 110).

In the remark below, Barab et al. (2003) discuss the importance of profiles in a community of practice called ILF[7].

In fact, the ILF encourages its members to create and edit their member profiles so other ILF members can learn more about one another. This enables ILF members to control how they are perceived by others within the community, and ideally, these profiles help ILF members to decide who they want to communicate with and how they might interpret statements or attitudes of others (Barab et al., 2003, p. 248).

Though social identity theory discourages anonymity, literature in MIS (especially in research on group decision support systems) has found anonymity to aid participation. However, an employee's trust in the content obtained from a KMS would be weaker if it were anonymous (Donath, 1999). Making authorship explicit adds legitimacy to the information. Literature in philosophy also discusses the importance of credentials in reliability on others for information (Hardwig, 1991). In addition, this makes the author responsible for making sure that the information he/she puts up is not erroneous. Investigating different levels of anonymity (such as using nicknames or real names) would be an interesting line of enquiry, as well.

Submissions to a KMS should not be anonymous.

The following statement supports this guideline.

An interesting metric developed by the specialists to assess data quality was their use of incident authorship as an indicator of quality. Each incident that is entered is automatically assigned a unique number, which includes a code identifying the particular specialist who entered it ... You tend to evaluate information differently from different people. So if you see 40 items from a search you go to the incidents of those folks you've gotten good information from in the past ... I know that Arthur has a reputation for writing shorts novels as resolutions. I mean, he's a wonderful source of information ... So when I get an incident from him, I'm very comfortable with that information. Whereas, some of the other people in the department will put in one or two sentence resolutions. And it tends to make it a little vaguer and more difficult to be confident about (Markus 2001, p. 68, 69).

Participation in a community needs to be encouraged (Cross et al., 2006; Davenport et al., 1998; Prandelli et al. 2006). Reluctance on behalf of employees to participate is a primary reason for the failure of KM efforts (Ardichvili, 2003; Alavi & Leidner, 1999). Specifically, perceived loss of power (Constant et al., 1994), fear of criticism and unintentionally misleading members (Ardichvili, 2003), and costs involved in terms of time and effort (Markus, 2001) are primary reasons why employees do not participate. While appropriate design of a system can help minimize the aforementioned costs (Bieber et al., 2002), the organization needs to make efforts to alleviate the other factors. Engendering an organizational culture which encourages pro-social behavior and knowledge sharing has been suggested in prior literature (Constant et al., 1994; Huber, 2001). At a community level, the moderator or community manager can play a vital role in increasing the level of participation. In particular, the community manager needs to ensure that knowledge is transferred from producers (experts) to consumers (users) (Markus, 2001; Cross et al., 2006) by ensuring that questions are answered and new knowledge is posted. These functions of a community manager maintain the value and usefulness of the community to its members (Preece et al., 2003). Hence:

A community manager should ensure that queries are answered.

Some of the responsibilities of community managers (Sysops, in this case) are shown in the statement below.

System operators (Sysops) were appointed to monitor the discussions in the forums, track requests and make sure they were answered. Sysops would try to get answers in 24 hours; if not they would contact people directly and ask them to respond. Additionally, they were to give positive feedback to those who did respond. Since there were likely to be cultural difference and sensitivities, Sysops were to monitor the content of messages ... Three translators were hired and Sysops would decide which messages were to be translated into English with technical replies to be translated back to the originator's own language. The goal for completion of translation was 48 hours (Markus 2001, p. 85).

A community manager should encourage experts to contribute tacit knowledge (experiences, ideas, etc.) to the community.

Some techniques for encouraging participation were observed by Jones and Rafaeli.

Where sustained interactive discourse is a goal, various techniques can be used to gain critical mass. Administrators can seed discussions by

systematically encouraging a group of key individuals to contribute. Economic incentives can be given and where a number of related virtual publics already exist, group segmentation can be used to gain instant critical mass for new virtual publics (Jones & Rafaeli 2000, p. 221).

In addition to voluntarily accessing the system for information, a KMS can be made a central repository for electronically storable information such as required documents, templates, and forms. Since knowledge sharing cannot be forced and extrinsic rewards may not work (Huber, 2001; Kankanhalli et al., 2005), making the KMS an obligatory passage point[8] (Callon, 1986) for explicit knowledge required for daily work would encourage employees who otherwise would not access the system, or make use of it. Most employees either do not think of using the KMS for posting such information, or may not want to spend the time and effort required to do so. More importantly, there should be a consistent format used for information posted. A community manager plays a key role in ensuring this consistency. Hence:

A community manager should encourage the posting of standardized documents, templates, forms, and other electronically storable information in a consistent manner.

The following shows some consequences of inconsistent postings.

We had to overhaul the [knowledge repository] after three years because nobody was following a consistent style [of classifying documents] ... [P]eople were building their case bases with different parameter settings, so it became like a soup of knowledge, and nobody could find anything (Markus 2001, p. 81).

We summarize the design guidelines and highlight our contributions in the following section.

DISCUSSION AND FUTURE RESEARCH

The design recommendations derived from theory and past research are summarized in Table 2. In broad terms, the recommendations can be seen to fall into four categories: technological, membership, content, and organizational. To achieve our objective, we study KMS from a perspective that incorporates technical, individual, as well as organizational level literature. These are reflected in the technological, membership, and organizational categories. We tie our recommendations to prior research by suggesting how each of them could be used to achieve a particular success factor. In Table 2, each recommendation is followed by success factor(s) from Table 1 which we believe to be applicable.

Employing the lens of communities of practice allowed us to add a fourth level of analysis, that of a community. The guidelines regarding the content of KM communities address this level. In practice, most KMS incorporate communities as part of their architecture. By drawing on literature on virtual communities, we add to research in KMS by proposing design guidelines for the content and management of such communities. These categories are not mutually exclusive. For example, membership profiles need the corresponding technological features to support them. Also, the list is not intended to be exhaustive. Further research studying actual participatory behavior in KMS would help determine if there are other design characteristics that facilitate community participation.

Prior research in IS has not systematically studied design guidelines for knowledge management communities. While the design features discussed have been implemented in current KM practices, they have emerged more from a process of trial-and-error; not grounded in research. Organizations are spending considerable resources, both in terms of time and money, on knowledge management

efforts, not all of which are successful. Results from this study could help managers concerned with KM increase their chances of success by designing sustainable communities.

Using a community of practice as a lens unifies the fragmented literature in knowledge management which, thus far, has studied the phenomenon separately at the individual, organizational, or

Table 2. Summary of design recommendations for sustainable communities of practice in knowledge management systems

Design Guidelines	Success Factor
I. Technological Features	
1. Site map	SF1, SF2, SF7
2. Search features	SF7
3. Document version control systems	SF1, SF2, SF7
II. Membership Features	
1. Non-anonymity	SF1
2. Maintaining usage statistics and profiles	SF1, SF5
3. Connectivity between members – Directories, Forums, IM capabilities	SF1
III. Content Features	
1. Making it an obligatory passage point (OPP) for documents, templates, forms, etc.	SF3, SF8
2. Home page to represent community – fresh, dynamic	SF7
3. Content on the home page for summary of new relevant topics, news, and highlights	SF2, SF7
4. Moderation to ensure topical focus and relevance	SF2, SF3, SF5, SF6, SF7
5. Moderation to ensure questions are answered	SF3, SF5
6. Moderation to encourage 'experts' to contribute	SF3
7. Moderation to ensure currency of documents	SF3, SF5, SF7
8. Splitting into sub-communities, if size grows	SF3, SF5, SF6, SF7
IV. Organizational Functions	
1. Creating and supporting role of moderator/ community manager	SF3, SF4

system level. A shift of perspective of community sustainability from one that retains more members to one that serves as an effective and efficient platform for knowledge management also stimulates new lines of inquiry.

We looked at existing research on knowledge management and virtual communities, synthesized the literature, and applied it in the context of organizational KMS. Our goal was to conceptually identify guidelines for successful design of communities in KMS which can be used by practitioners who wish to implement KMS. Hence, these guidelines are intended as a blueprint to design KMS communities. Furthermore, they can also be used as an evaluative tool for existing KMS. Future research would add value by evaluating and testing these recommendations in a representative setting through possibly an action research. An exploratory case study of existing KMS that implement these features would also provide validity. Each recommendation could be the subject of a separate study. For example, the role of a knowledge manager or the design of the search features could constitute substantial research agendas.

This chapter has the limitations associated with a conceptual study. It needs to be validated by empirical research. The suggestions presented for future research are intended to be a guide for a research program in this area. We draw on literature in virtual communities to inform us about the nature of communities in knowledge management systems. However, the design of these communities is contingent upon organizational factors such as organization culture, norms, practices, and structure. Hence, these guidelines may not be universally applicable but need to be tailored to the specific context and requirements. This research is intended to be a starting point for an inquiry on the topic.

REFERENCES

Akhavan, P., Jafari, M., & Fathian, M. (2005). Exploring failure-factors of implementing knowledge management systems in organizations. *Journal of Knowledge Management Practice.*

Alavi, M., & Leidner, D. (2001). Review: Knowledge management and knowledge management systems: Conceptual foundations and research issues. *Management Information Systems Quarterly*, *25*(1), 107–136. doi:10.2307/3250961

Alavi, M., & Leidner, D. E. (1999). Knowledge management systems: Issues, challenges, and benefits. *Communications of the AIS, 1*(7).

Appelt, W., & Mambrey, P. (1999), Experiences with the BSCW Shared Workspace System as the Backbone of a Virtual Learning Environment for Students. *In Proceedings of ED Media '99.'* Charlottesville, (pp. 1710-1715).

Ardichvili, A. (2003). Motivation and barriers to participation in virtual knowledge-sharing communities of practice. *Journal of Knowledge Management*, *7*(1), 64–77. doi:10.1108/13673270310463626

Argote, L., & Ingram, P. (2000). Knowledge transfer: A basis for competitive advantage in firms. *Organizational Behavior and Human Decision Processes*, *82*(1), 150–169. doi:10.1006/obhd.2000.2893

Balasubramanian, S., & Mahajan, V. (2001). The economic leverage of the virtual community. *International Journal of Electronic Commerce*, *5*(3), 103–138.

Barab, S. A. MaKinster, J. G., & Scheckler, R. (2003). Designing system dualities: Characterizing a web-supported professional development community. *The Information Society, 19*, 237-256.

Bieber, M., D., Engelbart, D., Furuta, R., Hiltz, S. R., Noll, J., Preece, J. et al. (2002). Toward virtual community knowledge evolution. *Journal of Management Information Systems, 18*(4), 11–35.

Blanchard, A. L., & Markus, M. L. (2004). The experienced 'sense' of a virtual community: Characteristics and processes. *The Data Base for Advances in Information Systems, 35*(1), 65–79.

Bock, G. W., Zmud, R. W., Kim, Y. G., & Lee, J. N. (2005). Behavioral intention formation in knowledge sharing: Examining the roles of extrinsic motivators, social-psychological forces, and organizational climate. *Management Information Systems Quarterly, 29*(1), 87–111.

Brown, J. S. & Duguid, P. (1999). Organizing knowledge. *The society for organizational learning, 1*(2), 28-44.

Brown, J. S., & Duguid, P. (2000). Balancing act: How to capture knowledge without killing it. *Harvard Business Review*, 73–80.

Brown, J. S., & Gray, E. S. (1995). The people are the company. *FastCompany*. Retrieved from http://www.fastcompany.com/online/01/people.html

Butler, B. (1999) *The dynamics of electronic communities*. Unpublished PhD dissertation, Graduate School of Industrial Administration, Carnegie Mellon University.

Butler, B. S. (2001). Membership size, communication activity and sustainability: A resource-based model of online social structures. *Information Systems Research, 12*(4), 346–362. doi:10.1287/isre.12.4.346.9703

Callon, M. (1986). Some elements of a sociology of translation: Domestication of the scallops and the fishermen of St. Brieuc Bay. In Law, J. (Ed.), *Power, action and belief* (pp. 196–233). London: Routledge & Kegan Paul.

Chae, B., Koch, H., Paradice, D., & Huy, V. (2005). Exploring knowledge management using network theories: Questions, paradoxes and prospects. *Journal of Computer Information Systems, 45*(4), 62–74.

Cheuk, B. W. (2006). Using social networking analysis to facilitate knowledge sharing in the British Council. *International Journal of Knowledge Management, 2*(4), 67–76.

Cohen, D. (2006). What's your return on knowledge? *Harvard Business Review, 84*(12), 28–28.

Constant, D., Keisler, S., & Sproull, L. (1994). What's mine is ours, or is it? A study of attitudes about information sharing. *Information Systems Research, 5*(4), 400–421. doi:10.1287/isre.5.4.400

Cross, R., Laseter, T., Parker, A., & Velasquez, G. (2006). Using social network analysis to improve communities of practice. *California Management Review, 49*(1), 32–60. doi:10.2307/41166370

Daconta, M., Orbst, L., & Smith, K. (2003). *The semantic web*. Indianapolis: Wiley Publishing.

Damodaran, L., & Olphert, W. (2000). Barriers and facilitators to the use of knowledge management systems. *Behaviour & Information Technology, 19*(6), 405–413. doi:10.1080/014492900750052660

Davenport, T. H., De Long, D. W., & Beers, M. C. (1997). *Building successful knowledge management projects*. Center for Business Innovation Working Paper, Ernst and Young.

Davenport, T. H., De Long, D. W., & Beers, M. C. (1998). Successful knowledge management projects. *MIT Sloan Management Review, 39*(2), 43–57.

Davenport, T. H., & Prusak, L. (1997). *Working knowledge: How organizations manage what they know*. Cambridge, MA: Harvard Business School Press.

Davis, F. D. (1989). Perceived usefulness, perceived ease of use, and user acceptance of information technology. *Management Information Systems Quarterly*, *13*(3), 319–341. doi:10.2307/249008

Donath, J. S. (1999). Identity and deception in the virtual community. In Kollock, M. A. S. P. (Ed.), *Communities in cyberspace* (pp. 29–59). London: Routledge.

Douglas, K. M., & McGarty, C. (2001). Identifiability and self-presentation: Computer-mediated communication and intergroup interaction. *The British Journal of Social Psychology*, *40*, 399–416. doi:10.1348/014466601164894

Dretske, F. (1991). *Explaining behavior: Reasons in a world of causes*. Cambridge, MA: MIT Press. doi:10.2307/2108238

Epple, D., & Argote, L. (1996). An empirical investigation of the microstructure of knowledge acquisition and transfer through learning by doing. *Operations Research*, *44*(1), 77–86. doi:10.1287/opre.44.1.77

Faraj, S., & Wasko, M. M. (Forthcoming). The web of knowledge: An investigation of knowledge exchange in networks of practice. *AMR*.

Garton, L., Haythornthwaite, C., & Wellman, B. (1997). Studying online social networks. *Journal of Computer-Mediated Communication*, *3*(1), 75–105.

Geroimenko, V., & Chen, C. (2003). *Visualizing the semantic web*. London: Springer.

Goel, L. & Mousavidin, E. (2007). vCRM: Virtual customer relationship management. forthcoming in DATABASE Special Issue on Virtual Worlds, November 2007.

Goodhue, D. L., & Thompson, R. L. (1995). Task-technology fit and individual performance. *Management Information Systems Quarterly*, *19*(2), 213–236. doi:10.2307/249689

Gordon, R., & Grant, D. (2005). Knowledge management or management of knowledge? Why people interested in knowledge management need to consider foucault and the construct of power. *Journal of Critical Postmodern Organization Science*, *3*(2), 27–38.

Granovetter, M. (1973). The strength of weak ties. *American Journal of Sociology*, *78*, 1360–1380. doi:10.1086/225469

Grant, R. M. (1996). Toward a knowledge-based theory of the firm. *Strategic Management Journal 17(Winter Special Issue)*, 109-122.

Habermas, J. (1984). *The theory of communicative action*. Boston: Beacon Press.

Hammer, M., Leonard, D., & Davenport, T. H. (2004). Why don't we know more about knowledge? *MIT Sloan Management Review*, *45*(4), 14–18.

Hansen, M. T. (1999). The search-transfer problem: The role of weak ties in sharing knowledge across organizational subunits. *ASQ*, *44*, 82–111. doi:10.2307/2667032

Hardwig, J. (1991). The role of trust in knowledge. *The Journal of Philosophy*, *88*(12), 693–708. doi:10.2307/2027007

Hart, D., & Warne, L. (2006). Comparing cultural and political perspectives of data, information, and knowledge sharing in organisations. *International Journal of Knowledge Management*, *2*(2), 1–15. doi:10.4018/jkm.2006040101

Hill, W., Stead, L., Rosenstein, M., & Furnas, G. (1995). Recommending and evaluating choices in a virtual community of use. *SIGCHI Conference on Human factors in Computing systems*, Denver, Colorado.

Hippel, E. V. (1994). Sticky information and the locus of problem solving: Implications for innovation. *Management Science*, *40*(4), 429–440. doi:10.1287/mnsc.40.4.429

Huber, G. (2001). Transfer of knowledge in knowledge management systems: unexplored issues and suggested studies. *European Journal of Information Systems (EJIS)*, *10*, 72–79. doi:10.1057/palgrave.ejis.3000399

Husted, K., & Michailova, S. (2002). Diagnosing and fighting knowledge-sharing hostility. *Organizational Dynamics*, *31*(1), 60–73. doi:10.1016/S0090-2616(02)00072-4

Inkpen, A. C. (2005). Learning through alliances: General motors and NUMMI. *California Management Review*, *47*(4), 114–136. doi:10.2307/41166319

Jarvenpaa, S. L., & Staples, D. S. (2000). The Use of Collaborative Electronic Media for Information Sharing: An Exploratory Study of Determinants. *The Journal of Strategic Information Systems*, *9*(2/3), 129–154. doi:10.1016/S0963-8687(00)00042-1

Jennex, M., & Olfman, L. (2000). Development recommendations for knowledge management/organizational memory systems. In *Proceedings of the Information Systems Development Conference*.

Jennex, M., & Olfman, L. (2004). Assessing Knowledge Management success/Effectiveness Models. *Proceedings of the 37th Hawaii International Conference on System Sciences*.

Jennex, M., & Olfman, L. (2005b). Assessing knowledge management success. *International Journal of Knowledge Management*, *1*(2), 33–49. doi:10.4018/jkm.2005040104

Jennex, M., Smolnik, S., & Croasdell, D. (2007). Knowledge management success. *International Journal of Knowledge Management*, *3*(2), i–vi.

Jennex, M. E. (2005a). What is knowledge management? *International Journal of Knowledge Management*, *1*(4), i–iv.

Jennex, M. E., & Olfman, L. (2006). A model of knowledge management success. *International Journal of Knowledge Management*, *2*(3), 51–68. doi:10.4018/jkm.2006070104

Jones, Q. (1997). Virtual-communities, virtual settlements and cyber archeology: A theoretical outline. *Journal of Computer-Mediated Communication*, *3*(3).

Jones, Q., & Rafaeli, S. (2000). Time to split, virtually: 'Discourse architecture' and 'community building' create vibrant virtual publics. *Electronic Markets*, *10*(4), 214–223. doi:10.1080/101967800750050326

Jones, Q., Ravid, G., & Rafaeli, S. (2004). Information overload and the message dynamics of online interaction spaces: A theoretical model and empirical exploration. *Information Systems Research*, *15*(2), 194–210. doi:10.1287/isre.1040.0023

Kankanhalli, A., Tan, B. C. Y., & Kwok-Kei, W. (2005). Contributing knowledge to electronic knowledge repositories: An empirical investigation. *Management Information Systems Quarterly*, *29*(1), 113–143.

King, W. R. (2006). The critical role of information processing in creating an effective knowledge organization. *Journal of Database Management*, *17*(1), 1–15. doi:10.4018/jdm.2006010101

Koeglreiter, G., Smith, R., & Torlina, L. (2006). The role of informal groups in organisational knowledge work: Understanding an emerging community of practice. *International Journal of Knowledge Management*, *2*(1), 6–23. doi:10.4018/jkm.2006010102

Kohlhase, M., & Anghelache, R. (2004). Towards collaborative content management and version control for structured mathematical knowledge. *In Proceedings Mathematical Knowledge Management: 2nd International Conference, MKM 2003*, Bertinoro, Italy

Lapre, M. A., & Van Wassenhove, L. N. (2003). Managing learning curves in factories by creating and transferring knowledge. *California Management Review*, *46*(1), 53–71. doi:10.2307/41166231

Lave, J., & Wenger, E. (1991). *Situated learning. Legitimate peripheral participation*. Cambridge: Cambridge University Press. doi:10.1017/CBO9780511815355

Leimeister, J. M., Ebner, W., & Krcmar, H. (2005). Design, implementation, and evaluation of trust-supporting components in virtual communities for patients. *Journal of Management Information Systems*, *21*(4), 101–135.

Licklider, J., & Taylor, R. (1968). The computer as a communication device. *Sciences et Techniques (Paris)*.

Lippman, S. A., & Rumelt, R. P. (1982). Uncertain imitability: An analysis of interfirm differences in efficiency under competition. *The Bell Journal of Economics*, *13*, 418–438. doi:10.2307/3003464

Malhotra, A., Gosain, S., & Hars, A. (1997). Evolution of a virtual community: Understanding design issues through a longitudinal study. *International Conference on Information Systems (ICIS), AIS.*

Markus, L. M. (1987). Towards a critical mass theory of interactive media: Universal access, interdependence and diffusion. *Communication Research*, *14*, 491–511. doi:10.1177/009365087014005003

Markus, L. M. (2001). Towards a theory of knowledge reuse: Types of knowledge reuse situations and factors in reuse success. *Journal of Management Information Systems*, *18*(1), 57–94.

Markus, L. M., Majchrzak, A., & Gasser, L. (2002). A design theory for systems that support emergent knowledge processes. *Management Information Systems Quarterly*, *26*(3), 179–212.

McDermott, R. (1999a). How to get the most out of human networks: Nurturing three-dimensional communities of practice. *Knowledge Management Review*, *2*(5), 26–29.

McDermott, R. (1999b). Why information inspired but cannot deliver knowledge management. *California Management Review*, *41*(4), 103–117. doi:10.2307/41166012

Nidumolu, S. R., Subramani, M., & Aldrich, A. (2001). Situated learning and the situated knowledge web: Exploring the ground beneath knowledge management. *Journal of Management Information Systems (JMIS)*, *18*(1), 115–150.

Nonaka, I. (1994). A dynamic theory of organizational knowledge creation. *Organization Science*, *5*(1), 14–37. doi:10.1287/orsc.5.1.14

Orlikowski, W. J. (2002). Knowing in practice: Enacting a collective capability in distributed organizing. *Organization Science*, *13*(3), 249–273. doi:10.1287/orsc.13.3.249.2776

Orr, J. E. (1989). Sharing knowledge, celebrating identity: War stories and community memory among service technicians. In Middleton, D. S., & Edwards, D. (Eds.), *Collective remembering: memory in society*. Newbury Park, CA: Sage Publications.

Pan, S. L., & Leidner, D. E. (2003). Bridging communities of practice with information technology in the pursuit of global knowledge sharing. *The Journal of Strategic Information Systems*, *12*, 71–88. doi:10.1016/S0963-8687(02)00023-9

Polanyi, M. (1958). *Personal knowledge, towards a post-critical philosophy*. Chicago, IL: University of Chicago Press.

Porra, J., & Goel, L. (2006, November 18-21). Importance of Power in the Implementation Process of a Successful KMS: A Case Study. *In 37th Annual Meeting of the Decision Sciences Institute*, San Antonio, TX.

Porra, J., & Parks, M.S. (2006). Sustaining virtual communities: Suggestions from the colonial model. *Information Systems and e-Business Management, 4*(4), 309-341.

Postmes, T., Spears, R., & Lea, M. (2000). The formation of group norms in computer-mediated communication. *Human Communication Research, 26*(3), 341–371. doi:10.1111/j.1468-2958.2000.tb00761.x

Prandelli, E., Verona, G., & Raccagni, D. (2006). Diffusion of web-based product innovation. *California Management Review, 48*(4), 109–135. doi:10.2307/41166363

Preece, J., Nonnecke, B., & Andrews, D. (2003). (in press). The top five reasons for lurking: improving community experiences for everyone. *Computers in Human Behavior.*

Prusak, L. (2001). Where did knowledge management come from? *IBM Systems Journal, 40*(4), 1002–1007. doi:10.1147/sj.404.01002

Rogers, E. M., & Agarwala-Rogers, R. (1975). Organizational communication. G. L. Hanneman, & W. J. McEwen, (Eds.), Communication behaviour (pp. 218–236). Reading, MA: Addision Wesley.

Ryle, G. (1949/1984). *The concept of mind.* Chicago, IL: University of Chicago Press.

Schultze, U., & Leidner, D. (2002). Studying knowledge management in information systems research: Discourses and theoretical assumptions. *MISQ, 26*(3), 213–242. doi:10.2307/4132331

Schultze, U., & Stabell, C. (2004). Knowing what you don't know? Discourses and contradictions in knowledge management research. *Journal of Management Studies, 41*(4), 549–573. doi:10.1111/j.1467-6486.2004.00444.x

Sherif, K., & Xing, B. (2006). Adaptive processes for knowledge creation in complex systems: the case of a global IT consulting firm. *Information & Management, 43*(4), 530–540. doi:10.1016/j.im.2005.12.003

Shin, M. (2004). A framework for evaluating economics of knowledge management systems. *Information & Management, 42,* 179–196. doi:10.1016/j.im.2003.06.006

Silva, L., Mousavidin, E., & Goel, L. (2006). Weblogging: Implementing Communities of Practice. *In Social Inclusion: Societal and Organizational Implications for Information Systems: IFIP TC8 WG 8.2,* Limirick, Ireland.

Stein, E. W. (2006). A qualitative study of the characteristics of a community of practice for knowledge management and its success factors. *International Journal of Knowledge Management, 1*(4), 1–24.

Tajfel, H., & Turner, J. C. (1986). The *social* identity theory of intergroup behavior. In Worchel, S., & Austin, W. G. (Eds.), *Psychology of intergroup relations* (pp. 7–24). Chicago: Nelson-Hall.

Usoro, A., & Kuofie, M. H. S. (2006). Conceptualisation of cultural dimensions as a major influence on knowledge sharing. *International Journal of Knowledge Management, 2*(2), 16–25. doi:10.4018/jkm.2006040102

Wasko, M. M., & Faraj, S. (2000). 'It is what one does:' Why people participate and help others in electronic communities of practice. *JSIS, 9*(2/3), 155–173.

Wasko, M. M., & Faraj, S. (2005). Why should I share? Examining knowledge contribution in networks of practice. *Management Information Systems Quarterly, 29*(1), 35–57.

Wasko, M. M., Faraj, S., & Teigland, R. (2004). Collective action and knowledge contribution in electronic networks of practice. *JAIS*, *5*(11-12), 493–513.

Wellman, B. (1996). *For a Social Network Analysis of Computer Networks: A Sociological Perspective on Collaborative Work and Virtual Community*. Denver, Colorado: SIGCPR/SIGMIS.

Wellman, B., & Gulia, M. (1997). *Net surfers don't ride alone: Virtual communities as communities. Communities and cyberspace*. New York: Routledge.

Wellman, B., Salaff, J., Dimitrova, D., Garton, L., Gulia, M., & Haythornthwaite, C. (1996). Computer networks as social networks: Collaborative work, telework, and virtual community. *Annual Review of Sociology*, *22*, 213–238. doi:10.1146/annurev.soc.22.1.213

Wenger, E. (1998). *Communities of practice: Learning, meaning, and identity*. Cambridge: Cambridge University Press.

Yue Wah, C., Menkhoff, T., Loh, B., & Evers, H. D. (2007). Social capital and knowledge sharing in knowledge-based organizations: An empirical study. *International Journal of Knowledge Management*, *3*(1), 29–38. doi:10.4018/jkm.2007010103

ENDNOTES

1. as opposed to theoretical knowledge
2. We adopt a broad definition of expertise. New hires can have tacit knowledge about certain technologies that more senior employees do not.
3. http://www.findarticles.com/p/articles/mi_m0NEW/is_2001_June_5/ai_75318288
4. California-based joint venture of GM and Toyota.
5. Signal is considered as relevant discussion.
6. In organizations, KMS typically consist of communities formed around functional areas or projects. For example, an engineering firm's KMS might have communities for piping, electrical engineering, structural and architectural design, process control, procurement, and for each client account.
7. Inquiry Learning Forum (ILF) is a "Web-based professional development system designed to support a community of practice (CoP) of in-service and preservice mathematics and science teachers who are creating, reflecting upon, sharing, and improving inquiry-based pedagogical practices" (Barab et al. 2003, p. 237).
8. An obligatory passage point would require employees use the system to access explicit work-related information. Using the system for explicit knowledge would make employees aware of the system and encourage them to refer to the KMS for tacit knowledge.

This work was previously published in the International Journal of Knowledge Management (IJKM) Volume 4, Issue 3, edited by Murray E. Jennex, pp. 82-100, copyright 2008 by IGI Publishing (an imprint of IGI Global).

Chapter 16

The Retaliatory Feedback Problem:
Evidence from eBay and a Proposed Solution

Ross A. Malaga
Montclair State University, USA

ABSTRACT

Online auctions are an increasingly popular avenue for completing electronic transactions. Many online auction sites use some type of reputation (feedback) system—where parties to a transaction can rate each other. However, retaliatory feedback threatens to undermine these systems. Retaliatory feedback occurs when one party in a transaction believes that the other party will leave them a negative feedback if they do the same. This chapter examines data gathered from E-Bay in order to show that retaliatory feedback exists and to categorize the problem. A simple solution to the retaliatory feedback problem—feedback escrow—is described.

INTRODUCTION

The past few years have seen the explosive growth of online auction transactions. In 2005, E-Bay listed 1.9 billion items for auction, representing a 33% increase over the previous year. Those listing were responsible for $44 billion in transactions (a 29.6%) increase over 2004 (E-Bay, 2006). While E-Bay is the major player in this area, it is not the only one. Many other companies, such as

Amazon, Yahoo, and Overstock, offer consumer-to-consumer (C2C) online auctions.

While online auction sites are an increasingly popular avenue for completing electronic transactions, they are characterized by a high degree of uncertainty. They face what Akerlof (1970) calls a "Lemons" market; that is, they have a high amount of uncertainty about the quality of the information and/or goods. Uncertainty primarily derives from the fact that buyers and sellers typically know little

DOI: 10.4018/978-1-4666-2803-8.ch016

about each other, are involved in one-time transactions, and pictures and descriptions of goods provide the only means for assessing the quality of goods available for bidding (Montano, Porter, Malaga, & Ord, 2005). This lack of information available to auction bidders, termed information asymmetry, leads to a higher level of uncertainty about potential outcomes from an auction transaction than if a bidder were able to learn more about the auction seller and his product prior to bidding (Liang & Huang, 1998).

In order to reduce information asymmetry and increase the level of trust between auction participants, reputation systems have been developed. Wilson (1985, pp. 27-28) states, "in common usage, reputation is a characteristic or attribute ascribed to one person, industry, and so forth, by another (e.g., A has a reputation for courtesy)." This is typically represented as a prediction about likely future behavior (e.g., "A is likely to be courteous"). It is, however, primarily an empirical statement (e.g., "A has been observed in the past to be courteous"). The predictive power of reputation depends on the supposition that past behavior is indicative of future behavior. Reputation systems (sometimes called feedback systems) allow the participants in a transaction to rate each other. Individuals' ratings are aggregated and are available for everyone to see. These systems promote trust between buyers and sellers because they serve as a benchmark for seller reliability. Trust has been shown to serve as a key factor in the success of online transactions, including electronic auctions (Brynjolfsson & Smith, 2000; Resnick, Zeckhauser, Friedman, & Kuwabara, 2000; Hoffman & Novak, 1999).

This chapter proceeds as follows. The next section discusses the existing literature on trust and reputations systems. Following that, the retaliatory feedback problem is further defined. The research methodology is then discussed. Finally, future trends and conclusions are detailed.

BACKGROUND

Trust and Reputation Systems

A large body of research has shown that trust plays an important role in a consumer's decision to purchase a product in an electronic market (e.g., Jarvenpaa, Tractinsky, & Vitale, 2000; Lim, Sia, Lee, & Benbasat, 2006; McKnight, Choudhury, & Kacmar, 2002). In many contexts such as off-line auctions and stores, buyers have multiple clues that help them determine their level of trust in a seller (e.g., branding, location, past experience, visual observation). However, in online auctions there is no such context, and there is no guarantee that any pictures or descriptions provided are accurate. Thus, performance on past transactions (reputation) is typically the only factor that can contribute to the development of trust.

An important aspect of reputation is the dissemination of reputation information. Landon and Smith (1997, p. 313) concluded, "results suggest that consumers place considerable value on mechanisms that disseminate information on the past quality performance of firms." One way in which reputation information can be generated and disseminated in electronic markets and online communities is through the use of a reputation system (Nielsen, 1998). These systems allow participants in a transaction to rate each other, and the ratings are aggregated in order to provide an overall score. The overall score is provided to potential participants in future transactions.

As E-Bay is the dominant online auction site, we focus on its reputation system as an exemplar. In the E-Bay reputation system (which E-Bay calls feedback), any completed transaction may be rated by the winning bidder of an item and the seller of that item. The feedback scores are +1, representing a positive experience; 0, representing "neutral" feedback; and -1, meaning the purchasing experience was negative for some reason. These ratings are then used to calculate an overall reputation score (feedback score) as well as a percent positive feedback rating.

A number of researchers (Houser & Wooders, 2001; Lucking-Reiley, Bryan, Prasad, & Reeves, 2005; Resnick & Zeckhauser, 2002) have shown a correlation between higher reputation (feedback) scores on E-Bay and higher prices. Therefore, sellers have a strong incentive to achieve and maintain a high reputation score. Some sellers will go to great lengths to ensure a high score. For example, individuals with a negative reputation can easily open a new account (although E-Bay has taken measures to curtail this behavior). In addition, some sellers enter into agreements to purchase each other's goods and provide positive feedback (Gormley, 2004). However, mechanisms can be put in place to prevent this. On E-Bay, for example, a seller's reputation score only uses feedback from unique buyers. Therefore, multiple feedback from the same buyer is not counted.

It should be noted that sellers have a much greater incentive to attain positive feedback than buyers. This is due to the fact that it is buyers who chose whom to transact with. Sellers merely post their items and wait for buyers to bid on them.

Problem Definition

According to Bunnell (2000, p. 55) the E-Bay system, was founded on the belief that most people are trustworthy and committed to honorable dealings with each other. We are encouraging our community to think that basically 99% of the people out there are doing the right thing.

In fact, Resnick and Zeckhauser (2002) found that 99.1% of feedback left on E-Bay was positive, only 0.6% negative, and 0.3% neutral. While the empirical evidence seems to support the basic tenant that most people will behave correctly, we must consider the possibility that the underlying data is skewed due the problem of retaliation.

A number of researchers (Dellarocas, 2003; Resnick & Zeckhauser 2002) have suggested that the high level of positive feedback observed on E-Bay is the result of retaliatory feedback. Retaliatory feedback occurs when one party in a transaction believes that the other party will leave them a negative feedback if they do the same. In order to better understand this phenomenon we turn to game theory and anecdotal data gathered from E-Bay's feedback discussion forum.

The problem of retaliatory feedback on E-Bay was recently noted by Brian Burke, E-Bay's Director of Global Feedback Policy, who stated, "if you take this trend to its extreme we'll have 100% universal positive Feedback—which would make the Feedback system valueless" (Ninad, 2007).

Retaliatory Feedback and Game Theory

The theoretical basis for reputation systems is game theory (Resnick & Zeckhauser, 2000). The game theoretical approach views online auction transactions as a continuous Prisoner' Dilemma game (Fudenberg & Tirole, 1991; Tullock, 1997). In this game, if both players cooperate (provide each other with positive feedback) they will both receive a payoff of +1. If both players defect (provide each other with negative feedback) they will both receive a payoff of -1. However, if one player defects (provides a negative feedback) and the other player does not (provides positive feedback) then the payoffs are -1 for the player receiving the negative feedback, and +1 for the user receiving the positive feedback. These payoffs are summarized in Table 1. Neutral scores are not considered due to their low incidence and for the sake of simplicity. Therefore, if one person defects (in this case meaning provides a negative feedback) the best option for the other transaction participant is to also to defect. This is particularly true under E-Bay's revised feedback system, where transaction participants can agree to mutually withdraw feedback. Thus, the best possible payoff over time for both participants, under Nash equilibrium, is to cooperate (provide

Table 1. Prisoner's dilemma payoffs

Player A Provides Feedback

	Positive	Negative
Positive	+1 / +1	-1 / +1
Negative	+1 / -1	-1 / -1

(left axis label: **Player B Provides Feedback**)

only positive feedback). This is in keeping with the findings of Bolton and Ockenfels (2000), who found that the second player in a Prisoner's Dilemma game will only cooperate when the first player has done likewise. With this in mind, many E-Bay participants have chosen to cooperate (never provide negative feedback) even when the circumstances dictate that they should not.

RESEARCH METHODOLOGY

We would not expect to see many instances of actual retaliation, since the fear of retaliation drives users to provide positive feedback. This tendency toward positive feedback has already been discussed above. Therefore, in order to find evidence to support the theory that users are concerned about retaliatory feedback, we monitored E-Bay's Feedback Discussion Forum during a one week period.

Data Analysis

During the period of observation there were 324 active threads. An active thread was defined a priori as either a new thread or a thread that contained a new reply. Each thread was reviewed to determine if the content was related to the retaliatory feedback problem. Of the 324 total threads, 51 or 15.7% concerned retaliatory feedback.

An example of a typical posting on these issues is I have to say this. I think it totally defeats the purpose of the feedback system if you don't leave feedback until the "other" person leaves it. In about fifty percent of my recent purchases, the sellers have said, "When I see you have left positive feedback, I will leave mine." No! No! No! I have held up my end of the bargain by paying you within 30 seconds via PayPal, if you are happy with that, then the rest of the deal is up to you. Honestly, you have nothing else to base your feedback on for me. This is not a trade; it is FEEDBACK on how the transaction went. What if I am unhappy with your end of the deal, now if I leave negative feedback I have to worry about receiving negative feedback, when I did nothing wrong. Something has got to change. Do other E-Bayers out there face this same problem?

The 51 threads of interest were further examined in order to determine the exact nature of the users' concerns. Based on this analysis, the retaliatory feedback problem can be broken into four main categories. Note, some threads fell into multiple categories resulting in the numbers below adding to more than 51.

First, 17 threads were primarily concerned with specific instances of retaliation. Many of the postings in this category discussed the problem of nonpaying winner bidders threatening to leave negative feedback for sellers who leave them a negative for nonpayment.

Second, a large number of threads (28) discussed the problem of who should leave feedback first after a completed transaction. Of course, most buyers believe that the seller should leave feedback first, and most sellers believe buyers are responsible for first feedback. The fact that both parties are waiting for the other to leave feedback first follows from game theory. If one party already knows that the other party either defected or cooperated then the best move is predefined. The impact of this waiting game is that many people do not leave feedback at all. This is in line with previous research by Resnick and Zeckhauser

(2002) who found that only 52.1% of buyers leave feedback and 60.6% of sellers.

Third, 8 threads sought advice on whether and how to leave negative feedback. A number of posts recommend maintaining two E-Bay accounts—one only to sell and the other only to buy. This allows an individual who is unhappy with a purchase to leave negative feedback for a seller without fear that it will impact his or her own seller account. Since E-Bay allows 90 days to leave feedback, the other common advice is to wait until the last minute to post negative feedback. This would not allow enough time for the other party to the transaction to retaliate. However, a number of posts warn that this approach may backfire, as E-Bay does not always lock out feedback at exactly 90 days.

Fourth, some threads (5) discussed the problem of feedback extortion. This occurs when one party to a transaction demands something from the other party and uses the threat of negative feedback as part of the demand. For example, in a recent posting a buyer complained that the seller had only shipped part of the purchase. After waiting 11 weeks for the rest of the purchase, the buyer posted negative feedback. At that point the seller told the buyer that he would not send the remainder of the purchase unless the negative feedback was withdrawn.

Retaliatory Feedback Solution

There is a very simple solution to the retaliatory feedback problem—feedback escrow. In a system with feedback escrow, parties to a transaction would be allowed to leave feedback only for a specified period of time (a few weeks perhaps). In addition, all feedback would be kept secret until both parties have left their feedback or until the time has expired. Finally, the option to mutually withdraw feedback would not exist.

In terms of game theory, feedback escrow changes the game from a Prisoner's Dilemma in which one party knows what the other party has already chosen, to one in which neither party has information about what the other has chosen. In this scenario there is essentially no first mover and thus no ability to retaliate. The elimination of mutual feedback withdrawal is an important aspect of this solution. If participants can withdraw feedback, then the optimal solution becomes defection.

It should be noted that feedback escrow addresses three of the categories of retaliation discussed above. Specific instances of retaliation would not occur. In addition, we would expect an increase in the amount of feedback and the speed with which it is left. Since feedback is kept secret until both parties have posted or until the time limit is reached, the concern about who should leave feedback first is eliminated.

The one area that feedback escrow cannot address is feedback extortion. If the system allows users to withdraw previously posted feedback, then some users will likely find ways to abuse the process.

FUTURE RESEARCH DIRECTIONS

While E-Bay is not the only auction site that uses a reputation system to provide user feedback, it is, by far, the largest. However, E-Bay has traditionally been reluctant to tinker with its feedback system. This is to be expected as millions of users have invested a considerable amount of time (and perhaps money) in building and protecting their E-Bay feedback scores. However, in 2007, it began pilot testing a revision to its feedback system, called Feedback 2.0. Under this new system auction, participants are able to leave detailed feedback. For example, a buyer will be able to rate a seller, using a five point Likert scale, on certain aspects of the transaction, such as accuracy of the listing and shipping (E-Bay, 2007). As these detailed ratings are delayed, it should prove very difficult for participants to determine exactly who left feedback. However, E-Bay has simply added detailed feedback to its original feedback system.

As Feedback 2.0 has only recently been rolled out, it provides a rich resource for future research in this area. For instance, how prevalent is the use of detailed feedback? Does the implementation of detailed feedback and the time delay increase the amount of negative feedback left on E-Bay?

CONCLUSION

The purpose of an online reputation system, such as E-Bay feedback, is to provide a level of trust between buyers and sellers. Game theory predicts, and comments from the E-Bay discussion board confirm, that retaliatory feedback likely occurs. In this situation, few users are likely to provide true feedback—guarding their own reputation against retaliation. Thus, less than one percent of all feedback left on E-Bay is negative (Resnick & Zeckhauser, 2002). If, as appears to be the case, almost everybody receives positive feedback, then the feedback system is ineffective in helping users determine with whom they should transact.

A simple solution to the retaliatory feedback problem, feedback escrow, has been proposed in this chapter. Feedback escrow alone would solve three of the four retaliation problems outlined above. Combined with the inability to retract previous feedback, the escrow system would also resolve the feedback extortion problem.

REFERENCES

Akerlof, G. A. (1970, August). The market for Lemons: Quality uncertainty and the market mechanism. *The Quarterly Journal of Economics*, *84*, 488–500. doi:10.2307/1879431

Bolton, G., & Ockenfels, A. (2000). ERC: A theory of equity, reciprocity, and cooperation. *The American Economic Review*, *90*(1), 166–193.

Brynjolfsson, E., & Smith, M. (2000). Frictionless commerce? A comparison of Internet and conventional retailers. *Management Science*, *46*(4), 563–585. doi:10.1287/mnsc.46.4.563.12061

Bunnell, D. (2000). *The E-bay phenomenon*. New York: John Wiley.

Dellarocas, C. (2003). The digitization of word of mouth: Promise and challenges of online feedback mechanisms. *Management Science*, *49*(10), 1407–1424. doi:10.1287/mnsc.49.10.1407.17308

E-Bay. (2006). E-bay form 10-K, Feb. 24, 2006. Retrieved June 14, 2008, from http://investor.E-Bay.com/secfiling.cfm?filingID=950134-06-3678

E-Bay. (2007). Detailed seller ratings. Retrieved June 14, 2008, from http://pages.E-Bay.com/help/feedback/detailed-seller-ratings.html

Fudenberg, D., & Tirole, J. (1991). *Game theory*. Cambridge: MIT Press.

Gormley, M. (2004, November 8). E-Bay sellers admit to phony bids. *Associated Press*.

Hoffman, D. L., Novak, T. P., & Peralta, M. (1999). Building consumer's trust online. *Communications of the ACM*, *42*(4), 80–86. doi:10.1145/299157.299175

Houser, D., & Wooders, J. (2001). Reputation in Internet auctions: Theory and evidence from E-Bay, Working Paper, University of Arizona. Retrieved June 14, 2008, from http://bpa.arizona.edu/~jwooders/E-Bay.pdf

Jarvenpaa, S., Tractinsky, N., & Vitale, M. (2000). Consumer trust in an Internet store. *Information Technology and Management*, *1*(1/2), 45–71. doi:10.1023/A:1019104520776

Landon, S., & Smith, C. E. (1997). The use of quality and reputation indicators by consumers: The case of Bordeaux wine. *Journal of Consumer Policy*, *20*, 289–323. doi:10.1023/A:1006830218392

Liang, T. P., & Huang, J. S. (1998). An empirical study on consumer acceptance of products in electronic markets: A transaction cost model. *Decision Support Systems, 24*, 29–43. doi:10.1016/S0167-9236(98)00061-X

Lim, K., Sia, C., Lee, M., & Benbasat, I. (2006). Do I trust you online, and if so, will I buy? An empirical study of two trust-building strategies. *Journal of Management Information Systems, 23*(2), 233–266. doi:10.2753/MIS0742-1222230210

Lucking-Reiley, D., Bryan, D., Prasad, N., & Reeves, D. (2005). *Pennies from E-Bay: The determinants of price in online auctions*. Working Paper, Vanderbilt University. Retrieved June 14, 2008, from http://www.u.arizona.edu/~dreiley/papers/PenniesFromE-Bay.pdf

McAfee, R. P., & McMillan, J. (1987). Auctions and bidding. *Journal of Economic Literature, 25*(2), 699–738.

McKnight, D. H., Choudhury, V., & Kacmar, C. (2002). Developing and validating trust measures for e-commerce: An integrative typology. *Information Systems Research, 13*(3), 334–359. doi:10.1287/isre.13.3.334.81

Montano, B. R., Porter, D. C., Malaga, R. A., & Ord, J. K. (2005). Enhanced reputation scoring for online auctions. In *Twenty-sixth International Conference on Information Systems, Proceedings.*

Nielsen, J. (1998, February 8). The reputation manager. Retrieved June 14, 2008, from http://www.useit.com/alertbox/980208.html

Ninad. (2007). *Feedback 2.0 – a conversation with Brian Burke*. Retrieved June 14, 2008, from http://www.E-Baychatter.com/the_chatter/2007/02/feedback_20_a_c.html

Resnick, P., & Zeckhauser, R. (2002). Trust among strangers in Internet transactions: Empirical analysis of E-Bay's reputation system. In M.R. Baye (Ed.), *The economics of the Internet and e-commerce* (Vol. 11 of Advances in Applied Microeconomics). Amsterdam: Elsevier Science.

Resnick, P., Zeckhauser, R., Friedman, E., & Kuwabara, K. (2000). Reputation systems: Facilitating trust in Internet interactions. *Communications of the ACM, 43*(12), 45–48. doi:10.1145/355112.355122

Tullock, G. (1997). Adam Smith and the prisoners' dilemma. In D.B. Klein (Ed.), *Reputation: Studies in the voluntary elicitation of good conduct* (pp. 21-28). Ann Arbor, MI: University of Michigan Press.

Wilson, R. (1985). Reputations in games and markets. In A. Roth (Ed.), *Game theoretic models of bargaining* (pp. 27-62). Cambridge: Cambridge University Press.

ADDITIONAL READING

Ba, S., & Pavlou, P. A. (2002). (. Evidence of the effect of trust building technology in electronic markets: Price premiums and buyer behavior. *MIS Quarterly, 26*(3), 1–26. doi:10.2307/4132332

Ba, S., Whinston, A. B., & Zhang, H. (2003). Building trust in online auction markets through an economic incentive mechanism. *Decision Support Systems, 35*, 273–286. doi:10.1016/S0167-9236(02)00074-X

Bajari, P., & Hortacsu, A. (2004). Economic insights from Internet auctions. *Journal of Economic Literature, 42*, 457–486. doi:10.1257/0022051041409075

Dellarocas, C. (2003). The digitization of word of mouth: Promise and challenges of online feedback mechanisms. *Management Science, 49*(10), 1407–1424. doi:10.1287/mnsc.49.10.1407.17308

Fan, M., Tan, Y., & Whinston, A. B. (2005). Evaluation and design of online cooperative feedback mechanisms for reputation management. *IEEE Transactions on Knowledge and Data Engineering, 17*(2), 244–254. doi:10.1109/TKDE.2005.26

Houser, D., & Wooders, J. (2001). Reputation in Internet auctions: Theory and evidence from E-Bay. Working Paper, University of Arizon. Retrieved June 14, 2008, from http://bpa.arizona.edu/~jwooders/E-Bay.pdf

Josang, A., Ismail, R., & Boyd, C. (2007). A survey of trust and reputation systems for online service provision. *Decision Support Systems*, *43*, 618–644. doi:10.1016/j.dss.2005.05.019

Lim, K., Sia, C., Lee, M., & Benbasat, I. (2006). Do I trust you online, and if so, will I buy? An empirical study of two trust-building strategies. *Journal of Management Information Systems*, *23*(2), 233–266. doi:10.2753/MIS0742-1222230210

Lucking-Reiley, D., Bryan, D., Prasad, N., & Reeves, D. (2005). Pennies from E-Bay:

Malaga, R. A. (2001). Web-based reputation management systems: Problems and suggested solutions. *Electronic Commerce Research*, *1*, 403–417. doi:10.1023/A:1011557319152

Melnik, M., & Alm, J. (2002). Does a seller's ecommerce reputation matter? Evidence from E-Bay auctions. *The Journal of Industrial Economics*, *50*, 337–349.

The determinants of price in online auctions. Working Paper, Vanderbilt University. Retrieved June 14, 2008, from http://www.u.arizona.edu/~dreiley/papers/PenniesFromE-Bay.pdf

Yu, B., & Singh, M. P. (2002). Distributed reputation management for electronic commerce. *Computational Intelligence*, *18*(4), 535–549. doi:10.1111/1467-8640.00202

This work was previously published in Information Systems Research Methods, Epistemology, and Applications, edited by Aileen Cater-Steel and Latif Al-Hakim, pp. 342-349, copyright 2009 by Information Science Reference (an imprint of IGI Global).

Chapter 17
Multiagents System Applied on a Cyberbullying Model for a Social Network

Carlos Alberto Ochoa Ortiz-Zezzatti
University of Ciudad Juárez, Mexico

Julio Cesar Ponce Gallegos
Autonomous University of Aguascalientes, Mexico

José Alberto Hernández Aguilar
Autonomous University of Morelos, Mexico

Felipe Padilla Diaz
Autonomous University of Aguascalientes, Mexico

ABSTRACT

The contribution of this chapter is to present a novel approach to explain the performance of a novel Cyberbullying model applied on a Social Network using Multiagents to improve the understanding of this social behavior. This approach will be useful to answer diverse queries after gathering general information about abusive behavior. These mistreated people will be characterized by following each one of their tracks on the Web and simulated with agents to obtain information to make decisions to improve their life's and reduce their vulnerability in different locations on a social network and to prevent its retort in others.

INTRODUCTION

Cyberbullying refers to repeated intimidation, over time, of a physical, verbal or psychological (including indirect and relational bullying) nature of a less powerful person by a more powerful person or group of persons through information and communication technologies, mediums such as mobile phone text messages, emails, phone calls, Internet chat rooms, instant messaging – and the latest trend – social networking websites such as Orkut, MySpace, Hi5, Facebook and video sharing

DOI: 10.4018/978-1-4666-2803-8.ch017

sites like YouTube. Cyberbullying is a fast growing trend that experts believe is more harmful than typical schoolyard bullying. Nearly all of us can be contacted via the Internet or our mobile phones. Victims can be reached anytime and anyplace. For many children, home is no longer a refuge from the bullies. "Children can escape threats and abuse in the classroom, only to find text messages and emails from the same tormentors when they arrive home" (Cyberbulling, 2010).

This chapter focuses on the social and cultural implications of cyber technologies. Identity, bullying and inappropriate use of communication are major issues that need to be addressed in relation to communication technologies for the security in the Web use. We present some concepts and tools of Artificial Intelligence that will be used to analyze information to know when someone can be victim of this type of crime, like Text Mining and Multiagent Systems.

First, in section 1 of this chapter, we explain the way to generate a specific behavior with this cyber bullying model on a Social Network, later we discuss social Blockade, Harassment, and Motivational Factors. In next sections, we show the Issues, Problems, and Trends of Cyberbullying (section 2), we approach different concepts related with Artificial Societies and Social Simulation using Multiagent Systems to analyze and model the necessary information to support the correct decisions for the proposed model (section 3) Social Manipulation and Coercion (section 4), and Social exclusion and intimidation (section 5). In section 6 we apply this Model in a social network over the Internet: Orkut (a popular social network in Brazil) and try to explain innovative perspectives of this Model. In section 7 we analyze a case of study: Social Blockade in a Social Networking based on Memory Alpha with 1000 societies represented by agents to demonstrate the concept of cyber bulling. In Section 8, we present a summary of the best practices to avoid cyber bulling by performing automatic text mining over top related Web sites.

Finally, in section 9 we provide our conclusions and our future research on this novel topic.

SOCIAL BLOCKADE AND HARASSMENT

There are different mechanisms of bulling a classmate. Bill Belsey, former president of www.bullying.org and www.cyberbullying.ca, considers next types of cyberbullying: by means of e-mail, cellular phone using text or multimedia messages: by means of instant messages, defamatory web blogs and personal Webs. According to Fante (2005) young people use web blogs, social networks and systems of instantaneous mail to intimidate their classmates, being the diffusion of altered photographs to ridicule to victims one of the preferred methods. These are distributed massively and sometimes indicating the identity of who is put under the humiliation to increase the impact. In the case of the virtual communities, many of them need an invitation to be able to enter to be member of a group, the harassment at school is based on isolating the victims of the humiliations.

Cyberbullying is a form of scholar harassment produced by means of virtual platforms and technological tools, like chats, blogs, fotologs, text messages of cellular phones, e-mail, forums, servers storing videos or photographs, telephone and other electronic media (Alcantara, 2007).

The Cyberbullying being an indirect and non actual form of harassment, the aggressor does not have contact with the victim, does not see his/her face, eyes, pain, suffering, consequently hardly the aggressor will be able to empathize or to wake up compassion on the other. The cyber aggressor obtains satisfaction in the elaboration of the violent act and in the imagination of the damage caused in the other, since is not possible to live it in situ (Hernandez, 2007).

Social Networks

Social network describes a group of social entities and the pattern of inter-relationships among them. What the relationship means varies, from those of social nature, such as values, visions, ideas, financial exchange, friendship, dislike, conflict, trade, kinship or friendship among people, to that of a transactional nature, such as trading relationship between countries. Despite the variability in semantics, social networks share a common structure in which social entities, generically termed actors, are inter-linked through units of relationship between a pair of actors known as: tie, link, or pair. By representing actors as nodes and ties as edges, social network can be represented as a graph (Ponce et al, 2009).

A social network is a social structure made of nodes (which are generally individuals or organizations) that are tied by one or more specific types of interdependency (See Figure 1).

Social Blockade

Also denominated social isolation, in which one the young person is not allowed to participate, ignoring his/her presence and avoid him/her in

Figure 1. Social network diagram

the normal activities between friends and class-mates (Alcantara, 2009). Sometimes, mainly at the beginning, the majority of the students that compose the class know the Web on which is making fun of a classmate, and, the victim does not even know it, although the purpose is that sooner or later this one has to find out. In the Web, also voting systems can also be included to choose to the ugliest, to the most idiotic, to the weakest, or contributing another way to stigmatise and to humiliate to the classmates.

Despite being able to denounce one harmful Web page, there is still no guarantee a new Web page won't arise shortly after. This is all taking place before the victim has even had a chance to recover from the social and psychological effects of the previous page.

Harassment

Bullying is a present reality in our scholar centers that contaminates coexistence, producing negative effects not only in those implied directly, but in the totality of the students and teaching staff. The scholar climate is deteriorated seriously, until the point, that for many students to go daily to the center, supposes a torture. All of this, together with the paper sensationalism that the mass media grant to this problem, contributes to untie the general anguish in parents, who until recently considered the schools like places of peace, security and well-being for the mature development of their children (Hernandez, 2007). According to this author there are two modalities of Cyber Bullying: One that acts as reinforcer of an already started bullying, and that way of harassment among equals by means of Information Technologies without antecedents.

For the first modality, cyber bullying is a more sophisticated way of harassment developed, generally, when traditional ways of harassment becomes unattractive or unsatisfactory. In this case the aggressor is easily identifiable, since it coincides with the actual bullier. The effects of this type of cyber bullying are additive to those the victim already suffers, but also amplify and increase the damages, given to the world-wide opening and generalization of the harassment through the pages Web.

Regarding to the second modality, they are forms of harassment between equals that do not present antecedents, so that without reason it pretends the boy begins to receive forms of harassment through the ITs. Sometimes, after a time to receive this type of harassment, the cyber aggressor decides to complete its work with an actual experience, giving the face. This type of harassment at network presents characteristics of similarity with other forms of harassment, like the fact of being premeditated and a highly deliberated violent conduct or harassment; that based on an asymmetric relation of control and power on the other, but with particular characteristics that differentiate it from other forms of harassment actually and directly:

- It requires of the use and domain of ITs.
- It is a form of indirect harassment.
- It is an act of violence.
- It is an act of camouflaged violence, on which the aggressor is complete unknown, unless he has been harassing actual to the victim before or who decides to be it after the cyber bullying.
- The ignorance of the aggressor superb the impotence feeling.
- It picks up diverse types or forms to indicate the harassment through the TIC's.
- Legal neglect of these forms of harassment, since although it is possible to close the Web, immediately can be opened another one.
- The harassment invades scopes of privacy and pretended security as it is the home, developing a feeling of total lack of protection.
- Harassment becomes public, is opened to more persons rapidly.

Sexual Harassment

On the other hand, sexual harassment also finds in Internet other forms of expression that accompany the actual harassment, for example: continue sending of threatening electronic mails, or the design of a Web page simulating that one woman is a sexual worker that offer her services and that includes her personal data (name, telephone, email, direction, etc.).

Profile of Victims

According to data of (Protégeles, 2009), most of the situations of cyber bullying took place between the 13 and 14 years, a 52% of the total; whereas a 10% of the cases took place before turning the ten years of age. Besides, most of the harassments were committed in the last quarterly of the course; a 35% of April to June, a 32% of October to December, a 13% of January to March and a 10% of July to September.

In the majority of the known cases the pursuers are women (19%) against a 10% of men. In 71% of the cases the gender of the aggressor was not determined.

In 60% of the cases, the girls are those that undergo harassment. The victims undergo cyber bullying through forums or social networks and programs of instantaneous mail in a 45%. In addition, 12% go through electronic mail and 19% through cellular phone. According this study, there are minors who undergo harassment by several of these means.

Finally, this study reflects that in almost half of the reviewed cases the situation of harassment takes place almost on a daily basis and that in 80% of the cases the parents finish knowing about the situation, and at the school in 74%.

Motivational Factors Associated with Cyberbullying

Bullying cannot be considered as part of the normal and common adolescent development, several studies have related it to emotional and behavioral problems (Rodríguez et al., 2006).

As part of an ongoing, statewide bullying prevention initiative in Colorado, 3,339 youth in Grades 5, 8, and 11 completed questionnaires in 78 school sites during the fall of 2005, and another 2,293 youth in that original sample participated in a follow-up survey in 65 school sites in the spring of 2006. Questionnaires included measures of bullying perpetration and victimization, normative beliefs about bullying, perceptions of peer social support, and perceptions of school climate. The highest prevalence rates were found for verbal, followed by physical, and then by Internet bullying. Physical and Internet bullying peaked in middle school and declined in high school. Verbal bullying peaked in middle school and remained relatively high during high school. Males were more likely to report physical bullying than females, but no gender differences were found for Internet and verbal bullying. All three types (verbal bullying, physical bullying and Internet Bullying) were significantly related to normative beliefs approving of bullying, negative school climate, and negative peer support (Williams & Guerra, 2007). This calls attention to the existence of little monitoring and consequently the few applied corrective measures in the educative centers and at home (Cabezas, 2007).

In the case of women, in a sample from Costa Rica (see (Cabezas, 2007)) it seems that confronting is the main reason for bullying ("to avoid menaces", "to avoid harassment", "she has something of mine", "she criticizes me", "they bother me", among others). In the case of males, motivation is diverse considering intention and pleasure to do it, they provide answers like: "to feel the adrenaline", "just to do it", "nothing special". Bullies have a great necessity to self impose

and dominate (Olweus, 1998). Those who bully repeatedly engage in conflicts that they are sure to win because of their superior power, and those who bully are merciless in their tactics (Bitney & Tile, 1997).

Situations like the above mentioned require paying special attention due to the fact that they could develop an antisocial disorder of personality that generates: A general pattern of scorn and violation of the rights of the others, that starts in childhood or at the beginning of the adolescence stage and continues during the adult age" (DSM-IV, 1995, p. 662).

Grouping phenomena is an answer to the established socio-cultural mechanisms that takes to the social contagion, to a weakening of the control, to the division of the responsibility, besides to graduated cognitive changes of the perception of the bullier and the victim, which simultaneously explains, why there are students who are not aggressive but also participate in passive form of the harassment (Olweus, 1998).

ISSUES, PROBLEMS, AND TRENDS OF CYBERBULLYING

Importance of Cyberbullying

In mathematical sociology, interpersonal ties are defined as information-carrying connections between people. Interpersonal ties, generally, come in three varieties: strong, weak, or absent. Weak social ties, it is argued, are responsible for the majority of the embeddedness and structure of social networks in society as well as the transmission of information through these networks. Specifically, more novel information flows to individuals through weak rather than strong ties. Because our close friends tend to move in the same circles that we do, the information they receive overlaps considerably with what we already know. Acquaintances, by contrast, know people that we do not, and thus receive more novel information.

Included in the definition of absent ties, according to Granovetter, are those relationships (or ties) without substantial significance, such as "nodding" relationships between people living on the same street, or the "tie", for example, to a frequent vendor one would buy from. Furthermore, the fact that two people may know each other by name does not necessarily qualify the existence of a weak tie. If their interaction is negligible the tie may be absent. The "strength" of an interpersonal tie is a linear combination of the amount of time, the emotional intensity, the intimacy (or mutual confiding), and the reciprocal services which characterize each tie.

In Granovetter's view, a similar combination of strong and weak bonds holds the members of society together. This model became the basis of his first manuscript on the importance of weak social ties in human life. He submitted his paper to the American Sociological Review in 1969, but it was rejected. Nevertheless, in 1972, Granovetter submitted a shortened version to the American Journal of Sociology, and it was finally published in May 1973. According to Current Contents, by 1986, the Weak Ties paper had become a citation classic, being one of the most cited papers in sociology.

In a related line of research, in 1969 anthropologist Bruce Kapferer, published "Norms and the Manipulation of Relationships in a Work Context" after doing field work in Africa. In the document, he postulated the existence of multiplex ties, characterized by multiple contexts in a relationship. In telecommunications, a multiplexer is a device that allows a transmission medium to carry a number of separate signals. In social relations, by extrapolation, "multiplexity" is the overlap of roles, exchanges, or affiliations in a social relationship (Granovetter, 1973)..

Problems of Cyberbullying

Risk perception is the subjective judgment that people make about the characteristics and severity of a risk. Several theories have been proposed to explain why different people make different estimates of the dangerousness of risks. Three major families of theory have been developed: psychology approaches (heuristics and cognitive), anthropology/sociology approaches (cultural theory) and interdisciplinary approaches (social amplification of risk framework).

Early Theories

The study of risk perception arose out of the observation that experts and lay people often disagreed about how risky various technologies and natural hazards were.

The mid 1960's saw the rapid rise of nuclear technologies and the promise for clean and safe energy. However, public perception shifted against this new technology. Fears of both longitudinal dangers to the environment as well as immediate disasters creating radioactive wastelands turned the public against this new technology. The scientific and governmental communities asked why public perception was against the use of nuclear energy when all of the scientific experts were declaring how safe it really was. The problem, from the perspectives of the experts, was a difference between scientific facts and an exaggerated public perception of the dangers.

A key early paper was written in 1969 by Chauncey Starr. He used a revealed preference approach to find out what risks are considered acceptable by society. He assumed that society had reached equilibrium in its judgment of risks, so whatever risk levels actually existed in society were acceptable. His major finding was that people will accept risks 1,000 greater if they are voluntary (e.g. driving a car) than if they are involuntary (e.g. a nuclear disaster).

This early approach assumed that individuals behave in a rational manner, weighing information before making a decision. Individuals have exaggerated fears due to inadequate or incorrect information. Implied in this assumption is that additional information can help people understand true risk and hence lessen their opinion of danger. While researchers in the engineering school did pioneer research in risk perception, by adapting theories from economics, it has little use in a practical setting. Numerous studies have rejected the belief that additional information, alone, will shift perceptions (Starr C. et al., 1969).

Psychology Approach

The psychology approach began with research in trying to understand how people process information. These early works maintain that people use cognitive heuristics in sorting and simplifying information which lead to biases in comprehension. Later work built on this foundation and became the psychometric paradigm. This approach identifies numerous factors responsible for influencing individual perceptions of risk, including dread, newness, stigma, and other factors (Starr C. et al., 1969).

Heuristics and Biases

The earliest psychometric research was done by psychologists Daniel Kahneman and Amos Tversky, who performed a series of gambling experiments to see how people evaluated probabilities. Their major finding was that people use a number of heuristics to evaluate information. These heuristics are usually useful shortcuts for thinking, but they may lead to inaccurate judgments in some situations – in which case they become cognitive biases.

- **Availability Heuristic:** Events that can be more easily brought to mind or imagined are judged to be more likely than events that could not easily be imagined.

- **Anchoring Heuristic:** People will often start with one piece of known information and then adjust it to create an estimate of an unknown risk – but the adjustment will usually not be big enough.

- **Asymmetry between Gains and Losses:** People are risk averse with respect to gains, preferring a sure thing over a gamble with a higher expected utility but which presents the possibility of getting nothing. On the other hand, people will be risk-seeking about losses, preferring to hope for the chance of losing nothing rather than taking a sure, but smaller, loss (e.g. insurance).

- **Threshold Effects:** People prefer to move from uncertainty to certainty over making a similar gain in certainty that does not lead to full certainty. For example, most people would choose a vaccine that reduces the incidence of disease A from 10% to 0% over one that reduces the incidence of disease B from 20% to 10%.

Another key finding was that the experts are not necessarily any better at estimating probabilities than lay people. Experts were often overconfident in the exactness of their estimates, and put too much stock in small samples of data.

PSYCHOMETRIC PARADIGM

Research within the psychometric paradigm turned to focus on the roles of affect, emotion, and stigma in influencing risk perception. Melissa Finucane and Paul Slovic have been among the key researchers here. These researchers first challenged Starr's article by examining expressed preference – how much risk people say they are willing to accept. They found that, contrary to Starr's basic assumption, people generally saw most risks in society as being unacceptably high. They also found that the gap between voluntary and involuntary risks was not nearly as great as Starr claimed.

Slovic and team found that perceived risk is quantifiable and predictable. People tend to view current risk levels as unacceptably high for most activities. All things being equal, the greater people perceived a benefit, the greater the tolerance for a risk. If a person derived pleasure from using a product, people tended to judge its benefits as high and its risks as low. If the activity was disliked, the judgments were opposite. Research in psychometrics has proven that risk perception is highly dependent on intuition, experiential thinking, and emotions.

Psychometric research identified a broad domain of characteristics that may be condensed into three high order factors: (1) the degree to which a risk is understood, (2) the degree to which it evokes a feeling of dread, and (3) the number of people exposed to the risk. A dread risk elicits visceral feelings of terror, uncontrollable, catastrophe, inequality, and uncontrolled. An unknown risk is new and unknown to science. The more a person dreads an activity, the higher its perceived risk and the more that person wants the risk reduced.

ANTHROPOLOGY/ SOCIOLOGY APPROACH

The anthropology/sociology approach posits risk perceptions as produced by and supporting social institutions. In this view, perceptions are socially constructed by institutions, cultural values, and ways of life.

CULTURAL THEORY

"The Cultural Theory of risk" (with capital C and T). Cultural Theory is based on the work of anthropologist Mary Douglas and political scientist Aaron Wildavsky first published in 1982.

In Cultural Theory, Douglas and Wildavsky outline four "ways of life" in a grid/group arrangement. Each way of life corresponds to a specific social structure and a particular outlook on risk. Grid categorizes the degree to which people are constrained and circumscribed in their social role. The tighter binding of social constraints limits individual negotiation. Group refers to the extent to which individuals are bounded by feelings of belonging or solidarity. The greater the bonds, the less individual choice are subject to personal control. Four ways of life include: Hierarchical, Individualist, Egalitarian, and Fatalist.

Risk perception researchers have not widely accepted Cultural Theory. Even Douglas says that the theory is controversial; it poses a danger of moving out of the favored paradigm of individual rational choice of which many researchers are comfortable.

INTERDISCIPLINARY APPROACH

Social Amplification of Risk Framework

The Social Amplification of Risk Framework (SARF), combines research in psychology, sociology, anthropology, and communications theory. SARF outlines how communications of risk events pass from the sender through intermediate stations to a receiver and in the process serve to amplify or attenuate perceptions of risk. All links in the communication chain, individuals, groups, media, etc., contain filters through which information is sorted and understood.

The theory attempts to explain the process by which risks are amplified, receiving public attention, or attuned, receiving less public attention. The theory may be used to compare responses from different groups in a single event, or analyze the same risk issue in multiple events. In a single risk event, some groups may amplify their perception of risks while other groups may attune, decrease, and their perceptions of risk (Valdez et al., 2009).

The main thesis of SARF states that risk events interact with individual psychological, social and other cultural factors in ways that either increase or decrease public perceptions of risk. Behaviors of individuals and groups then generate secondary social or economic impacts while also increasing or decreasing the physical risk itself.

These ripple effects caused by the amplification of risk include enduring mental perceptions, impacts on business sales, and change in residential property values, changes in training and education, or social disorder. These secondary changes are perceived and reacted to by individuals and groups resulting in third-order impacts. As each higher-order impacts are reacted to, they may ripple to other parties and locations. Traditional risk analyses neglect these ripple effect impacts and thus greatly underestimate the adverse effects from certain risk events. Public distortion of risk signals provides a corrective mechanism by which society assesses a fuller determination of the risk and its impacts to such things not traditionally factored into a risk analysis.

Trends of Cyberbullying

The main explanation for social loafing is that people feel unmotivated when working with a team, because they think that their contributions will not be evaluated or considered.

According to the results of a meta-analysis study, social loafing is a pervasive phenomenon, but it does not occur when team members feel that the task or the team itself is important. It can occur when the person feels underappreciated within their team or group.

Social loafing occurs in a group situation in which the presence of others causes relaxation instead of arousal. When individuals relax their performance, they are able to fade into the crowd, which is especially appealing to people when they know they are not going to be accountable for their actions or performance. In easier, less demanding tasks, such as singing happy birthday or giving applause, one is likely to exert less effort due to the concept of diffusion of responsibility. This occurs when people think that they can "get a free ride" because someone else will surely pick up the slack.

Social loafing is associated with poor performance on easy tasks. However, people tend to exert more effort on challenging or rewarding tasks. If a group is completing a task for some kind of reward, such as money or a good grade, then members are more likely to try harder. Generally, a greater reward results in more motivation to perform well, and therefore, more effort. People will also work harder when they feel their particular tasks or efforts are indispensable to the group's success (Karau & Williams, 1993).

ARTIFICIAL SOCIETIES AND SOCIAL SIMULATION USING MULTIAGENT SYSTEMS

Social simulation is the modeling or simulation, usually with a computer, of a social phenomena (e.g., cooperation, competition, markets, social networks dynamics, among others). A subset within social simulations is Agent Based Social Simulations (ABSS) which are an amalgam of computer simulations, agent-based modeling, and social sciences.

History and Development

The history of the agent based model can be traced back to Von Neumann, with his theoretical machine capable of reproduction. The device that Von Neumann proposed would follow detailed instructions to build a copy of it. The concept was then improved by Von Neumann's friend Stanislaw Ulam, also a mathematician, whom suggested that the machine were built on paper, as a collection of cells on a grid. The idea intrigued Von Neumann, who drew it up—creating the first of the devices later named as cellular automata. Another improvement was brought by mathematician (Conway, 2004). He constructed the well-known Game of Life. Unlike the Von Neumann's machine, Conway's Game of Life operated by tremendously simple rules in a virtual world in the form of a 2-dimensional checkerboard.

The birth of agent based model as a model for social systems was primarily brought by computer scientist (Reynolds, 1987). He tried to model the reality of lively biological agents, known as the artificial life, developed the first large-scale agent model, the Sugarscape, to simulate and explore the role of social phenomena such as seasonal migrations, pollution, sexual reproduction, combat, transmission of diseases, and even culture, more recently. Ron Sun developed methods for basing agent based simulation on models of human cognition, known as cognitive social simulation.

Multiagent Systems

The goal of multiagent systems research is to find methods that allow us to build complex systems composed of autonomous agents who, while operating on local knowledge and possessing only limited abilities, are nonetheless capable of enacting the desired global behaviors. We want to know how to take a description of what a system of agents should do and break it down into individual agent behaviors. At its most ambitious, multiagent systems aims at reverse-engineering emergent phenomena as typified by ant colonies. Multiagent systems approaches the problem using the well proven tools from game theory, Economics, and Biology. It supplements these with ideas and algorithms from artificial intelligence research, namely planning, reasoning methods, search methods, and machine learning (Vidal, 2007).

Types of Simulation and Modeling

Social simulation can refer to a general class of strategies for understanding social dynamics using computers to simulate social systems. Social simulation allows for a more systematic way of viewing the possibilities of outcomes.

The history of the agent based model can be traced back to Von Neumann, with his theoretical machine capable of reproduction. The device that Von Neumann proposed would follow detailed instructions to build a copy of it. The concept was then improved by Von Neumann's friend Stanislaw Ulam, also a mathematician, whom suggested that the machine were built on paper, as a collection of cells on a grid. The idea intrigued Von Neumann, who drew it up—creating the first of the devices later named as cellular automata. Another improvement was brought by mathematician (Conway, 2004). He constructed the well-known Game of Life. Unlike the Von Neumann's machine, Conway's Game of Life operated by tremendously simple rules in a virtual world in the form of a 2-dimensional checkerboard.

The birth of agent based model as a model for social systems was primarily brought by computer scientist (Reynolds, 1987). He tried to model the reality of lively biological agents, known as the artificial life, developed the first large-scale agent model, the Sugarscape, to simulate and explore the role of social phenomena such as seasonal migrations, pollution, sexual reproduction, combat, transmission of diseases, and even culture, more recently. Ron Sun developed methods for basing agent based simulation on models of human cognition, known as cognitive social simulation.

There are four major types of social simulation:

1. System level simulation
2. Agent based simulation
3. System level modeling
4. Agent based modeling

A social simulation may fall within the rubric of computational sociology which is a recently developed branch of sociology that uses computation to analyze social phenomena. The basic premise of computational sociology is to take advantage of computer simulations in the construction of social theories. It involves the understanding of social agents, the interaction among these agents, and the effect of these interactions on the social aggregate. Although the subject matter and methodologies in social science differ from those in natural science or computer science, several of the approaches used in contemporary social simulation originated from fields such as physics and artificial intelligence.

System Level Simulation

System Level Simulation (SLS) is the oldest level of social simulation. System level simulation looks at the situation as a whole. This theoretical outlook on social situations uses a wide range of information to determine what should happen to society and its members if certain variables are present. Therefore, with specific variables presented, society and its members should have a certain response to the situation. Navigating through this theoretical simulation will allow researchers to develop educated ideas of what will happen under some specific variables.

AGENT BASED SIMULATION

Agent Based Social Simulation (ABSS) consists of modeling different societies after artificial agents, (varying on scale) and placing them in a computer simulated society to observe the behaviors of the agents. From this data it is possible to learn about the reactions of the artificial agents and translate them into the results of non-artificial agents and simulations. Three main fields in ABSS are agent based computing, social science, and computer simulation. Agent based computing is the design of the model and agents, while the computer simulation is the simulation of the agents in the model as well as the outcomes. The social science is a mixture of sciences and social part of the model. It is the social phenomena that is developed and theorized. The main purpose of ABSS is to provide models and tools for agent based simulation of social phenomena. With ABSS which can explore different outcomes for phenomena where it might not be able to view the outcome in real life. It can provide us valuable information on society and the outcomes of social events or phenomena.

SYSTEM LEVEL MODELING

System Level Modeling (SLM) aims to specifically (unlike system level simulation's generalization in prediction) predict and convey any number of actions, behaviors, or other theoretical possibilities of nearly any person, object, construct, etcetera within a system using a large set of mathematical equations and computer programming in the form of models. A model is a representation of a specific thing ranging from objects and people to structures and products created through mathematical equations and are designed, using computers, in such a way that they are able to stand-in as the aforementioned things in a study. Models can be either simplistic or complex, depending on the need for either; however, models are intended to be simpler than what they are representing while remaining realistically similar in order to be used accurately. They are built using a collection of data that is translated into computing languages that allow them to represent the system in question. These models, much like simulations, are used to help us better understand specific roles and actions of different things so as to predict behavior and the like.

AGENT BASED MODELING

Agent Based Modeling (ABM) is a system in which a collection of agents independently interact on networks. Each individual agent is responsible for different behaviors that result in collective behaviors. These behaviors as a whole help to define the workings of the social networking. ABM focuses on human social interactions and how people work together and communicate with one another without having one, single "group mind". This essentially means that it tends to focus on the consequences of interactions between people (the agents) in a population. Researchers are better able to understand this type of modeling by modeling these dynamics on a smaller, more localized level. Essentially, ABM helps to better understand interactions between people (agents) who, in turn, influence one another (in response to these influences). Simple individual rules or actions can result in coherent group behavior. Changes in these individual acts can affect the collective group in any given population.

Agent-based modeling is simply just an experimental tool for theoretical research. It enables one to deal with more complex individual behaviors, such as adaptation. Overall, through this type of modeling, the creator, or researcher, aims to model behavior of agents and the communication between them in order to better understand how these individual interactions impact an entire population. In essence, ABM is a way of modeling and understanding different global patterns.

CURRENT RESEARCH

Agent based modeling is most useful in providing a bridge between micro and macro levels, which is a large part of what sociology studies. Agent based models are most appropriate for studying processes that lack central coordination, including the emergence of institutions that, once established, impose order from the top down. The

models focus on how simple and predictable local interactions generate familiar but highly detailed global patterns, such as emergence of norms and participation of collective action. (Macy & Willer, 2002) researched a recent survey of applications and found that there were two main problems with agent based modeling the self-organization of social structure and the emergence of social order. Below is a brief description of each problem:

1. **Emergent Structure:** In these models, agents change location or behavior in response to social influences or selection pressures. Agents may start out undifferentiated and then change location or behavior so as to avoid becoming different or isolated (or in some cases, overcrowded). Rather than producing homogeneity, however, these conformist decisions aggregate to produce global patterns of cultural differentiation, stratification, and hemophiliac clustering in local networks. Other studies reverse the process, starting with a heterogeneous population and ending in convergence: the coordination, diffusion, and sudden collapse of norms, conventions, innovations, and technological standards.

2. **Emergent Social Order:** These studies show how egoistic adaptation can lead to successful collective action without either altruism or global (top down) imposition of control. A key finding across numerous studies is that the viability of trust, cooperation, and collective action depends decisively on the embeddings of interaction.

These examples simply show the complexity of our environment and that agent based models are designed to explore the minimal conditions, the simplest set of assumptions about human behavior, required for a given social phenomenon to emerge at a higher level of organization (see Figure 2).

Researchers working in social simulation might respond that the competing theories from the

Figure 2. Social simulation using a dyoram to represent a social networking

social sciences are far simpler than those achieved through simulation and therefore suffer the afore-mentioned drawbacks much more strongly. Theories in social science tend to be linear models that are not dynamic and which are inferred from small laboratory experiments. The behavior of populations of agents under these models is rarely tested or verified against empirical observation.

As we will see in next section, Social Blockade is a process that involves the skills of a number of individuals. The creation of Social Blockade begins with a concept, an argumentation and a situation. For its social modeling, a dyoram is proposed to display this behavior inside a social networking, that is to say, in this sequence of scenarios, each one of elements who conforms the network according to their roll and society's parameters that they have within the same one. The development of this behavior inside a social networking requires of the concept and diverse similar measures that allow actions of a separated part of a society about a sustained concept in the data. But it is necessary to prioritize the conceptual development and of the categories of the system that conforms the social model, and think simultaneously about the mathematical model. We proposed the next model to represent this behavior (See Figure 3).

Social Manipulation and Coercion

Cyberbullying is a form of violence, social isolation; public humiliation and malicious gossip have long been the stock in trade of bullies. With the advent of modern communications the bully's reach and powers of social manipulation have been increased exponentially, for intimidate an individual in the attempt to gain power and control over them. The new technologies are advancing more and more every day. People can never know what will be the technologies in the future - a few years ago, in the past we carried a phone with a big size and now – there are small phones with tons of features. Who would have thought? And of course, as the world advances, the bullies do not lag behind and some people said that something needs to be done against cyber-bullies. I would have never thought that it would get to this, but it did. Now, the worst part about this new technique is that it doesn't stop or start with school. You can stay at home and be cyber-bullied from a distance. This activity includes - but is not limited to - threats, intimidation harassment and of course unauthorized publication of personal info in different mediums. So now, not just the whole school will know who the wimps are, but the whole web. This is a lot worse than normal bullying. In any case, there is a good part about

Figure 3. Social blockade model to separate a society of an organized cluster

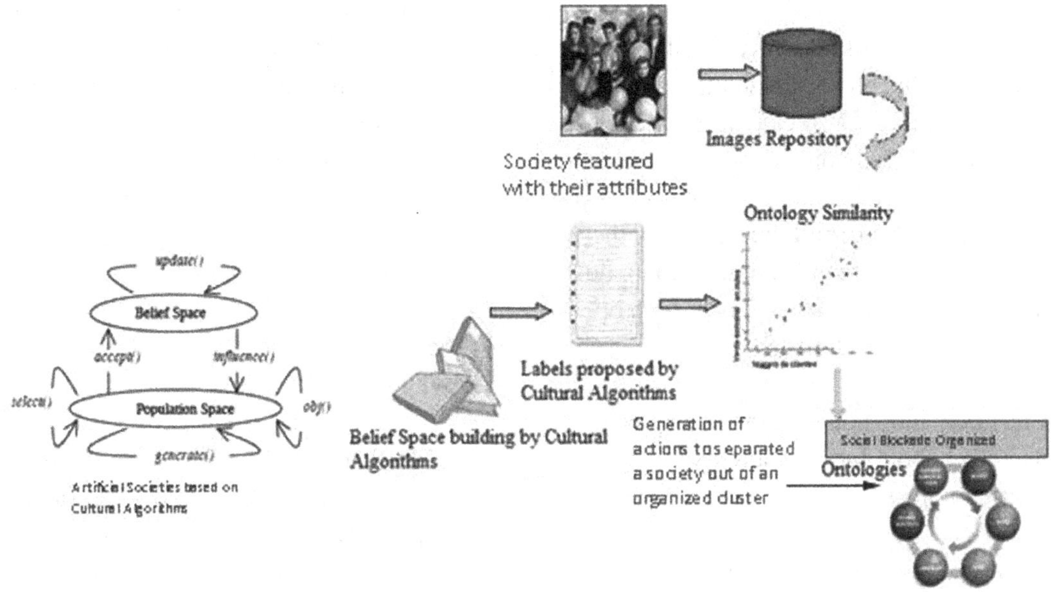

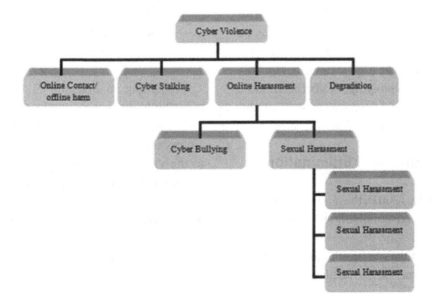

this. It will lead to children getting informed on how to stay safe on the web. Some types or cyber violence are showed in Figure 4.

"Cyberbullying is the use of modern communication technologies to embarrass, humiliate, threaten, or intimidate an individual in the attempt to gain power and control over them." (Stutzky, 2010)

Parents are well advised to pay close attention to how this new threat can impact their children. We have already seen too many cases of children subjected to a cyberbullying attack who have been

Figure 4. Overview of cyber violence

so traumatized that they have committed suicide as a direct result.

As the trial in the first-ever cyberbullying suicide case began, the prosecutors in the case said the federal court that the suicide of the Missouri teen could have been avoided had she not been tormented online by the woman.

Some signs of cyberbullying (none of these by themselves are a sure indication of being bullied, but several taken together are cause to open a discussion) are the following:

- Unusually long hours on the computer
- Secretive about Internet activity
- Lack of appetite
- Fear
- Closes windows when parents enter his/her room
- Getting behind in school work
- Stomachache

No type of bullying is harmless. In extreme incidents, cyber bullying has led teenagers to suicide. Most victims, however, suffer shame, embarrassment, anger, depression and withdrawal. Cyber bullying is often seen as anonymous, and the nature of the Internet allows it to spread quickly to hundreds and thousands of people. In some cases, it can constitute criminal behavior. The inappropriate use of the Internet, telephones or text messages is a criminal offense. "Bullycide" is the term that has been used to describe suicides caused by relentless bullying. "Cyberbullycide", to coin a phrase, would describe someone driven to suicide following a cyber bullying attack.

Social Exclusion and Intimidation

According to Silva (2001), the intimidation forms that can be presented on bullying are the following:

1. Verbal intimidations (insults, motes, to speak badly of somebody, to seed rumors).

2. Psychological intimidations (threats to provoke fear, to obtain some object or money, or simply to force the victim to make things). The same author refers regarding social isolation as:

3. Social isolation, preventing the young person to participate, ignoring its presence and not having him in the normal activities between friends and classmates.

According (Aviles, 2006) main types of abusive behavior can be classified on:

- **Physical:** like pushes, kicks, and aggressions with objects. This type of mistreat occurs with more frequency in the primary school than in the secondary school.
- **Verbal:** here appear the insults and motes mainly. Contempts in public or being emphasizing and mocking on frequent basis of a physical defect or action are frequent.
- **Psychological:** are directed actions to mine the self-esteem of the individual and to foment on his/her a sensation of insecurity and fear. The psychological component can be found in all the forms of mistreat.
- **Social:** they pretend to separately locate to the individual with respect to the group in a bad status and sometimes to make other individuals contributors of this action. These actions are considered indirect bullying.

According to Faustino & Oliveira (2008), cyber bullying, in the virtual community, is characterized, not only for verbal and psychological aggression, besides with diverse insults directed to the victim, as well as the nicknames that are conferred to this person. And also:

- Threats, as in: "I'm waiting to see her in the street to kick her", "Lets go together everybody and give her a …. in the street" and "ahhh and also in her living room… smart loafer… that I will finish her any day….";

- Calumny and defamation. A topic was created by a boy who said that B. had taken a naked photo of a girl in a barbecue, but, when they had asked for showed it, a test of what he was saying, he contradicted the fact to have seen her completely naked and placed doubts in relation to the quality of the photo: "It was not a very clear one, but she wears white panties and with half of it inside her" …
- Provoking or empowering to commit suicide.

Application of the Proposed Social Model on Orkut (A Popular Social Network in Brazil)

Orkut is a social networks used in Brazil by 13 million users, many create more of a profile, and generate different relationships, to be able to establish communications with people of different life styles, and they doing to believe other users that they are different people.

Nowadays it is demonstrated the feasibility of large-scale, passive deanonymization of real world social networks (Narayanan & Shmatikov, 2009). See Figure 5.

Some problems in Orkut that dislike much to their users are: the loss of privacy, the lack of materialization of the relations established through the network. False profiles are created to: make a joke, harass other users, or see who visualizes its profile. Now the users can make denunciations against those false profiles, but without a clear profile of Orkut, the original author (aggressor) can create a new profile.

A serious event in Orkut is the creation of communities with concepts of racism, xenophobia, sales of drugs and paedophilia. Unfortunately, the users who denounce these facts to be eliminated

Figure 5. Traffic on Orkut by country (http://en.wikipedia.org/wiki/Orkut)

Traffic of Orkut on March 31, 2004			Traffic of Orkut on May 13, 2009		
	United States	51.56%		Brasil	50%
	Japan	7.76%		India	15%
	Brasil	5.16%		United States	8.9%
	Netherlands	4.10%		Japan	8.8%
	United Kingdom	3.72%		Pakistan	6.9%
	Other	27.92%		Other	29.6%

do not reach their objective: the criminals create these data again. Actuality there are a controversy around the use of Orkut by various groups of hate. Some fanatic users, racists and religious, allegedly have a solid following there.

In Orkut there are several hate communities focused on racism, Nazism, and other kinds of specific hate. In 2005, various cases of racism were brought to police attention and reported on in the Brazilian media. According (Racism in Brazil, 2009) in 2006, a judicial measure was opened by the Brazil federal justice denouncing a 20-year-old student accused of racism against those of Black African ancestry and spreading defamatory content on Orkut. In (*Racismo na Internet*, 2006) the Brazilian Federal Justice subpoenaed Google on March 2006 to explain the crimes that had occurred in Orkut (Ministerio, 2006).

The next numbers about some crimes in Orkut was finding in: (http://www.theregister.co.uk/2008/04/12/google_brazil_pledge/, last quest December 2009). To others Statistics and Tips of cyberbullying in general see I-SAFE website.

- Google is to give Brazilian police access to 3,261 private photo albums on social networking website Orkut, which may contain child pornography.
- Orkut has more than 60 million users, most of them in Brazil. The federal prosecutor for Sao Paulo, in the last two years nearly 90 per cent of the 56,000 complaints in Brazil about net-based paedophilia were linked to the website.
- Until September 2007, all requests for information about users suspected of crimes such as racism or paedophilia were sent to Google US to be examined.
- The private photo albums, considered by some as a safe haven for criminals, were introduced in November 2007, and allow users to block access to the pictures by anyone outside their direct network.

Using the tool of Data Mining denominated WEKA, it was come to develop a denominated Model "Ahankara" of prediction of profiles in users of Orkut, which allows to understand the motivations of this type of profile and determine if the user has a generated Bipolar Syndrome (Ponce et al., 2009).

SOCIAL BLOCKADE IN A SOCIAL NETWORKING REPRESENTED BY AGENTS TO DEMONSTRATE THE CONCEPT OF CYBERBULLYING

A social networking is a social structure that can be represented making use of different types of diagrams. Both more common types are the graph and dyorams. The graph is a collection of objects called vertex or nodes that are connected by lines, edges or arcs. The nodes represent individuals which sometimes are denominated actors and the edges represent the relations that exist between these actors. Consisting of one or more graphs where these graphs conceptualize the network, which is to say, the representation is made mainly on the basis of the relations that exist between the actors who conform the network. In this research, we focused our attention in a practical problem of the Literature related to Modeling of Societies, the social blockade of a society organized for several societies, which allows to understand the behavior that keeps a society with respect to others, the capacity to establish this behavior, allows to establish "the actions to defense" for the given set of societies. The solution to this problem could be given by a sequence of generations of agents, denoted like "community." The agents can only be reassigned a position to blockade with respect to the other societies, determined this according to previous behaviors, shown in each one of them (Ochoa et al., 2007).

Social Blockade in a Social Networking Represented by Multi Agents

From the point of view of the agents, this problem is very complex, on account that the group of agents using strategies to separate another individuals of a society, with respect to the behavior. In the algorithm proposed for the social blockade, the individuals in the space of beliefs (beliefscape) through their better paradigm (BestParadigm) are set to zero to represent the fact that the diversity increases the amount of expectations associated with the separation of a society with respect to other, giving an incentive to the behavior associated with the best paradigm (BestParadigm). For this purpose, we selected 1000 societies described in (Memory Alpha, 2009) and characterized their social behavior with base in seven attributes: emotional control, ability to fight, intelligence, agility, force, resistance, and speed (see Figure 6), these characteristics allow to describe so much to the society as to the individual. The development of the tool was based on our desire to share the intuitive understanding about the treatment of a new class of systems, individuals able to have restlessness, a reserved characteristic of people alive.

We identify twelve features which increase social blockade: Ethnicity, Use of Technology, Cultural Identity, Commerce, Architecture, Textile Heritage, Language, Environment Protection, Spatial Travels and Exo-collections, Telepathy and Psycho Abilities, The Art of war, Holographic Societies and Living in Class K Planets. Each feature is adjusted according highest interclass similarity and highest intraclass similarity.

We propose seven different similarity equations to organize the separation in societies:

1. Language
   Ordered Symbol and None ordered Symbol and Hamming Similarity to realize disambiguation:

   $$d(Q,C) = b+c$$

2. Use of Technology
   Simple Matching Coefficient:

   $$sim(Q,C) = 1 - \frac{b+c}{n} = \frac{a+d}{a+b+c+d}$$

3. Spatial Location
   Euclidean Distance (see Figure 7).

4. Anthropometry
   None ordered Symbol to Qualitative Parts and Ordered Symbol to Quantitative parts

5. Dyoram (Social Representation)
   Nearest Neighborhood:

   $$sim(Q,C) = \sum_{i=1}^{n} f(Qi, Ci) * wi$$

6. Commerce
   Strategies Optimist and Pessimist:
   If sim(x,y) is a similarity measure.
   If has a value of $\alpha = 1 - \alpha \leftrightarrow \alpha > 1/2$, this is optimist.
   With $\alpha > 1 - \alpha \leftrightarrow \alpha < \frac{1}{2}$, the strategy turn pessimist

Figure 6. Features related with a society in memory alpha (Memory Alpha, 2009)

Specie: Kelemane
Quadrant: Delta
Musical ability (1..7): 6   Logical ability (1..7): 5
Understanding emotion (1..7): 5   Creativity (1..7): 6
Narrative ability (1..7): 7   Spatial awareness (1..7): 4
Physical ability (1..7): 6

Figure 7. Formula of Euclidean distance

$$d(Q, C) = \sqrt{\sum_{i=1}^{n} w_i (q_i - c_i)^2}$$

7. Social Negotiation
    Asymmetric Similarity Measures:

$$sim(Q, C) = \frac{a}{a + b + c}$$

With these seven specific equations is possible to organize the separation of societies and identify the societies which are involved in this behavior of social blockade.

Experimentation

In order to be able to similar the most efficient arrangement of individuals in a social network, we developed an atmosphere that is able to store the data of each one of the representing individuals of each society; this was done with the purpose of evaluating these societies. One of the most interesting characteristics observed in this experiment was the separation of cultural patterns established by each community. The structured scenes associated with the agents cannot be reproduced in general, since they only represent a little dice in the space and time of the different societies. These represent a unique form and innovating approach of adaptive behavior which explains a computational problem that is separating societies only with a factor associated with his external appearance (genotype), trying to solve a computational problem that involves a complex change between the existing relations. The generated configurations can be metaphorically related to the knowledge of the behavior of the community with respect to an optimization problem (to conform a cluster culturally similar to other societies). The main experiment consisted of detailing each one of the 1000 communities, with 500 agents, and one condition of unemployment of 50 époques (a moment in the history), this allowed us to generate different scenes of Social Blockade in a specific cluster, which was obtained after comparing the different cultural and social similarities from each community, and to determine the existing relations between each one of them (Ustaoglu, 2009). The developed tool represents social blockade, this permits to identify changes over the time respect to other societies (See Figure 8).

Best Practices to Avoid Cyber Bullying

Currently, people are not informed about cyber-security in school, neither about cyber bullying though they should be and now this pops up. We hope people are going to be better informed about cyber-security in near future. The situation will change, on this matter there are anti-bullying experts working on it. They're cooperating with a lot of experts to thwart cyber-bullying and all we can say is that this is a great initiative, not only for the victims, but also for cyber-security advance. In this section we present a proposal to answer objectively and in automatically manner the: How to avoid cyber bullying? Question.

Problem at Hand

Is our desire to develop a intelligent tool to provide automatic answers to the question: How to avoid cyber bullying? After considering several options, we decided to use automatic text mining by means of intelligent agents to retrieve responses to answer this question from top related Websites.

Figure 8. Environment constructed by means of the use of multiagents system (cultural algorithms) describing the concept of Social Blockade, each color represented a different society. Green Society generates Social Blockade to the Red Society.

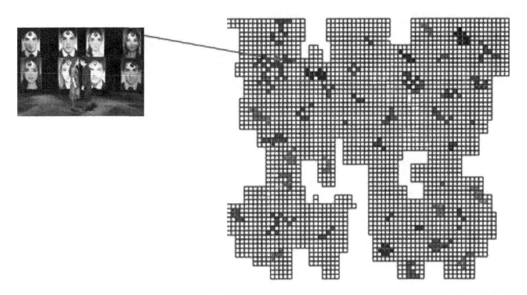

Web Based Text Mining

According to Pérez & Santin(2006), knowledge discovering in databases is a process, a set of stages or tasks, among them are: establishing a relevant problem, selection of proper data to problem solution, exploration and cleaning of data, data processing and transformation, application of modeling techniques, generation and interpretation of models proposed, use of obtained knowledge, and generation of new data by means of its application on the real world. The Web based text mining process we proposed is shown in Figure 9. Determining research objectives is the first stage on this process and determines the success or failure of research, researcher must have a clear idea of which specific information does he want to obtain and the level of depth required to satisfy information requirements (See Figure 9).

Considering objectives, web based data (text) is retrieved from top sites by means of an intelligent agent which explores (executes queries on a web site) and process data (transforms meta data on HTML o XML code on valuable answers for the question), which is stored in a.csv file for further process. Resulting data is transformed and modeled by means of a visual data mining tool (i.e. rapid miner, enterprise miner or WEKA), based on findings a visual model is generated and proved, finally resulting answers become part of new knowledge.

Methodology

Our proposal simplifies this process to three stages:

1. Selection and data base creation from Web Sites using a query with keywords: "How to avoid cyber bullying?" executed on Web search engine.
2. Automatic retrieving of answers using intelligent agents from selected web sites.
3. Text mining of data using a visual tool.

Figure 9. Web based text mining to answer the question: How can we avoid cyber bullying?

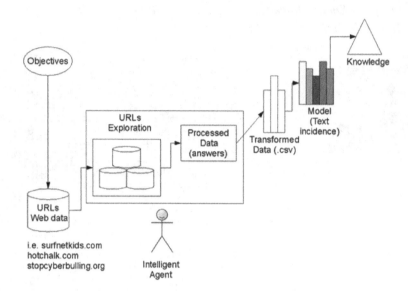

The Experiment

We created a URL database based on Web Sites retrieved from the Google search engine using the proposed keywords. When URLs were stored, we respected the order provided by Google engine. Each one of the URLs in database was processed by means of our prototype Intelligent Agent written on Java to retrieve automatic answers based on keywords (i.e. Avoid, prevent, don't, refuse, etc.). We limited the number of possible answers to five, once individual URL was processed, we send the id, the url, and the automatically retrieved answers to a.csv file for further analysis using rapid miner.

Preliminary Results and Discussion

On this early stage, our prototype system provided some recommendations to avoid cyber bullying: Do not give/provide private information to anyone (p.e passwords, telephone, address); Never share/post personal information; Report incidents to law enforcements/ school officials/to a trusted adult; Learn to recognize types of cyber bullying; never participate/never transmit cruel/harmful messages or images; and stop communications with cyber

bullies/tell friends to stop cyber bullying. Despite our system does not provide a complete set of measures to avoid cyber bullying, is summarizes some guidelines to avoid this undesirable practice.

CONCLUSION AND FUTURE RESEARCH

Using Cultural Algorithms, we improved the understanding substantially to obtain the change of "best paradigm", because we appropriately separated the agent communities basing to us on an approach to the relation that keep their attributes. This allowed us to understand that the concept of "Social Blockade" exists with base in the time of époques of interaction with their relationships. This technique allows including the possibility of generating experimental knowledge created by the community of agents for an organized social blockade. The analysis of the level and degree of cognitive knowledge for each community is an aspect that is desired to evaluate for the future work. Understanding the true similarities that have different societies with base in the characteristics that make them contributors of cluster as well as it allows them to keep their own identity, demonstrates that the small variations go beyond phenotype characteristics and are mainly associate to tastes and similar characteristics developed through the time (Ochoa et al, 2009). A new modeling of Artificial Societies can take care of analyzing individually the complexities that each society keeps, without forgetting that they still need methods to understand the original features and particular things for each society.

ACKNOWLEDGMENT

The authors would like to thank to Eliza Coli, and Anna Tordanelli for their insightful comments on the preparation of this chapter.

REFERENCES

Alcantara, M. (2009). El Bullying, Acoso Escolar. Innovación y experiencias Educativas. *CSI-CSIF Revista Digital, 55*.

Aviles, J. M. (2006). Bulling. Intimidación y maltrato entre alumnado. Retrieved May 1, 2010, From: www.educacionenvalores.org/ Bullying-Intimidacion-y-maltrato.html

Bellsey, B. (2009). *Where you are not alone*. Retrieved May 1, 2010, from: http://www.bullying.org

Bitney, J., & Title, B. (1997). *No bullying program: Preventing bullying/victim violence at school (programs directors manual)*. Central City, MN: Hazelden and Johnson Institute.

Cabezas, H. (2007). Detección de Conductas Agresivas "Bullyings" en escolares de Sexto a Octavo Año en Una Muestra Costarricense. *Revista Educación, 31*(1), 123–133.

Conway, J. (2004). Theory of games to improve artificial societies. *International Journal of Intelligent Games & Simulation, 2*(1).

Cyberbulling (2010). *What is cyberbulling?* Obtained May 1, 2010, from: http://www.cyberbullying.info/ whatis/whatis.php

DSM-IV. (1995). *Manual diagnóstico y estadístico de los transtornos mentales (DSM-IV)*. Madrid, Spain: Masson, S.A.

Fante, C. (2005). *Fenómeno Bullying. Como prevenir a violência nas escolas e educar para a paz*. Florianopolis, Brasil: Verus.

Google. (2009). *Orkut help forum*. Retrieved May 1 2010 from http://www.google.com/support/forum/p/ orkut/thread?tid=1792d1c9901741fe&hl=en

Granovetter, M. S. (1973). The strength of weak ties. *American Journal of Sociology, 78*(6), 1360–1380. doi:10.1086/225469

Hernández, M., & Solano, I. (2007). CiberBullying, Un Problema de Acoso Escolar. *RIED*, 17-36.

I-SAFE. (2010). *Cyber Bullying: Statistics and Tips*. Retrieved May 1 2010, from http://www.isafe.org/channels/sub.php? ch=op&sub_id=media_cyber_bullying

Karau, S. J., & Williams, K. D. (1993). Social loafing: A meta-analytic review and theoretical integration. *Journal of Personality and Social Psychology, 65*, 681–706. doi:10.1037/0022-3514.65.4.681

Macy, M., & Willer, R. (2002). From factors to actors. England. *Annual Review of Sociology,28*.

Memory Alpha. (2009). *Memory Alpha wiki (Star Trek World)*. Retrieved May 1 2010, from: http://memory-alpha.org/ wiki/Portal:Main

Ministerio (2009). Ministério público pede que Google explique crimes no Orkut (in Portuguese). *Folha Online*. March 10, 2006.

Narayanan, A., & Shmatikov, V. (2009). De-anonymizing social networks. *In IEEE Symposium on Security and Privacy.*

Ochoa, A., et al. (2007). Baharastar – simulador de algoritmos culturales para la minería de datos social. *In Proceedings of COMCEV'2007.*

Ochoa, A. (2008). Social data mining to improve bioinspired intelligent systems. In Giannopoulou, E. G. (Ed.), *Data Mining in Medical and Biological Research*. Vienna, Austria: In-Tech.

Ochoa, A., et al. (2009) Six degrees of separation in a graph to a social networking. *In Proceedings of ASNA'2009*, Zürich, Switzerland.

Olweus, D. (1998). *Bullying at school: What we know and what we can do* [Spanish edition: Conductas de acoso y amenaza entre escolares]. Madrid, Spain: Ediciones Morata.

Pérez, C., & Santín, D. (2006). *Data mining. Soluciones con Enterprise Manager*. Alfa Omega Grupo Editor, S.A. de C.V., Mexico

Ponce, J., Hernández, A., Ochoa, A., Padilla, F., Padilla, A., Alvarez, F., & Ponce de León, E. (2009). Data mining in web applications. In Ponce, J., & Karahoka, A. (Eds.), *Data Mining and Discovery Knowledge in Real Life Application*. Vienna, Austria: In-Tech.

Protégeles. (2009). *Protegeles línea de denuncia*. Retrieved May 1, 2010, from http://www.protegeles.com/

Racism in Brazilian. (2009). *Racism in Brazilian Orkut*. Retrieved May 1, 2010, from http://zonaeuropa.com/20050326_2.htm

Racismo na internet (2006). *Racismo na internet chega à Justiça* (in Portuguese). Estadão. Retrieved July 10, 2007, from http://www.estadao.com.br/tecnologia/internet/noticias/2006/fev/01/97.htm.

Reynolds, R. (1998). An *introduction to cultural algorithms. Cultural algorithms repository*. Retrieved May 1, 2010, from http://www.cs.wayne.edu/~jcc/car.html

Rodríguez, R., Seoane, A., & Pereira, J. (2006). Niños contra niños: El bullying como transtorno emergente. *Anal Pediátrico, 64*(2), 162–166. doi:10.1157/13084177

Silva, J. A. (2001). Do hipertexto ao algo mais: usos e abusos do conceito de hipermídia pelo jornalismo on-line. In: A. Lemos, & M. Palacios (orgs.) *Janelas do Ciberespaço - Comunicação e Cibercultura*. Sao Paulo, Brasil: Editora Sulina, pp. 128-139.

Starr, C. (1969). Social benefits versus technological risks. *Science, 165*(3899), 1232–1238. doi:10.1126/science.165.3899.1232

Ustaoglu, Y. (2009). Simulating the behavior of a minority in Turkish society. *In Proceedings of ASNA'2009.*

Valdez, S. I., Hernández, A., & Botello, S. (2009). Approximating the search distribution to the selection distribution. *In Proceedings of GECCO 2009*, 461-468.

Vidal, J. (2007). *Fundamentals of multiagent systems with NetLogo examples*. Retrieved May 1, 2010, from: http://jmvidal.cse.sc.edu/papers/mas.pdf

Williams, K. R., & Guerra, N. G. (2007). Prevalence and predictors of internet bullying. *The Journal of Adolescent Health, 41*, S14–S21. doi:10.1016/j.jadohealth.2007.08.018

ADDITIONAL READING

Blumen, W. J. (2005). Cyberbullying: A New Variation on an Old Theme. *Proceedings of Abuse: The darker side of Human-Computer Interaction.* An INTERACT workshop, Rome.

Bontempo, R. N., Bottom, W. P., & Weber, E. U. (1997). Cross-cultural differences in risk perception: A model-based approach. *Risk Analysis, 17*(4), 479–488. doi:10.1111/j.1539-6924.1997.tb00888.x

Cunningham, S. (1967). The major dimensions of perceived risk. In Cox, D. F. (Ed.), *Risk taking and information handling in consumer behavior.* Cambridge, MA: Harvard University Press.

Kowalski, R. M., Limber, S. P., & Agatston, P. W. (2008). *Cyber bullying: Bullying in the digital age.* Boston: Blackwell Publishing. doi:10.1002/9780470694176

McGrath, H. (2009). *Young people and technology. A review of the current literature (2nd edition). The Alannah and Madeline Foundation.* Retrieved March 25, 2010, from http://www.amf.org.au/Assets/Files/ 2ndEdition_Youngpeopleandtechnology_LitReview_June202009.pdf

McKenna, K. Y. A., & Bargh, J. A. (2000). Plan 9 from cyberspace: The implications of the internet for personality and social psychology. *Personality and Social Psychology Review, 4*(1), 57–75. doi:10.1207/S15327957PSPR0401_6

Shariff, S. (2009). *Confronting Cyber-Bullying: What Schools Need to Know to Control Misconduct and Avoid Legal Consequences.* New York: Cambridge University Press. doi:10.1017/CBO9780511551260

Stacey, E. (2009). Research into cyberbullying: Student perspectives on cybersafe learning environments. Informatics in Education -. *International Journal (Toronto, Ont.), 8*(1), 115–130.

Stutzky, G. R. (2010). *What is Cyberbulling?* Retrieved May 1, 2010, from: http://www.opheliaproject.org /main/ra_cyberbullying.htm

Szoka B. & Thierer A. (2009). Cyberbullying Legislation: Why Education is Preferable to Regulation. *Progress on Point, 16*(12).

Weber, E. U., & Hsee, C. (1998). Cross-cultural differences in risk perception, but cross-cultural similarities in attitudes towards perceived risk. *Management Science, 44*(9), 1205–1217. doi:10.1287/mnsc.44.9.1205

KEY TERMS AND DEFINITIONS

Artificial Societies: The representation of a group of individuals using Multiagents Systems.

Cyberbullying: A type of bullying (embarrass, humiliate, threaten, or intimidate an individual) through information and communication technologies such as mobile phone text messages, emails, phone calls, Internet chat rooms, instant messaging.

Intelligent Agent (IA): An autonomous entity which observes and acts upon an environment (i.e. it is an agent) and directs its activity towards achieving goals (i.e. it is rational). Intelligent agents may also learn or use knowledge to achieve their goals.

Intelligent Social Modeling: Technique which permits analyzing diverse social behaviors to its interpretation.

Social Network: A social structure made of nodes (which are generally individuals or organizations) that are tied by one or more specific types of interdependency, the relationship means varies, from those of social nature, such as values, visions, ideas, financial exchange, friendship, dislike, conflict, trade, kinship or friendship among people, to that of transactional nature, such as trading relationship between countries.

Social Blockade: Action that tries to separate a person or group of persons of a society, with base in a different attribute for example: race, religion, studies or status. This behavior is near to segregation.

Text Mining: Sometimes alternately referred to as text data mining, refers generally to the process of deriving high-quality information from text. Text mining usually involves the process of structuring the input text (usually parsing, along with the addition of some derived linguistic features and the removal of others, and subsequent insertion into a database), deriving patterns within the structured data, and finally evaluation and interpretation of the output.

Social Simulation: The modeling, usually with a computer, of a social phenomena.

This work was previously published in Technology for Facilitating Humanity and Combating Social Deviations: Interdisciplinary Perspectives, edited by Miguel Vargas Martin, Miguel A. Garcia-Ruiz, and Arthur Edwards, pp. 69-92, copyright 2011 by Information Science Reference (an imprint of IGI Global).

Chapter 18
Trust Modeling and Management:
From Social Trust to Digital Trust

Zheng Yan
Nokia Research Center, Finland

Silke Holtmanns
Nokia Research Center, Finland

ABSTRACT

This chapter introduces trust modeling and trust management as a means of managing trust in digital systems. Transforming from a social concept of trust to a digital concept, trust modeling and management help in designing and implementing a trustworthy digital system, especially in emerging distributed systems. Furthermore, the authors hope that understanding the current challenges, solutions and their limitations of trust modeling and management will not only inform researchers of a better design for establishing a trustworthy system, but also assist in the understanding of the intricate concept of trust in a digital environment.

INTRODUCTION

Trust plays a crucial role in our social life. Our social life is characterized by the trust relationships that we have. Trust between people can be seen as a key component to facilitate coordination and cooperation for mutual benefit. Social trust is the product of past experiences and perceived trustworthiness. We constantly modify and upgrade our trust in other people based on our feelings in response to changing circumstances. Often, trust is created and supported by a legal framework, especially in business environments or when financial issues are involved. The framework ensures that misbehavior can be punished with legal actions and increases the incentive to initiate a trust relationship. The legal framework decreases the risk of misbehavior and secures the financial transactions. With the rapid growth of global digital computing and networking

DOI: 10.4018/978-1-4666-2803-8.ch018

technologies, trust becomes an important aspect in the design and analysis of secure distributed systems and electronic commerce. However, the existing legal frameworks are often focused on local legislation and are hard to enforce on a global level. The most popular examples are email spam, software piracy, and a breach of warranty. Particularly, because legal regulation and control cannot keep pace with the development of electronic commerce, the extant laws in conventional commerce might not be strictly enforceable in electronic commerce. In addition, resorting to legal enforcement in electronic commerce might be impracticably expensive or even impossible, such as in the case of micro payment transactions (Ba, Whinston, & Zhang 1999). This raises the importance of trust between interacting digital entities. People cannot assume that the legal framework is able to provide the needed trustworthiness for their digital relationships, for example, for an electronic transaction purpose. It has been a critical part of the process by which trust relationships are required to develop in a digital system. In particular, for some emerging technologies, such as MANET (Mobile Ad Hoc Networks), P2P (Peer-to-Peer) computing, and GRID virtual systems, trust management has been proposed as a useful solution to break through new challenges of security and privacy caused by the special characteristics of these systems, such as dynamic topology and mobility (Lin, Joy, & Thompson, 2004; Yan, 2006; Yan, Zhang, & Virtanen 2003).

Trust is a very complicated phenomena attached to multiple disciplines and influenced by many measurable and non-measurable factors. It is defined in various ways for different purposes and cultures, even though in information technology area. Thereby, it is difficult to have a common definition for this comprehensive concept.

Establishing a trust relationship in digital networking environment involves more aspects than in the social world. This is because communica-tions in the computing network rely on not only relevant human beings and their relationships, but also digital components. On the other hand, the visual trust impression is missing and need somehow to be compensated. Moreover, it is more difficult to accumulate accurate information for trust purposes in remote digital communications where information can be easily distorted or faked identities can be created. The mapping of our social understanding of trust into the digital world and the creation of trust models that are feasible in practice are challenging. Trust is a special issue beyond and will enhance a system security and personal privacy. Understanding the trust relationship between two digital entities could help selecting and applying feasible measures to overcome potential security and privacy risk. From social trust to digital trust, how can trust theory help in designing and implementing a trustworthy digital system? The literature suggests the usage of trust modeling and management. This book chapter aims to help readers understanding trust modeling and management in the emerging technologies. The reader is guided from the creation of a digital trust concept to the deployment of trust models for trust management in a digital environment.

BACKGROUND

The problem to create trust in the digital environment has led to a number of approaches, for example, expressing and modeling trust in a digital way, evaluating trust in a reputation system, rating agencies, certificate authorities that equip trading partners with certificates as trusted providers and brokers, trustworthy user interface design, and so on. Current academic and industrial work related to trust covers a wide area of interest ranging from such aspects as perception of trust, cryptographic-security enforced trust solutions, trust modeling, and trust management to trusted computing activities.

Perception of Trust Concept (from Social Trust towards Digital Trust)

What is trust and how is trust influenced? We will now examine the most common definitions of trust and start from a classical social trust concept that is supported by a legal framework to a concept for digital processing. Through the study of various definitions of trust, we explain the properties of trust relationships and classify the factors that influence trust.

Definitions of Trust. The concept of trust has been studied in disciplines ranging from economic to psychology, from sociology to medicine, and to information science. It is hard to say what trust exactly is because it is a multidimensional, multidiscipline and multifaceted concept. We can find various definitions of trust in the literature. For example, it can be loosely defined as a state involving confident positive expectations about another's motives with respect to oneself in situations entailing risk (Boon & Holmes, 1991). This definition highlights three main characteristics of trust. First, a trust relationship involves at least two entities: a trustor and a trustee, reliant on each other for mutual benefit. Second, trust involves uncertainty and risk. There is no perfect guarantee to ensure that the trustee will live up to the trustor's expectation. Third, the trustor has faith in the trustee's honesty and benevolence, and believes that the trustee will not betray his/her risk-assuming behavior.

Gambetta (1990) defined trust as trust (or, symmetrically, distrust) is a particular level of the subjective probability with which an agent will perform a particular action, both before [we] can monitor such action (or independently of his capacity of ever to be able to monitor it) and in a context in which it affects [our] own action. Mayer, Davis, and Schoorman (1995) provided the definition of trust as the willingness of a party to be vulnerable to the actions of another party based on the expectation that the other party will perform a particular action important to the trustor, irrespective of the ability to monitor or control that other party. These definitions point out another important main characteristic of trust: Trust is subjective. The level of trust considered sufficient is different for each individual in a certain situation. Trust may be affected by those actions that we cannot (digitally) monitor. The level of trust depends on how our own actions are in turn affected by the trustee's actions. Grandison and Sloman (2000) hold an opinion that trust is a qualified belief by a trustor with respect to the competence, honesty, security and dependability of a trustee within a special context.

McKnight and Chervany (2000, 2003) conducted analysis on the trust definitions and noted that trust is a concept hard to define because it is itself a vague term. Looking up the term "trust" in a dictionary may reveal many explanations since it is a cross-disciplinary concept. For example, from the sociologists' point of view, it is related to social structure. From the psychologists' point of view, it concerns personal trait. From the economists' point of view, it is a mechanism of economic choice and risk management. The definitions of trust can be classified based on the consideration of structural, disposition, attitude, feeling, expectancy, belief, intention, and behavior. There are suggestions to evaluate trust with regard to competence, benevolence, integrity, and predictability. But generally, these attributes are only applicable to very narrow scenarios and hard to measure.

Other expressions of trust are targeting at different context and technology areas, for example:

- **Online System:** Online trust is an attitude of confident expectation in an online situation of risk that one's vulnerabilities will not be exploited (Corritore, Kracher, & Wiedenbeck, 2003).
- **Multi Agent System:** In a multi-agent system, trust is a subjective expectation an agent has about another agent's future behavior (Mui, 2003).

- **Software Engineering:** From a software engineering point of view, trust is accepted dependability (Avizienis, Laprie, Randell, & Landwehr, 2004).
- **Ad-Hoc Networks:** For an ad hoc network, trust could be defined as the reliability, timeliness, and integrity of message delivery to a node's intended next-hop (Liu, Joy, & Thompson, 2004).

Denning (1993) emphasizes the importance of assessment for trust in a system, which is of particular importance in the digital environment, where the entities often just have digital artifacts to base their trust judgment on. The current paradigm for trusted computing systems holds that trust is a property of a system. It is a property that can be formally modeled, specified, and verified. It can be "designed into" a system using a rigorous design methodology. Trust is an assessment that a person, organization, or object can be counted on to perform according to a given set of standards in some domain of action. In particular, a system is trusted if and only if its users trust it. Trust itself is an assessment made by users based on how well the observed behavior of the system meets their own standards.

Common to many definitions are the notions of confidence, belief, faith, hope, expectation, dependence, and reliance on the goodness, strength, reliability, integrity, ability, or character of a person or entity. Generally, a trust relationship involves two entities: a trustor and a trustee. The trustor is the person or entity who holds confidence, belief, faith, hope, expectation, dependence, and reliance on the goodness, strength, reliability, integrity, ability, or character of another person or entity, which is the object of trust - the trustee.

Although the richness of the concept, we can still summarize the subjective and objective factors that are relevant to a decision of trust, as shown in Table 1.

Factors Influencing Trust. Trust is subjective because the level of trust considered sufficient is different for each entity. It is the subjective expectation of the trustor on the trustee related to the trustee's behaviors that could influence the trustor's belief in the truestee. Trust is also dynamic as it is affected by many factors that are difficult to measure and monitor. It can be further developed and evolved due to good experiences about the trustee. Or it is sensitive to be decayed caused by one or several bad experiences. From the digital system point of view, trust is a kind of assessment on the trustee based on a number of referents, for example, competence, security, reliability, and so on. Trust is influenced by a number of factors. Those factors can be classified into five categories, as shown in Figure 1.

From the digital system point of view, we pay more attention to the objective properties of both the trustor and the trustee. For social human interaction, we consider more about the trustee's subjective and objective properties and the trustor's subjective properties. For economic transactions, we study more about the context for risk

Table 1. Factors influencing trust extracted from some definitions of trust

Factors related to trustee's objective properties	competence; ability; security; dependability; integrity; predictability; reliability; timeliness; (observed) behavior; strength
Factors related to trustee's subjective properties	honesty; benevolence; goodness
Factors related to trustor's objective properties	assessment; a given set of standards; trustor's standards
Factors related to trustor's subjective properties	confidence; (subjective) expectations or expectancy; subjective probability; willingness; belief; disposition; attitude; feeling; intention; faith; hope; trustor's dependence and reliance
Context	situations entailing risk; structural; risk; domain of action

Figure 1. Classification of factors that influence trust

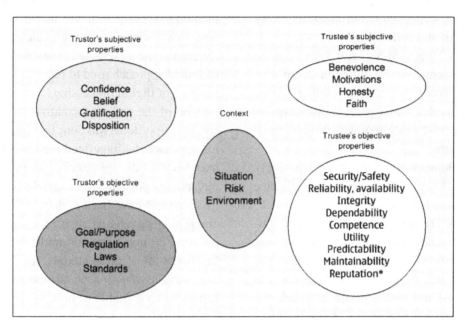

management. The context of trust is a very important factor that influences a trust relationship, e.g. the why and when to trust. It specifies any information that can be used to characterize the background or situation of the involved entities. Trust is influenced by many factors, but the impact of different factors could be various in dissimilar contexts.

Trust Modeling Technology (Modeling Trust in a Digital Approach)

We introduce the existing trust modeling work in the area of distributed systems, e-commerce and Web services to understand the methodologies used for trust modeling. We will answer questions like: What are the characteristics of trust? What is a trust model? What is trust modeling? What kinds of trust models have been developed and applied in the emerging technologies? Which factors related to trust are considered in the modeling in the literature?

Characteristics of Trust. Despite the diversity among the existing definitions of trust, and despite that a precise definition is missing in the literature, there is a large commonality on the properties of trust. Rousseau, Sitkin, Burt, and Camerer (1998) also observed considerable overlap and synthesis in contemporary scholarship on trust. Particularly, the most common characteristics of trust, which play as the important guidelines for trust modeling are:

- **Trust is Directed:** This property says that trust is an oriented relationship between the trustor and the trustee.
- **Trust is Subjective:** Trust is inherently a personal opinion. According to the survey conducted by Grandison and Sloman (2000), trust is considered a personal and subjective phenomenon that is based on various factors or evidence and that some of those may carry more weight than others. Trust is different for each individual in a certain situation.

- **Trust is Context-Dependent:** In general, trust is a subjective belief about an entity in a particular context.

- **Trust is Measurable:** Trust values can be used to represent the different degrees of trust an entity may have in another. "Trust is measurable" also provides the foundation for trust modeling and computational evaluation.

- **Trust Depends on History:** This property implies that past experience may influence the present level of trust.

- **Trust is Dynamic:** Trust is usually non-monotonically changed with time. It may be refreshed or revoked periodically, and must be able to adapt to the changing conditions of the environment in which the trust decision was made. Trust is sensitive to be influenced due to some factors, events, or changes of context. In order to handle this dynamic property of trust, solutions should take into account the notion of learning and reasoning. The dynamical adaptation of the trust relationship between two entities requires a sophisticated trust management approach (Grandison and Sloman, 2000).

- **Trust is Conditionally Transferable:** Information about trust can be transmitted/received along a chain (or network) of recommendations. The conditions are often bound to the context and the trustor's objective factors.

- **Trust Can Be a Composite Property:** "Trust is really a composition of many different attributes: reliability, dependability, honesty, truthfulness, security, competence, and timeliness, which may have to be considered depending on the environment in which trust is being specified" (Grandison and Sloman, 2000, pp. 3). Compositionality is an important feature for making trust calculations.

Trust Modeling. What is a trust model? The method to specify, evaluate and set up trust relationships amongst entities for calculating trust is referred as the trust model. Trust modeling is the technical approach used to represent trust for the purpose of digital processing.

One of the earliest formalizations of trust in computing systems was done by Marsh (1994). In his approach, he integrated the various facets of trust from the disciplines of economics, psychology, philosophy and sociology. Since then, many trust models have been constructed for various computing paradigms such as ubiquitous computing, P2P networks, and multi-agent systems. In almost all of these studies, trust is accepted as a subjective notion by all researchers, which brings us to a problem: how to measure trust? Translation of this subjective concept into a machine readable language is the main objective needed to be solved. Rahman and Hailes (2000) proposed a trust model based on the work done by Marsh (1994). Their trust model focuses on online virtual communities where every agent maintained a large data structure representing a version of global knowledge about the entire network. Gil and Ratnakar (2002) described a feedback mechanism that assigns credibility and reliability values to sources based on the averages of feedback received from individual users.

Regarding the trust modeling, there are various methodologies can be applied for different purposes. Some trust models are based on cryptographic technologies, for example Public Key Infrastructure (PKI) played as the foundation in a trust model (Perlman, 1999). A big number of trust models are developed targeting at some special trust properties, such as reputations, recommendations and risk studied by Xiong and Liu (2004) and Liang and Shi (2005). Seldom, they support multi-property of trust that is needed to take into account the factors like multiple objective factors of the trustee and context. Many trust models have been constructed for various computing paradigms such as GRID computing, ad hoc networks, and P2P systems. Those models use computational, linguistic or graphical methods. For example, Maurer (1996) described an entity's opinion about the trustworthiness of a certificate as a value in the scope of [0, 1]. Theodorakopoulos and Baras (2006) used a two-tuple in [0, 1]^2 to describe a trust opinion. In Jøsang (1999), the metric is a triplet in [0, 1]^3, where the elements in the triplet represent belief, disbelief, and uncertainty, respectively. Abdul-Rahman and Hailes (2000) used discrete integer numbers to describe the degree of trust. Then, simple mathematics, such as minimum, maximum, and weighted average, is used to calculate unknown trust values through concatenation and multi-path trust propagation. Jøsang and Ismail (2002) and Ganeriwal and Srivastava (2004) used a Bayesian model to take binary ratings as input and compute reputation scores by statistically updating beta probability density functions. Linguistic trust metrics were used for reasoning trust with provided rules by Manchala (2000). In the context of the "Web of Trust," many trust models (e.g., Reiter and Stubblebine, 1998) are built upon a graph where the resources/entities are nodes and trust relationships are edges.

One promising approach of trust modeling aims to conceptualize trust based on user studies through a psychological or sociological approach (e.g., a measurement scale). This kind of research aims to prove the complicated relationships among trust and other multiple factors in different facets. Two typical examples are the initial trust model proposed by McKnight, Choudhury, and Kacmar (2002) that explained and predicted customer's trust towards an e-vender in an e-commerce context, and the Technology Trust Formation Model (TTFM) studied by Li, Valacich, and Hess (2004) to explain and predict people's trust towards a specific information system. Both models used the framework of the theory of reasoned action (TRA) created by Fishbein and Ajzen (1975) to explain how people form initial trust in an

unfamiliar entity, and both integrated important trusting antecedents into this framework in order to effectively predict people's trust. For other examples, Gefen (2000) proved that familiarity builds trust; Pennington, Wilcox, and Grover (2004) tested that one trust mechanism, vendor guarantees, has direct influence on system trust; Bhattacherjee (2002) studied three key dimensions of trust: trustee's ability, benevolence and integrity; Pavlou and Gefen (2004) explained that institutional mechanisms engender buyer's trust in the community of online auction sellers. The trust models generated based on this approach are generally linguistic or graphic. They do not quantify trust for machine processing purposes. Therefore, the achieved results could only help people understanding trust more precisely in order to work out a design guideline or an organizational policy towards a trustworthy digital system or a trustworthy user interface. Although little work has been conducted to integrate psychological, sociological and technological theories together, we believe, however, the psychological and sociological study results could further play as practical foundations of computational trust—modeling trust for a digital processing purpose,.

Modeling trust in a digital way is important in a distributed digital system in order to automatically manage trust. Although a variety of trust models are available, it is still not well understood what fundamental criteria the trust models must follow. Without a good answer to this question, the design of trust models is still at an empirical stage and can never reach the expectation to simulate social trust to a satisfying degree. Current work focuses on concrete solutions in special systems. We would like to advocate that the trust model should reflect the characteristics of trust, consider the factors that influence the trust, and thus support the trust management in a feasible way.

Despite the variety of trust modeling methods, a common approach can be found in a number of publications regarding computational trust, for example, Xiong and Liu (2004); Theodorakopoulos and Baras (2006); Song, Hwang, Zhou, and Kwok (2005), Liu, Joy, and Thompson (2004), and Sun, Yu, Han, and Liu (2006). This approach is applied by firstly presenting an understanding of the characteristics of trust, principles or axioms, then modeling them in a mathematical way, and further applying the model into trust evaluation or trust management for a specific issue.

Taxonomy of Trust Models. The trust model aims to process and/or control trust using digital methods. Most of the modeling work is based on the understanding of trust characteristics and considers some factors influencing trust. The current work covers a wide area including ubiquitous computing, distributed systems, multi-agent systems, Web services, e-commerce, and component software. The discussed trust models can be classified into categories according to different criteria, as shown in Table 2.

Current Research Status. Trust is influenced by reputations (i.e., the public evidence of the trustee), recommendations (i.e., a group of entities' evidence on the trustee), the trustor's past experiences, and context (e.g., situation, risk, time,

etc.). Most of work focused on a singular trust value or level calculation by taking into account the previous behavior of the trustee. The reputations, the recommendations and the trustor's own experiences are assessed based on the quality attributes of the trustee, the trust standards of the trustor, and the local context for making a trust or distrust conclusion. A number of trust models support the dynamics of trust. So far, some basic elements of context are considered, such as time, context similarity, and so on. The time element has been considered in many pieces of work, such as Wang and Varadharajan (2005) and Xiong and Liu (2004). For peer-to-peer systems, Sarkio and Holtmanns (2007) proposed a set of functions to produce a tailored trustworthiness estimation, which takes into account factors like age of reputation, value of transaction, frequency of transactions, reputation of the reputation giver, and so on. However, no existing work gives a common consideration on all factors that influence trust in a generic digital environment, especially those subjective factors of the trustor and the trustee, as shown in Figure 1. It is still a challenge to

Table 2. Taxonomy of trust models

Criteria of Classification	Categories		Examples
Based on the method of modeling	Models with linguistic description		Blaze, Feigenbaum, and Lacy (1996) and Tan and Thoen (1998)
	Models with graphic description		Reiter and Stubblebine (1998)
	Models with mathematic description		Xiong and Liu (2004) and Sun, Yu, Han, and Liu (2006)
Based on modeled contents	Single-property modeling		Xiong and Liu (2004) and Sun, Yu, Han, and Liu (2006)
	Multi-property modeling		Zhou, Mei, and Zhang (2005), Wang and Varadharajan (2005), and Yan and MacLaverty (2006)
Based on the expression of trust	Models with binary rating		
	Models with numeral rating	Continuous rating	Maurer (1996) and Xiong and Liu (2004)
		Discrete rating	Liu, Joy, and Thompson (2004)
Based on the dimension of trust expression	Models with single dimension		Maurer (1996) and Xiong and Liu (2004)
	Models with multiple dimensions		Theodorakopoulos and Baras (2006) and Jøsang (1999)

digitally model the subjective factors related to the trustor and the trustee, especially when the trustor or the trustee is a person.

Mechanisms for Trust Management (Applying and Deploying Trust Models)

About the following questions will be addressed in this part: "What is trust management? What does trust management do? Why is trust management needed in the emerging technologies? What is the current research status of trust management?"

As defined by Grandison and Sloman (2000), trust management is concerned with: collecting the information required to make a trust relationship decision; evaluating the criteria related to the trust relationship as well as monitoring and reevaluating existing trust relationships; and automating the process. Due to the amount of data collected and processed in the digital environment, the definition should be extended to accommodate support for automatic processing in order to provide a system's trustworthiness. Yan and MacLaverty (2006) proposed that autonomic trust management includes four aspects and these four aspects are processed in an automatic way:

- **Trust Establishment:** The process for establishing a trust relationship between a trustor and a trustee.
- **Trust Monitoring:** The trustor or its delegate monitors the performance or behaviour of the trustee. The monitoring process aims to collect useful evidence for trust assessment of the trustee.
- **Trust Assessment:** The process for evaluating the trustworthiness of the trustee by the trustor or its delegate. The trustor assesses the current trust relationship and decides if this relationship is changed. If it is changed, the trustor will make decision which measure should be taken.

- **Trust Control and Re-Establishment:** If the trust relationship will be broken or is broken, the trustor will take corresponding measures to control or re-establish the trust relationship.

As we can see from the above, trust management can be achieved through trust modeling and evaluation.

Reputation Systems. There are various trust management systems in the literature and practice. A category of large practical importance is reputation based trust management system. Trust and reputation mechanisms have been proposed in various fields such as distributed computing, agent technology, grid computing, economics, and evolutionary biology. Reputation-based trust research stands at the crossroads of several distinct research communities, most notably computer science, economics, and sociology.

As defined by Aberer and Despotovic (2001), *reputation* is a measure that is derived from direct or indirect knowledge on earlier interactions of entities and is used to assess the level of trust an entity puts into another entity. Thus, reputation based trust management (or simply reputation system) is a specific approach to evaluate and control trust.

Reputation schemes can be classified into two different categories depending on what sort of reputation they utilize. Global reputation is the aggregation of all available assessments by other entities that have had interactions with the particular entity, and thus it has an *n-to-1 relationship*. On the other hand, the local reputation of an entity is each entity's own assessment based on past history of interaction with the particular entity, thus it is a *1-to-1 relationship*. This reflects the social situation that a person trusts another one, because "they are good friends."

Several representative P2P reputation systems currently exist, although the list we present is by no means exhaustive. The eBay and PeerTrust systems focus on trust management in securing commodity exchanges in e-commerce applications, as does the FuzzyTrust system by Song et al. (2005). Other systems focus on generic P2P applications such as P2P file sharing and Web service-based sharing platforms.

The eBay (www.ebay.com) user feedback system described by Resnick and Zeckhauser (2002) is by far the simplest and most popular trust-management system, and is specifically tailored for e-auction applications. It applies a centralized database to store and manage the trust scores. Data is open to the general public, so a newcomer can easily obtain a peer score. It's a hybrid P2P system using both distributed client resources and centralized servers. This system tries to be user friendly by providing a limited amount of data to a user, but on the other hand the provided and processed information is not complete and does not provide a "full picture."

Singh and Liu (2003) presented Trustme, a secure and anonymous protocol for trust. The protocol provides mutual anonymity for both a trust host and a trust querying peer. Guha and Kumar (2004) developed an interesting idea about the propagation of distrust. In addition to maintaining positive trust values for peers, the system also allows the proactive dissemination of some malicious peers' bad reputations. Buchegger and Le Boudec (2004) designed a distributed reputation system using a Bayesian approach, in which the second-hand reputation rating is accepted only when it is compatible with the primary rating.

Several universities are working on the research projects involving trust management in P2P applications. Xiong and Liu (2004) developed the PeerTrust model. Their model is based on a weighted sum of five peer feedback factors: *peer records, scope, credibility, transaction context,* and *community context.* PeerTrust is fully distributed, uses overlay for trust propagation, public-key infrastructure for securing remote scores, and prevents peers from some malicious abuses.

Kamvar, Schlosser, and Garcia-Molina (2003) proposed the EigenTrust algorithm, which captures peer reputation in the number of satisfactory transactions and then normalizes it over all participating peers. The algorithm aggregates the scores by a weighted sum of all raw reputation scores. The fully distributed system assumes that pre-trusted peers exist when the system is initiated. It uses majority voting to check faulty reputation scores reported.

Liang and Shi (2005) proposed the TrustWare system (retrieved from http://mist.cs.wayne.edu/trustware.html), a trusted middleware for P2P applications. Their approach consists of two models: the Multiple Currency Based Economic model (M-CUBE) and the Personalized Trust model (PET). The M-CUBE model provides a general and flexible substrate to support high-level P2P resource management services. PET derives peer trustworthiness from long-term reputation evaluation and short-term risk evaluation.

Sherwood and Bhattacharjee (2003) proposed in the Nice project a scheme for trust inference in P2P networks. The trust inference consists of two parts for local trust inference and distributed search. After each transaction, the system generates cookies to record direct trust between peers. It also uses trust graphs to infer transitive trust along a peer chain.

Credence is a robust and decentralized system for evaluating the reputation of files in a P2P file sharing system (Retrieved from http://www.cs.cornell.edu/people/egs/credence/index.html). Its goal is to enable peers to confidently gauge *file authenticity*, the degree to which the content of a file matches its advertised description. At the most basic level, Credence employs a simple, network-wide voting scheme where users can contribute positive and negative evaluations of files. On top of this, a client uses statistical tests to weight the importance of votes from other peers. It allows the clients to share selected information with other peers. Privacy is ensured by not collecting or using any personally identifiable information in any way in the protocol. Each Credence-equipped client is supplied with a unique, randomly generated key pair that is not bound to any personal information for use in cryptographic operations.

Meanwhile, European Union (EU) project SECURE investigated the design of security mechanisms for pervasive computing based on the human notion of trust. It addresses how entities in unfamiliar pervasive computing environments can overcome initial suspicion to provide secure collaboration (Cahill et al., 2003). Another EU project Trust4All aims to build up trustworthy middleware architecture in order to support easy and late integration of software from multiple suppliers and still have dependable and secure operations in the resulting system (Retrieved from https://nlsvr2.ehv.compus.philips.com/).

Requirements of Trust for Ad Hoc Networks. More dimensions are needed to secure the communications in wireless mobile ad hoc networks. Balakrishnan and Varadharajan (2005) demonstrated the issues that might creep out in the security design, when a cryptographic technique alone is involved. They also suggested how to counter those issues through the combination of trust management with cryptographic mechanisms. Moreover, they proposed the need to introduce the notion of heterogeneity resource management in the security design to address the divergence among the nodes, which can be taken advantage to diminish the packet drop attacks. To handle the dynamic nature of the medium, the authors proposed that the design of secure mobile ad hoc networks should envisage including trust management as another dimension apart from the cryptographic mechanisms. In addition, inclusion of trust management alone cannot guarantee secure communication due to some persisting issues such as packet dropping. Therefore, the resource should be also considered in order to provide a trustworthy system.

Trust Evaluation Mechanisms (Methodologies for Trust Decision)

Trust evaluation is a technical approach of representing trustworthiness for digital processing, in which the factors influencing trust will be evaluated by a continuous or discrete real number, referred to as a trust value. A trust evaluation mechanism aims to provide supporting information for the actual trust decision of an entity for managing trust. Embedding a trust evaluation mechanism is a necessity to provide trust intelligence in future computing devices.

The trust evaluation is the main aspect in the research for the purpose of digitalizing trust for computer processing. A number of theories about trust evaluation can be found in the literature. For example, Subjective Logic was introduced by Jøsang (2001). It can be chosen as trust representation, evaluation and update functions. The Subjective Logic has a mathematical foundation in dealing with evidential beliefs rooted in the Shafer's theory and the inherent ability to express uncertainty explicitly. The trust valuation can be calculated as an instance of the opinion in the Subjective Logic. An entity can collect the opinions about other entities both explicitly via a recommendation protocol and implicitly via limited internal trust analysis using its own trust base. It is natural that the entity can perform an operation in which these individual opinions can be combined into a single opinion to allow relatively objective judgment about other entity's trustworthiness. It is desirable that such a combination operation shall be robust enough to tolerate situations where some of the recommenders may be wrong or dishonest. Other situation with respect to trust valuation combination includes combining the opinions of different entities on the same entity together; aggregation of an entity's opinion on two distinct entities together with logical AND support or with logical OR support. (The description and demo about the Subjective Logic can be retrieved from http://sky.fit.qut.edu.au/~josang/sl/demo/Op.html.)

But the Subjective Logic is a theory about opinion that can be used to represent trust. Its operators mainly support the operations between two opinions. It doesn't consider context support, such as time based decay, interaction times or frequency, and trust standard support like importance weights of different trust factors. Concretely, how to generate an opinion on a recommendation based on credibility and similarity and how to overcome attacks of trust evaluation are beyond the Subjective Logic theory. The solutions of these issues need to be further developed in practice.

Fuzzy Cognitive Maps (FCM) developed by Kosko (1986) could be regarded as a combination of Fuzzy Logic and Neural Networks. In a graphical illustration, the FCM seems to be a signed directed graph with feedback, consisting of nodes and weighted arcs. Nodes of the graph stand for the concepts that are used to describe the behavior of the system and they are connected by signed and weighted arcs representing the causal relationships that exist between the concepts, as depicted in Figure 2.

The FCM can be used for evaluating trust. In this case, the concept nodes are trustworthiness and factors that influence trust. The weighted arcs

represent the impact of the trust influencing factors to the trustworthiness. These weighted arcs allow putting weight on the trust influencing factors. The FCM is convenient and practical for implementing and integrating trustworthiness and its influencing factors. In addition, Song et al. (2005) made use of fuzzy logic approach to develop an effective and efficient reputation system.

Theodorakopoulos and Baras (2006) introduced Semiring. It views the trust inference problem as a generalized shortest path problem on a weighted directed graph $G(V, E)$ (*trust graph*). The vertices V of the graph are the users/entities in the network. A weighted edge that belongs to E from vertex i to vertex j corresponds to the *opinion* that the trustor i has about the trustee j. The weight function is $l(i,j)$: $V \times V$ S, where S is the opinion space. Each opinion consists of two numbers: the *trust* value, and the *confidence* value. The former corresponds to the trustor's estimate of the trustee's trustworthiness. The confidence value corresponds to the accuracy of the trust value assignment. Since opinions with a high confidence value are more useful in making trust decisions, the confidence value is also referred to as the *quality or reliability* of the opinion. The space of opinions can be visualized as a rectangle (*ZERO_TRUST, MAX_TRUST*) $\times$ (*ZERO_CONF, MAX_CONF*) in the Cartesian plane. They don't

Figure 2. A simple fuzzy cognitive map

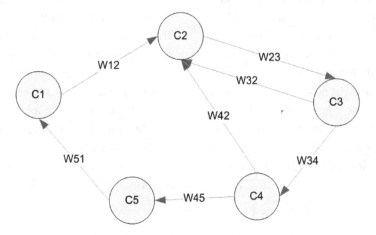

treat distrust or negative confidence, but these could be accommodated by rescaling S.

A *Semiring* is an algebraic structure $(S, \oplus, \otimes)$, where S is a set, and $\oplus$, $\otimes$ are binary operators. $\oplus$, $\otimes$ are associative and $\oplus$ is commutative. $\oplus$ and $\otimes$ can be used to aggregate opinions along the paths from the trustor to the trustee together. Concretely, $\otimes$ is used to calculate the opinion along a path from the trustor to the trustee, while $\oplus$ is applied to compute the opinion as a function of all paths from the trustor to the trustee. Theodorakopoulos and Baras (2006) gave the formula of $\oplus$ and $\otimes$ regarding path Semiring and distance Semiring.

Sun et al. (2006) presented an information theoretic framework to quantitatively measure trust and model trust propagation in the ad hoc networks. In the proposed framework, trust is a measure of uncertainty with its value represented by entropy. The authors develop four Axioms that address the basic understanding of trust and the rules for trust propagation. Based on these axioms two trust models are introduced: entropy-based model and probability-based model, which satisfy all the axioms. Xiong and Liu (2004) introduced five trust parameters in PeerTrust. By formalizing these parameters, they presented a general trust metric that combines these parameters in a coherent scheme. This model can be applied into a decentralized P2P environment. It is effective against dynamic personality of peers and malicious behaviors of peers.

DIGITAL MANAGEMENT OF TRUST

Issues, Controversies and Problems

The rapid evolution of the digital environment and emerging technologies creates a number of issues related to trust.

E-Commerce. Electronic commerce and services are revolutionizing the way we conduct our personal and organizational business. And this trend will be extended to mobile domain. But it

is very hard to build up a long-term trust relationship between all involved parties: manufactures, service providers, application providers, access providers and end users. Different legislation areas and distrust in the applicable legal frameworks of transaction partners make the creation of a trust foundation necessary for the electronic transaction quite challenging. This could be a major obstacle that retards the further development of e-commerce.

Hoffman, Novak, and Peralta (1999) pointed out that the reason more people have yet to shop online or even provide information to Web providers in exchange for access to information is the fundamental lack of faith between most businesses and consumers on the Web. Almost 95 percent of Web users have declined to provide personal information to Web sites when asked because they are not clear how the personal data will be used and they feel there is no way for them to control over secondary use of their personal information (Hoffman et al., 1999). In addition, differently from traditional commerce, uncertainty about product quality, that is, information asymmetry is always a problem for consumers in an online environment (Ba & Pavlou, 2002). Lack of consumer trust is a critical impediment to the success of e-commerce. Ultimately, the most effective way for commercial Web providers to develop profitable exchange relationships with online customers is to earn their trust. Trust thus becomes a vital factor influencing the final success of e-commerce.

Digital Distributed Networking and Communications. On the other hand, new networking is raising with the fast development of Mobile Ad hoc Networks (MANET) and local wireless communication technologies. Boundaries between traditional computers, laptops, mobile phones, Personal Digital Assistants (PDA), and consumer electronics devices dissolve. It is more convenient for mobile users to communicate in their proximity to exchange digital information in various circumstances. However, the special characteristics of the new networking paradigms (e.g., dynamically changed topology) introduce additional challenges on security. The ad hoc networks are generally more prone to physical security threats than conventional networks, due to their lack of any security infrastructure support. Security approaches used for fixed networks are not feasible due to the salient characteristics of the ad hoc networks. The root of the threats is originated from the lack of trust among network nodes.

In addition, new P2P computing technology has emerged as a significant paradigm for providing distributed services, in particular collaboration for content sharing and distributed computing. Generally, a P2P system consists of a decentralized and self-organizing network of autonomous devices that interact as peers. Each peer acts as both client and server to share its resources with other peers. However, this computing paradigm suffers from several drawbacks that obstruct its wide adoption. Lack of trust between peers is one of the most serious issues, which causes security challenges in the P2P systems. Building up trust collaboration among the system peers is a key issue to overcome, especially in the mobile environment.

GRID computing systems have attracted research communities in recent years. This is due to the unique ability of marshalling collections of heterogeneous computers and resources, enabling easy access to diverse resources and services that could not be possible without a GRID model. The context of GRID computing does introduce its own set of security challenges, as user(s) and resource provider(s) can come from mutually distrusted administrative domains and either participant can behave maliciously.

User-Device or User-System Interaction. Most current digital systems are designed based on the assumptions that the user trusts his/her device fully; or the user has to trust a service provider. Generally, the current systems are not designed to be configured by user with regard to their trust preferences. As can be seen from the above study, trust evaluation based management technology

has been proposed to overcome the challenge of trust and security in distributed systems and Internet e-commerce. However, the human-machine interaction in order to support trust management, especially in mobile domain, has still many open research questions. Embedding personal criteria of trust regarding different events into the device or system requires interaction between the end user and his/her devices. This would require friendly user interface for the device to collect useful information for trust evaluation and present the evaluation results in a comprehensive manner to the user. It also provides a technical challenge to design an effective trust management system that is light weight with regard to memory management, process observation and data mining.

Privacy Support. User data privacy is hard to control once the personal data is released into digital format. New customized services require detailed user information like location, preferences or general behavior profiles. People expect such a privacy control service that only trusted parties can get specified personal information. The user-friendly administration of this control is a tough challenge especially for the mobile systems with a limited user interface even through taking into account the latest privacy visualization approaches for the mobile environment (Hjelm & Holtmanns, 2006).

How to solve these issues described above depends on better understanding of trust, trust modeling and management technologies.

Solutions and Recommendations

Solutions for the Trust Issues of E-Commerce. Trust management is introduced to evaluate trust for reduced risk. Tan and Thoen (1998) specified a generic model of transaction trust for e-commerce as party trust supplemented with control trust. It provides a generic model of transaction trust for e-commerce. This model is based on separating the mostly subjective party trust and mostly objective control trust. The transaction trust consists of the sum of the party trust and the control trust. If the level of the transaction trust is not sufficient, then the party trust should be possibly complemented by the control trust in order to reach a required level. This theory was further developed to build up online trust between trading parties in a first trade situation through a trust matrix model.

Manchala (2000) described metrics and models for the measurement of trust variables and fuzzy verification of transactions. The trust metrics help preserve system availability by determining risk on transactions. Several variables (such as cost of transaction, transaction history, customer loyalty, indemnity, spending pattern, and system specific) on which trust depends are used to define trust. These variables in turn influence actions taken by a transacting entity (e.g., verification, security level decision, and authorization). Certain parameters, such as time and location, modify trust actions. In addition, Manchala pointed out that the existing e-commerce protocols have not been equipped with mechanisms to protect a vendor from a customer who makes a fraudulent payment or a customer from a vendor who supplies low quality or garbage goods. In other words, these protocols need to be equipped with suitable trust mechanisms and they should be strengthened by adding a non-reputable context to the transaction protocol. In an e-commerce transaction, mutual trust should exist between a vendor and a customer with several intermediaries involved in the transaction. In practical scenarios, eBay, for example, offers a safeguard service, that ensures the payment and the satisfactory delivery of the goods, but this service is not without costs.

The game theory-based research lays the foundation for online reputation systems research and provides interesting insight into the complex behavioral dynamics. Most of the game theoretic models assume that stage game outcomes are publicly observed. Online feedback mechanisms, in contrast, rely on private (pair-wise) and subjec-

tive ratings of stage game outcomes. Dellarocas (2003) introduced two important considerations, the incentive for providing feedback and the credibility or the truthfulness of the feedback.

Khare and Rifkin (1998) presented pragmatic details of Web-based trust management technology for identifying principals, labeling resources, and enforcing policies. It sketched how trust management might be integrated into Web applications for document authoring and distribution, content filtering, and mobile code security. By measuring today's Web protocols, servers, and clients, the authors called for stakeholders' support in bringing automatable trust management to the Web.

The reputation systems play an important role to evaluate the trustworthiness of transaction parties. A number of reputation systems and mechanisms were proposed for online environments and agent systems. Pujol, Sanguesa, and Delgado (2002) applied network flow techniques and proposed a generalized algorithm that extracts the reputation in a general class of social networks. Jøsang and Ismail (2002) and Jøsang and Tran (2003) developed and evaluated the beta reputation system for electronic markets by modeling reputation as posterior probability given a sequence of experiences. Among other things, they showed that a market with limited duration rather than infinite longevity of transaction feedback provides the best condition. Sabater and Sierra (2002) proposed the Regret system and showed how social network analysis can be used in the reputation system. Sen and Sajja (2002) proposed a word-of-mouth reputation algorithm to select service providers. Their focus was on allowing querying agent to select one of the high-performance service providers with a minimum probabilistic guarantee. Yu and Singh (2000) developed an approach for social reputation management and their model combines agents' belief ratings using combination schemes similar to certainty factors. The reputation ratings are propagated through neighbors. Zacharia and Maes (2000) proposed an approach that is an approximation of game-theoretic models and studied the effects of feedback mechanisms on markets with dynamic pricing using simulation modeling.

A few proposals specifically attempted to address the issue of quality or credibility of the feedback. Chen and Singh (2001) differentiated the ratings by the reputation of raters that is computed based the majority opinions of the rating. Adversaries who submit dishonest feedback can still gain a good reputation as a rater in a method simply by submitting a large amount of feedback and becoming the majority opinion. Dellarocas (2000) proposed mechanisms to combat two types of cheating behaviors when submitting feedbacks. The basic idea is to detect and filter out the feedbacks in certain scenarios using cluster-filtering techniques. The technique can be applied into feedback-based reputation systems to filter out the suspicious "fake" ratings before an actual aggregation. Miller, Resnick, and Zeckhauser (2002) proposed a mechanism, based on budget balanced payments in exchange for feedbacks, which provides strict incentives for all agents to tell the truth. This provides yet another approach to the problem of feedback trustworthiness. However, such a mechanism is vulnerable to malicious collusion. The development of effective mechanisms for dealing with collusive manipulations of online reputations systems is currently an active area of research.

On the other side, Salam, Iyer, Palvia, and Singh (2005) explored a framework to highlight the importance of nurturing consumer trust in the context of e-commerce. In particular, the authors pointed that the technical approaches to establish credibility and integrity are necessary but not sufficient for creating the long-term trusting relationships between consumers and online businesses. Web vendors must align both their long-term and short-term relationships with consumers and develop interventions to inspire consumer beliefs that affect their attitudes, intentions, and dependence, and ultimately their willingness to spend money. The Web vendors must address the factors affecting different belief classes to establish the trustworthiness of their organizations. They need a long-term approach to manage trust and generate a positive consumer experience from each and every Internet transaction. Quite a number of studies attempted to seek trust building factors and their relationships in order to propose guidelines or policies for designing a trustworthy e-commerce solution, for example, Jones, Wilikens, Morris, and Masera (2000); Rutter (2001); McKnight et al. (2002); Li et al. (2004); Gefen (2000); Pennington, Wilcox, and Grover (2004); Bhattacherjee (2002); and Pavlou and Gefen (2004). As we can see from the above, technologies and policies are in parallel influencing and enhancing trust management.

Solutions for Distributed Systems. Trust modeling and management can be applied to enhance security, dependability and other quality attributes of a system or a system entity. Main usage of trust management can be summarized below:

- Detecting malicious entity in a system.
- Helping in decision making in system processing.
- Selecting the best entity from a number of candidates, for example, selecting the best route node or the best path in MANET.
- Benefiting on system optimization. For example, if an entity is trusted, some procedures can be saved, which could benefit the efficiency of a system.
- Improving Quality of Services through applying trust management technology in a system.

So far, the trust evaluation and management technology have been studied and developed in many areas, such as distributed systems: P2P, Ad hoc, GRID, e-commerce, Web services, and software engineering. For example, P2P library, resource management in GRID, security enhancement in GRID computing, trusted Ad hoc routing and component software system configurations. We have introduced many of them in Section 2.3.

Tajeddine, Kayssi, Chehab, and Artail (2005) proposed a comprehensive reputation based trust model for distributed system. This approach requires that a host asks about the reputation of a target host that it wants to interact with. It calculates a reputation value based on its previous experiences and the gathered reputation values from other hosts, and then it decides whether to interact with the target host or not. The initiator also evaluates the credibility of hosts providing reputation values by estimating the similarity, the activity, the popularity, and the cooperation of the queried host. Moreover, each host uses different dynamic decay factors that depend on the consistency of the interaction results of a certain host.

Common opinions can be summarized based on the literature study. For a distributed system, trust modeling and evaluation can be used for improving system security and reliability. The trust of a trustor on a trustee is based on the trustor's past experience in the same or similar context, and recommendations or reputations generated from the experiences of other system entities. The contributions of the recommendations and the trustor's experiences to the trust value calculation are influenced by their age or time. The recommendations' contribution is also influenced by such factors as the trustor's opinion on the recommenders, distance (e.g., hops between the trustor and the recommender in ad hoc networks), and so on. Taking some detailed examples of policies for trust evaluation:

- Trust value is no bigger than trust value generated by trustor's experiences or recommendations.
- Latest information from experiences and recommendations will contribute more the calculation of trust value.

Furthermore, we summarize the factors considered in trust modeling:

- Recommendations, reputations, feedback from other entities (based on their experiences).
- Personal or local experience and its influencing factor (e.g. the time of the experience).
- Trust (credibility) on recommendations/ reputations of the trustor.
- Context factors (e.g. time, distance, transaction context, community context).
- Policy factors, i.e. the trustor's standards with regard to trust (e.g. accepted level of recommendations).

The above factors are generally aggregated through weighting. The weighting has some dependencies, for example, similarity, activity, popularity, and cooperation of a certain entity.

However, most modeling aims to support trust evaluation for decision making. Little considers trust control or management, for example, how to maintain trust for a period of time based on the evaluation. Trust management is more than trust evaluation, especially in an open computing platform, where autonomic trust management is becoming an important research topic.

Solutions for User-Device or User-System Interaction. A number of trusted computing projects have been conducted in the literature and industry. For example, Trusted Computing Group (TCG) defines and promotes open standards for hardware-enabled trusted computing and security technologies, including hardware building blocks and software interfaces, across multiple platforms, peripherals, and devices. TCG specified technology enables more secure computing environments without compromising functional integrity, privacy, or individual rights. It aims to build up a trusted computing device on the basis of a secure hardware chip—Trusted Platform Module (TPM). In short, the TPM is the hardware that controls the boot-up process. Every time the computer is reset, the TPM steps in, verifies the Operating System (OS) loader before letting boot-up continue. The OS loader is assumed to verify the Operating System. The OS is then assumed to verify every bit of software that it can find in the computer, and so on. The TPM allow all hardware and software components to check whether they have woken up in trusted states. If not, they should refuse to work. It also provides a secure storage for confidential information. In addition, it is possible for the computer user to select whether to boot his/her machine in a trusted computing mode or in a legacy mode.

eBay, Amazon, and other famous Internet services show the recommendation level of sellers and products based on the feedback accumulated. Credence employs a simple, network-wide voting scheme where users can contribute positive and negative evaluations of files. On top of this, a client uses statistical tests to weight the importance of votes from other peers. And finally, Credence allows clients to extend the horizon of information by selectively sharing information with other peers.

In order to design a trustworthy system, a number of user studies provide results on how to design a trustworthy user interfaces, especially for the recommendation systems. For example, Herlocker, Konstan, and Riedl (2000) studied explanation's aid on user trust regarding Automated Collaborative Filtering (ACF)—a technological recommendation approach based on the similarity of interest). It addressed explanation interfaces for the ACF systems—*how* they should be implemented and *why* they should be implemented. It presented that experimental evidence shows that providing explanations can improve the acceptance of the ACF systems.

As we have discussed, trust is a subjective topic. A system is trusted if and only if its users trust it. But little work practiced to formalize user-device or user-system interaction in order to extract the user's standards, as well as adaptively provide information about trust status to the user, particularly if the device is portable with a limited screen.

Privacy Support. Trust modeling and management can not only enhance security, but also support privacy. The term *privacy* denotes the ability of an entity to determine whether, when, and to whom information is to be released. With the trust modeling and evaluation, it is helpful to determine above "whether," "when," and "to whom." However, it lacks discussion on how to enhance privacy through trust management in the literature.

Limitations and Further Discussion

Trust modeling and management remain an active research area in recent years. Trust is today's fashion in security (Gollmann, 2007). However, many interesting research issues are yet to be fully explored. We summarize some of them for interested readers.

As discussed above, it lacks a widely accepted trust definition across multiple related disciplines. This could cause a comprehensive trust model missing in the literature. Perhaps it is impossible to have such a model that can be applied in various situations and systems. Diverse definitions could make normal people confused. This would make it hard for them to understand a trust management solution.

Secondly, it lacks general criteria to assess the effectiveness of a trust model and a trust management solution in the literature. Why is a trust model trustworthy? Why and how is a trust management system effective? Most of the existing work overlooked these issues and missed discussions about them. In addition, new attacks or malicious behaviors could also destroy a trust management system. Current prove is still based on empirical and experimental study.

Thirdly, the literature lacks discussions on the competence of trust management. That is when and in which situation trust is possibly or impossibly managed. This is a very interesting and important issue worth our attention, especially for autonomic trust management.

Finally, from the practical point of view, some important issues such as trust/reputation value storage (Li & Singhal, 2007), usability of a trust management system, what is the proper settings of a user's policies for trust decision, and how to extract these policies in a user-friendly way need further exploration.

FUTURE TRENDS

Herein, we provide insights about future and emerging trends on trust modeling and management.

An Integrated "Soft Trust" and "Hard Trust" Solution

Theoretically, there are two basic approaches for building up a trust relationship. We name them "soft trust" solutions and "hard trust" solutions. The "soft trust" solution provides trust based on trust evaluation according to subjective trust standards, facts from previous experiences and history. The "hard trust" solution builds up the trust through structural and objective regulations, standards, as well as widely accepted rules, mechanisms and sound technologies. Possibly, both approaches are applied in a real system. They can cooperate and support with each other to provide a trustworthy system. The "hard trust" provides a guarantee for the "soft trust" solution to ensure the integrity of its functionality. The "soft trust" can provide a guideline to determine which "hard trust" mechanisms should be applied and at which moment. It provides intelligence for selecting a suitable "hard trust" solution.

An integrated solution is expected to provide a trust management framework that applies both the "hard trust" solution and the "soft trust" solution. This framework should support data collection and management for trust evaluation, trust standards extraction from the trustor (e.g., a system or device user), and experience or evidence

dissemination inside and outside the system, as well as a decision engine to provide guidelines for applying effective "hard trust" mechanisms for trust management purposes.

In addition, how to store, propagate and collect information for trust evaluation and management is seldom considered in the existing theoretical work, but is a relevant issue for deployment. Human-device interaction is also crucial to transmit a user's trust standards to the device and the device needs to provide its assessment on trust to its user. These factors influence the final success of trust management.

Autonomic Trust Management

There is a trend that all the processing for trust management is becoming autonomic. This is benefited from the digitalization of trust model. Since trust relationship is dynamically changed, this requires the trust management should be context-aware and intelligent to handle the context changes. Obviously, it does not suffice to require the trustor (e.g., most possibly a digital system user) to make a lot of trust related decisions because that would destroy any attempt at user friendliness. For example, the user may not be informed enough to make correct decisions. Thus, establishing trust is quite a complex task with many optional actions to take. Rather trust should be managed automatically following a high level policy established by the trustor. We call such trust management autonomic. Autonomic trust management automatically processes evidence collection, trust evaluation, and trust (re-)establishment and control. We need a proper mechanism to support autonomic trust management not only on trust establishment, but also on trust sustaining. In addition, the trust model itself should be adaptively adjusted in order to match and reflect real system situation. Context-aware trust management and adaptive trust model optimization for autonomic trust management are developing research topics (Campadello, Coutand, Del Rosso, Holtmanns,

Kanter, Räck, Mrohs, & Steglich, 2005; Yan & Prehofer, 2007).

Cross-Domain Benefit

We can estimate that trust management will not only benefit security, but also other properties of the system, such as privacy, usability, dependability and Quality of Services. Combining trust management with other management tasks (e.g., resource management, power management, identity management, risk management and fault management) or applying it into other areas could produce cross-domain benefits. The outcome system will be a more intelligent system to help users managing their increasing amount of digital trust relationships, while providing also good performance.

CONCLUSION

This chapter firstly introduced the perception of trust in the literature. Based on the various definitions of trust, we summarized the factors influencing trust and the characteristics of trust. The trust modeling for digital processing is actually based on the understanding of trust, the influencing factors of trust and its characteristics. From a social concept, trust has become a digital object that can be processed. The research on trust modeling, trust evaluation and trust management that have been conducted in the area of distributed systems and e-commerce was presented. The latest trust management systems model trust using mathematical approaches. Thus, it is possible to conduct digital trust management for the emerging technologies.

Current research on trust modeling mostly focuses on the theoretic study based on empirical and experimental results. It lacks experiences in practice. Most of existing deployed solutions are special system driven. The support of a generic solution for trust management, which also benefits other system properties, is usually not considered. In addition, the user-device interaction with regard to trust management is a topic that needs further study. There are still many interesting research issues requiring full investigation.

Regarding the future trends, we believe an integrated solution is very promising that combines traditional security solution with newly developed trust evaluation based management together. This integrated solution should handle trust management in an automatic way and cooperate with other technologies to offer a better system performance.

REFERENCES

Abdul-Rahman, A., & Hailes, S. (2000). Supporting trust in virtual communities. *Proceedings of the 33rd Hawaii International Conference on System Sciences, 6007.*

Aberer, K., & Despotovic, Z. (2001). Managing trust in a peer-to-peer information system. *Proceedings of the ACM Conference on Information and Knowledge Management (CIKM),* USA, 310–317.

Avizienis, A., Laprie, J. C., Randell, B., & Landwehr, C. (2004). Basic concepts and taxonomy of dependable and secure computing. *IEEE Transactions on Dependable and Secure Computing, 1*(1), 11–33. doi:10.1109/TDSC.2004.2

Azzedin, F., & Muthucumaru, M. (2004). A trust brokering system and its application to resource management in public-resource grids. Proceedings of the 18th International Symposium on Parallel and Distributed Processing, April, 22.

Ba, S., & Pavlou, P. A. (2002). Evidence of the effects of trust building technology in electronic markets: Price premiums and buyer behavior. *MIS Quarterly, 26*(3), 243–268. doi:10.2307/4132332

Ba, S., Whinston, A., & Zhang, H. (1999). Building trust in the electronic market through an economic incentive mechanism. *Proceedings of the International Conference on Information Systems*, 208–213.

Balakrishnan, V., & Varadharajan, V. (2005). Designing secure wireless mobile ad hoc networks. Proceedings of the 19th International Conference on Advanced Information Networking and Applications (AINA 2005), March, Vol. 2, 5–8.

Bhattacherjee, A. (2002). Individual trust in online firms: scale development and initial test. *Journal of Management Information Systems*, *19*(1), 211–241.

Blaze, M., Feigenbaum, J., & Lacy, J. (1996). Decentralized trust management. *Proceedings of IEEE Symposium on Security and Privacy*, May, 164–173.

Boon, S., & Holmes, J. (1991). The dynamics of interpersonal trust: Resolving uncertainty in the face of risk. In R. Hinde & J. Groebel (Eds.), *Cooperation and prosocial behavior*, pp. 190-211. Cambridge, UK: Cambridge University Press.

Buchegger, S., & Le Boudec, J. Y. (2004). A robust reputation system for P2P and mobile ad-hoc networks. *Proceedings of the 2nd Workshop Economics of Peer-to-Peer Systems*.

Cahill, V. (2003). Using trust for secure collaboration in uncertain environments. *IEEE Pervasive Computing / IEEE Computer Society [and] IEEE Communications Society*, *2*(3), 52–61. doi:10.1109/MPRV.2003.1228527

Campadello, S., Coutand, O., Del Rosso, C., Holtmanns, S., Kanter, T., Räck, C., et al. (2005). Trust and privacy in context-aware support for communication in mobile groups. *Proceedings of Context Awareness for Proactive Systems (CAPS) 2005,* Helsinki, Finland.

Chen, M., & Singh, J. P. (2001). Computing and using reputations for internet ratings. *Proceedings of 3rd ACM Conference on Electronic Commerce*.

Corritore, C. L., Kracher, B., & Wiedenbeck, S. (2003). Online trust: Concepts, evolving themes, a model. *International Journal of Human-Computer Studies. Trust and Technology*, *58*(6), 737–758.

Dellarocas, C. (2000). Immunizing online reputation reporting systems against unfair ratings and discriminatory behavior. *Proceedings of the 2nd ACM Conference on Electronic Commerce*.

Dellarocas, C. (2003). The digitization of word-of-mouth: promise and challenges of online reputation mechanism. *Management Science*, *49*(10), 1407–1424. doi:10.1287/mnsc.49.10.1407.17308

Denning, D. E. (1993). A new paradigm for trusted systems. *Proceedings of the IEEE New Paradigms Workshop*.

Fishbein, M., & Ajzen, I. (1975). *Beliefs, attitude, intention, and behavior: An introduction to theory and research*. Addison-Wesley, Reading, MA.

Gambetta, D. (1990). *Can we trust Trust? Trust: Making and breaking cooperative relations*. Oxford: Basil Blackwell.

Ganeriwal, S., & Srivastava, M. B. (2004). Reputation-based framework for high integrity sensor networks. *Proceedings of the ACM Security for Ad-Hoc and Sensor Networks*, pp. 66–67.

Gefen, D. (2000). E-commerce: The role of familiarity and trust. *Omega*, *28*(6), 725–737. doi:10.1016/S0305-0483(00)00021-9

Gil, Y., & Ratnakar, V. (2002). Trusting information sources one citizen at a time. *Proceedings of the 1st International Semantic Web Conference*, Italy, June.

Gollmann, D. (2007). Why trust is bad for security. Retrieved May 2007, from http://www.sics.se/policy2005/Policy_Pres1/dg-policy-trust.ppt

Grandison, T., & Sloman, M. (2000). A survey of trust in Internet applications. *IEEE Communications and Survey, 4th Quarter, 3*(4), 2–16.

Guha, R., & Kumar, R. (2004). Propagation of trust and distrust. *Proceedings of the 13th international conference on World Wide Web*, ACM Press, pp. 403–412.

Guo, W., Xiong, Z., & Li, Z. (2005). Dynamic trust evaluation based routing model for ad hoc networks. Proceedings of International Conference on Wireless Communications, Networking and Mobile Computing, Vol. 2, September, pp. 727–730.

Herlocker, J. L., Konstan, J. A., & Riedl, J. (2000). Explaining collaborative filtering recommendations. *Proceedings of the 2000 ACM conference on Computer supported cooperative work.*

Hjelm, J., & Holtmanns, S. (2006). Privacy and trust visualization. Computer Human Interaction - CHI 2006 Conference Proceedings and Extended Abstracts, ACM Press, Montréal, Canada.

Hoffman, D. L., Novak, T. P., & Peralta, M. (1999). Building consumer trust online. *Communications of the ACM, 42*(4), 80–87. doi:10.1145/299157.299175

Jones, S., Wilikens, M., Morris, P., & Masera, M. (2000). Trust requirements in e-business - A conceptual framework for understanding the needs and concerns of different stakeholders. *Communications of the ACM, 43*(12), 80–87. doi:10.1145/355112.355128

Jøsang, A. (1999). An algebra for assessing trust in certification chains. *Proceedings of the Networking Distributed System Security Symposium.*

Jøsang, A. (2001). A logic for uncertain probabilities. *International Journal of Uncertainty. Fuzziness and Knowledge-Based Systems, 9*(3), 279–311.

Jøsang, A., & Ismail, R. (2002). The beta reputation system. *Proceedings of the 15th Bled Electronic Commerce Conference*, June.

Jøsang, A., & Tran, N. (2003). Simulating the effect of reputation systems on e-markets. *Proceedings of the First International Conference on Trust Management.*

Kamvar, S. D., Schlosser, M. T., & Garcia-Molina, H. (2003). The EigenTrust algorithm for reputation management in P2P networks. *Proceedings of the 12th International World Wide Web Conference.*

Khare, R., & Rifkin, A. (1998). Trust management on the World Wide Web. *First Monday, 3*(6).

Kosko, B. (1986). Fuzzy cognitive maps. *International Journal of Man-Machine Studies, 24*, 65–75. doi:10.1016/S0020-7373(86)80040-2

Lee, S., Sherwood, R., & Bhattacharjee, B. (2003). Cooperative peer groups in NICE. *Proceedings of the IEEE Conf. Computer Comm. (INFOCOM 03)*, IEEE CS Press, pp. 1272–1282.

Li, H., & Singhal, M. (2007). Trust management in distributed systems. *Computer, 40*(2), 45–53. doi:10.1109/MC.2007.76

Li, X., Valacich, J. S., & Hess, T. J. (2004). Predicting user trust in information systems: a comparison of competing trust models. Proceedings of the 37th Annual Hawaii International Conference on System Sciences

Liang, Z., & Shi, W. (2005). PET: a PErsonalized Trust model with reputation and risk evaluation for P2P resource sharing. Proceedings of the 38thAnnual Hawaii International Conference on System Sciences, January, pp. 201b–201b.

Lin, C., Varadharajan, V., Wang, Y., & Pruthi, V. (2004). Enhancing GRID security with trust management. Proceedings of the IEEE International Conference on Services Computing (SCC 2004), pp. 303–310.

Liu, Z., Joy, A. W., & Thompson, R. A. (2004). A dynamic trust model for mobile ad hoc networks. Proceedings of the 10th IEEE International Workshop on Future Trends of Distributed Computing Systems (FTDCS 2004), May, pp. 80–85.

Manchala, D. W. (2000). E-commerce trust metrics and models. *IEEE Internet Computing, 4*(2), 36–44. doi:10.1109/4236.832944

Marsh, S. (1994). *Formalising trust as a computational concept.* (Doctoral dissertation, University of Stirling, 1994).

Maurer, U. (1996). Modeling a public-key infrastructure. *Proceedings of the European Symposium of Research on Computer Security, LNCS, 1146*, 325–350.

Mayer, R. C., Davis, J. H., & Schoorman, F. D. (1995). An integrative model of organizational trust. *Academy of Management Review, 20*(3), 709–734. doi:10.2307/258792

McKnight, D. H., & Chervany, N. L. (2000). What is Trust? a conceptual analysis and an interdisciplinary model. *Proceedings of the 2000 Americas Conference on Information Systems (AMCI2000). AIS,* August, Long Beach, CA.

McKnight, D. H., & Chervany, N. L. (2003). The meanings of trust. UMN university report. Retrieved December 2006, from http://misrc.umn.edu/wpaper/WorkingPapers/9604.pdf

McKnight, D. H., Choudhury, V., & Kacmar, C. (2002). Developing and validating trust measures for e-commerce: An integrative typology. *Information Systems Research, 13*(3), 334–359. doi:10.1287/isre.13.3.334.81

Miller, N. H., Resnick, P., & Zeckhauser, R. J. (2002). *Eliciting honest feedback in electronic markets.* KSG Working Paper Series RWP02-039.

Mui, L. (2003). *Computational models of trust and reputation: agents, evolutionary games, and social networks.* Doctoral dissertation, Massachusetts Institute of Technology.

Pavlou, P., & Gefen, D. (2004). Building effective online marketplaces with institution-based trust. *Information Systems Research, 15*(1), 37–59. doi:10.1287/isre.1040.0015

Pennington, R., Wilcox, H. D., & Grover, V. (2004). The role of system trust in business-to-consumer transactions. *Journal of Management Information Systems, 20*(3), 197–226.

Perlman, R. (1999). An overview of PKI trust models. *IEEE Network, 13*(6), 38–43. doi:10.1109/65.806987

Pujol, J. M., Sanguesa, R., & Delgado, J. (2002). Extracting reputation in multi-agent systems by means of social network topology. *Proceedings of the First International Joint Conf. Autonomous Agents and Multiagent Systems.*

Reiter, M. K., & Stubblebine, S. G. (1998). Resilient authentication using path independence. *IEEE Transactions on Computers, 47*(12), 1351–1362. doi:10.1109/12.737682

Resnick, P., & Zeckhauser, R. (2002). Trust among strangers in Internet transactions: Empirical analysis of eBay's reputation system. In M. Baye, (Ed.), *Advances in applied microeconomics: The economics of the Internet and e-commerce, 11*, 127–157.

Rousseau, D. M., Sitkin, S. B., Burt, R. S., & Camerer, C. (1998). Not so different after all: A cross-discipline view of trust. *Academy of Management Review, 23*(3), 393–404.

Rutter, J. (2001). From the sociology of trust towards a sociology of "e-trust". *International Journal of New Product Development & Innovation Management, 2*(4), 371–385.

Sabater, J., & Sierra, C. (2002). Reputation and social network analysis in multi-agent systems. *Proceedings of the First International Joint Conference on Autonomous Agents and Multiagent Systems*.

Salam, A. F., Iyer, L., Palvia, P., & Singh, R. (2005). Trust in e-commerce. *Communications of the ACM*, *48*(2), 73–77. doi:10.1145/1042091.1042093

Sarkio, K., & Holtmanns, S. (2007). Tailored trustworthiness estimations in peer to peer networks. To appear in International Journal of Internet Technology and Secured Transactions IJITST, Inderscience.

Sen, S., & Sajja, N. (2002). Robustness of reputation-based trust: Boolean case. *Proceedings of the First International Joint Conference on Autonomous Agents and Multiagent Systems*.

Singh, A., & Liu, L. (2003). TrustMe: Anonymous management of trust relationships in decentralized P2P systems. *Proceedings of the IEEE International Conference on Peer-to-Peer Computing*, pp. 142–149.

Song, S., Hwang, K., Zhou, R., & Kwok, Y. K. (2005). Trusted P2P transactions with fuzzy reputation aggregation. *IEEE Internet Computing*, *9*(6), 24–34. doi:10.1109/MIC.2005.136

Sun, Y., Yu, W., Han, Z., & Liu, K. J. R. (2006). Information theoretic framework of trust modeling and evaluation for ad hoc networks. *IEEE Journal on Selected Areas in Communications*, *24*(2), 305–317. doi:10.1109/JSAC.2005.861389

Tajeddine, A., Kayssi, A., Chehab, A., & Artail, H. (2005, September). A comprehensive reputation-based trust model for distributed systems. Proceedings of the Workshop of the 1st International Conference on Security and Privacy for Emerging Areas in Communication Networks, pp. 118–127.

Tan, Y., & Thoen, W. (1998). Toward a generic model of trust for electronic commerce. *International Journal of Electronic Commerce*, *5*(2), 61–74.

TCG. Trusted Computing Group, Trusted Platform Module - TPM Specification v1.2, 2003. Retrieved May, 2006, from https://www.trustedcomputing-group.org/specs/TPM/

Theodorakopoulos, G., & Baras, J. S. (2006). On trust models and trust evaluation metrics for ad hoc networks. *IEEE Journal on Selected Areas in Communications*, *24*(2), 318–328. doi:10.1109/JSAC.2005.861390

Wang, Y., & Varadharajan, V. (2005). Trust2: developing trust in peer-to-peer environments. *Proceedings of the IEEE International Conference on Services Computing*, *1*(July), 24–31.

Xiong, L., & Liu, L. (2004). PeerTrust: Supporting reputation-based trust for peer-to-peer electronic communities. *IEEE Transactions on Knowledge and Data Engineering*, *16*(7), 843–857. doi:10.1109/TKDE.2004.1318566

Yan, Z. (2006). A conceptual architecture of a trusted mobile environment. *Proceedings of IEEE SecPerU06*, France, pp. 75–81.

Yan, Z., & MacLaverty, R. (2006). Autonomic trust management in a component based software system. *Proceedings of 3rd International Conference on Autonomic and Trusted Computing (ATC06)*, *LNCS*, Vol. 4158, September, pp. 279–292.

Yan, Z., & Prehofer, C. (2007). An adaptive trust control model for a trustworthy software component platform. *Proceedings of the 4th International Conference on Autonomic and Trusted Computing (ATC2007)*, *LNCS*, Vol. 4610, pp. 226-238.

Yan, Z., Zhang, P., & Virtanen, T. (2003). Trust evaluation based security solution in ad hoc networks. *Proceedings of the Seventh Nordic Workshop on Secure IT Systems (NordSec03)*, Norway.

Yu, B., & Singh, M. P. (2000). A social mechanism of reputation management in electronic communities. *Proceedings of the Seventh International Conference on Cooperative Information Agents.*

Zacharia, G., & Maes, P. (2000). Trust management through reputation mechanisms. *Applied Artificial Intelligence*, *14*(8), 881–908.

Zhang, Z., Wang, X., & Wang, Y. (2005). A P2P global trust model based on recommendation. *Proceedings of the 2005 International Conference on Machine Learning and Cybernetics,* Vol. 7, August, pp. 3975–3980.

Zhou, M., Mei, H., & Zhang, L. (2005). A multi-property trust model for reconfiguring component software. Proceedings of the Fifth International Conference on Quality Software QAIC2005, September, pp. 142–149.

Compilation of References

Abdul-Rahman, A., & Hailes, S. (2000). Supporting trust in virtual communities. *Proceedings of the 33rd Hawaii International Conference on System Sciences, 6007.*

Aberer, K., & Despotovic, Z. (2001). Managing trust in a peer-to-peer information system. *Proceedings of the ACM Conference on Information and Knowledge Management (CIKM),* USA, 310–317.

Acquisti, A., & Gross, R. (2005). *Information revelation and privacy in online social networks (The Facebook case).* Pittsburgh, PA: Carnegie Mellon University.

Acquisti, A., Carrara, E., Stutzman, F., Callas, J., Schimmer, K., Nadjm, M., et al. (2007, October). *Security issues and recommendations for online social networks* (ENISA Position Paper No. 1). Crete, Greece: ENISA.

Adam, B., Beck, U., & van Loon, J. (Eds.). (2000). *The risk society and beyond: Critical issues in social theory.* London: Sage.

Adamic, L. A., Buyukkokten, O., & Adar, E. (2003). A social network caught in the web. *First Monday, 8*(6). Retrieved December 28, 2005, from http://www.firstmonday.rog/issues/issues8_6/adamic/

Adar, E., & Huberman, B. (2000). *Free riding on Gnutella.* Technical report, Xerox PARC.

Agraz, D. (2004). *Making friends through social networks: A new trend in virtual communication.* Palo Alto, CA: Stanford University.

Aizpurua, I., Ortix, A., Oyarzum, D., Arizkuren, I., Ansrés, A., Posada, J., & Iurgel, I. (2004). Adaption of mesh morphing techniques for avatars used in Web applications. In: F. Perales & B. Draper (Eds.), *Articulated motion and deformable objects: Third international workshop.* London: Springer.

Akerlof, G. A. (1970). The market for lemons: Quality uncertainty and the market mechanism. *The Quarterly Journal of Economics, 84*(3), 488–500. doi:10.2307/1879431

Akhavan, P., Jafari, M., & Fathian, M. (2005). Exploring failure-factors of implementing knowledge management systems in organizations. *Journal of Knowledge Management Practice.*

Alavi, M., & Leidner, D. E. (1999). Knowledge management systems: Issues, challenges, and benefits. *Communications of the AIS, 1*(7).

Alavi, M., & Leidner, D. (2001). Review: Knowledge management and knowledge management systems: Conceptual foundations and research issues. *Management Information Systems Quarterly, 25*(1), 107–136. doi:10.2307/3250961

Alcantara, M. (2009). El Bullying, Acoso Escolar. Innovación y experiencias Educativas. *CSI-CSIF Revista Digital, 55.*

Alexander, C. (1964). Notes on the synthesis of form. Cambridge, Ma: Harvard University Press.

Alexander, D. (2005). An interpretation of disaster in terms of changes in culture, society and international relations. In R.W. Perry & E.L. Quarantelli (Eds.), *What is a disaster? New answers to old questions* (pp. 25-38). Philadelphia: XLibris.

Alexander, D. E. (2006). Globalization of disaster: trends, problems and dilemmas. *Journal of International Affairs, 59*(2), 1-22. Retrieved August 3, 2009, from http://www.policyinnovations.org/ideas/policy_library/data/01330/_res/id=sa_File1/alexander_globofdisaster.pdf

Allen, C. (2004). *My advice to social networking services.* Retrieved December 28, 2005, from http://www.lifewith-alacrity.com/2004/02/my_advice_to_so.html

Allen, I. E., & Seaman, J. (2003). *Sizing the opportunity: The quality and extent of online education in the United States, 2002-2003.* Alfred P. Sloan Foundation.

Amazon glitch outs authors reviewing own books. (2004). Retrieved December 28, 2005, from http://www.ctv.ca/serv-let/ArticleNews/story/CTVNews/1076990577460_35

Anon. (2003). *Net grief for online 'suicide'.* Retrieved from http://news.bbc.co.uk/1/hi/technology/2724819.stm

Appelt, W., & Mambrey, P. (1999), Experiences with the BSCW Shared Workspace System as the Backbone of a Virtual Learning Environment for Students. *In Proceedings of ED Media '99.'* Charlottesville, (pp. 1710-1715).

Ardichvili, A. (2003). Motivation and barriers to participation in virtual knowledge-sharing communities of practice. *Journal of Knowledge Management, 7*(1), 64–77. doi:10.1108/13673270310463626

Argote, L., & Ingram, P. (2000). Knowledge transfer: A basis for competitive advantage in firms. *Organizational Behavior and Human Decision Processes, 82*(1), 150–169. doi:10.1006/obhd.2000.2893

Armstrong, A., & Hagel, J. (1997). *Net gain: Expanding markets through virtual communities.* Boston, MA: Harvard Business School Press.

Armstrong, A., & Hagel, J. III. (1996). The real value of online communities. *Harvard Business Review, 74*(3), 134–141.

Aschbacher, P. R. (2003). Gender differences in the perception and use of an informal science learning Web site. *Journal of the Learning Sciences, 9.*

ATA News (2009). *RCMP and CTF join forces to fight cyberbullying, 43*(10), 6.

Attewell, P., Suazo-Garcia, B., & Battle, J. (2003). Computers and Young Children: Social Benefit or Social Problem? *Social Forces, 82*(1), 277–296. doi:10.1353/sof.2003.0075

Ausubel, D. P. (1963). *The psychology of meaningful verbal learning.* New York: Grune and Stratton.

Aviles, J. M. (2006). Bulling. Intimidación y maltrato entre alumnado. Retrieved May 1, 2010, From: www.educacionenvalores.org/ Bullying-Intimidacion-y-maltrato.html

Avizienis, A., Laprie, J. C., Randell, B., & Landwehr, C. (2004). Basic concepts and taxonomy of dependable and secure computing. *IEEE Transactions on Dependable and Secure Computing, 1*(1), 11–33. doi:10.1109/TDSC.2004.2

Aytes, K., & Connolly, T. (2004). Computer security and risky computing practices: A rational choice perspective. *Journal of Organizational and End User Computing, 16*(3), 22–40.

Azzedin, F., & Muthucumaru, M. (2004). A trust brokering system and its application to resource management in public-resource grids. Proceedings of the 18th International Symposium on Parallel and Distributed Processing, April, 22.

Ba, S., Whinston, A., & Zhang, H. (1999). Building trust in the electronic market through an economic incentive mechanism. *Proceedings of the International Conference on Information Systems,* 208–213.

Bagchi, K., & Udo, G. (2003). An analysis of the growth of computer and internet security breaches. *Communications of the Association for Information Systems, 12*(46), 129.

Bakardjieva, M., Feenberg, A., & Goldie, J. (2004). Usercentered Internet research: The ethical challenge. In Buchanan, E. (Ed.), *Readings in virtual research ethics: Issues and controversies* (pp. 338–350). Hershey, PA: Idea Group Publishing. doi:10.4018/978-1-59140-152-0.ch018

Baker, G. (2002). The effects of synchronous collaborative technologies on decision making: A study of virtual teams. *Information Resources Management Journal, 15*(4), 79–93. doi:10.4018/irmj.2002100106

Baker, J. R., & Moore, S. M. (2008). Blogging as a social tool: A psychosocial examination of the effects of blogging. *Cyberpsychology & Behavior, 11*(6), 747–749. doi:10.1089/cpb.2008.0053

Balakrishnan, V., & Varadharajan, V. (2005). Designing secure wireless mobile ad hoc networks. Proceedings of the 19th International Conference on Advanced Information Networking and Applications (AINA 2005), March, Vol. 2, 5–8.

Balasubramanian, S., & Mahajan, V. (2001). The economic leverage of the virtual community. *International Journal of Electronic Commerce, 5*(3), 103–138.

Bandura, A. (1986). *Social foundations of thought & action.* Englewood Cliffs, NJ: Prentice Hall.

Barab, S. A. MaKinster, J. G., & Scheckler, R. (2003). Designing system dualities: Characterizing a web-supported professional development community. *The Information Society, 19*, 237-256.

Baroudi, J. J., & Orlikowski, W. J. (1988). A short form measure of user information satisfaction: A psychometric evaluation and notes. *Journal of Management Information Systems, 4*, 45–59.

Barron, C. (2000). *Giving Youth a Voice: A Basis for Rethinking Adolescent Violence.* Halifax, Nova Scotia: Fernwood Publishing.

Barton, A. H. (2005). Disaster and collective stress. In R.W. Perry & E.L. Quarantelli (Eds.), *What is a disaster? New answers to old questions* (pp.125-152). Philadelphia: XLibris.

Ba, S., & Pavlou, P. A. (2002). Evidence of the effects of trust building technology in electronic markets: Price premiums and buyer behavior. *MIS Quarterly, 26*(3), 243–268. doi:10.2307/4132332

Bassi, L.J., Ludwig, J., McMurrer, D.P., & Van Buren, M. (2000, September). *Profiting from learning: Do firms' investments in education and training pay off?* American Society for Training and Development (ASTD) and Saba.

Baym, N. (1998). The emergence of online community. In Jones, S. G. (Ed.), *Cybersociety 2.0* (pp. 35–68). Thousand Oaks, CA: Sage.

BBC. (2005). *ITV buys Friends Reunited Web site.* London: BBC Online. Retrieved from http://news.bbc.co.uk/1/hi/business/4502550.stm.

Beale, A., & Hall, K. (2007). Cyberbullying: What school administrators (and parents) can do. *Clearing House (Menasha, Wis.), 81*(1), 8–12. doi:10.3200/TCHS.81.1.8-12

Beaty, J., Hunter, P., & Bain, C. (1998). *The Norton introduction to literature.* New York, NY: W.W. Norton & Company.

Beck, U. (1992). *Risk Society: Towards a new modernity.* London: Sage.

Beck, U. (2009). *World at risk.* Cambridge: Polity.

Beenen, G., Ling, K., Wang, X., Chang, K., Frankowski, D., Resnick, P., & Kraut, R. E. (2004). Using social psychology to motivate contributions to online communities. *Proceedings of the 2004 ACM Conference on Computer Supported Cooperative Work,* (p. 221).

Beike, D., & Wirth-Beaumont, E. (2005). Psychological closure as a memory phenomenon. *Memory (Hove, England), 13*(6), 574–593. doi:10.1080/09658210444000241

Belilos, C. (1998). *Networking on the net.* Professionalism, ethics and courtesy on the net. Retrieved November, 2004, from www.easytraining.com/networking.htm

Bell, J. (2004). Why software training is a priority? *Booktech the Magazine, 7*, 8.

Bellsey, B. (2009). *Where you are not alone.* Retrieved May 1, 2010, from: http://www.bullying.org

Berne, E. (1961). *Transactional analysis in psychotherapy.* New York, NY: Evergreen.

Berne, E. (1964). *Games people play: The psychology of human relationships.* New York, NY: Deutsch.

Berners-Lee, T. (2000). Weaving the Web: The original design and ultimate destiny of the world wide Web. New York: Harper-Collins.

Bernstein, T., & Wagner, J. (1976). *Reverse dictionary.* London: Routledge.

Bhat, C. (2008). Cyber Bullying: Overview and Strategies for School Counsellors Guidance Officers and All School Personnel. *Australian Journal of Guidance & Counselling, 18*(1), 53–66. doi:10.1375/ajgc.18.1.53

Bhattacherjee, A. (2002). Individual trust in online firms: scale development and initial test. *Journal of Management Information Systems, 19*(1), 211–241.

Bickart, B., & Schindler, R. M. (2001). Internet forums as influential sources of consumer information. *Journal of Interactive Marketing, 15*(3), 31–41. doi:10.1002/dir.1014

Bieber, M., D., Engelbart, D., Furuta, R., Hiltz, S. R., Noll, J., Preece, J. et al. (2002). Toward virtual community knowledge evolution. *Journal of Management Information Systems, 18*(4), 11–35.

Bielefield, A., & Cheeseman, L. (1997). *Technology and copyright law.* New York: Neal-Schuman.

Bielen, M. (1997). Online discussion forums are the latest communication tool. *Chemical Market Reporter, 252*(7), 9.

Biocca, F., Burgoon, J., Harms, C., & Stoner, M. (2001). Criteria and scope conditions for a theory and measure of social presence. *Proceedings of Presence 2001,* Philadelphia, PA.

Bishop, J. (2002). *Development and evaluation of a virtual community.* Unpublished dissertation. http://www.jonathanbishop.com/ publications/display.aspx?Item=1.

Bishop, J. (2005). The role of mediating artefacts in the design of persuasive e-learning systems. In: *Proceedings of the Internet Technology & Applications 2005 Conference.* Wrexham: North East Wales Institute of Higher Education.

Bishop, J. (2006). Social change in organic and virtual communities: An exploratory study of bishop desires. *Paper presented to the Faith, Spirituality and Social Change Conference.* University of Winchester.

Bishop, J. (2007). The psychology of how Christ created faith and social change: Implications for the design of e-learning systems. *Paper presented to the 2ⁿᵈ International Conference on Faith, Spirituality, and Social Change.* University of Winchester.

Bishop, J. (2007). Ecological cognition: A new dynamic for human computer interaction. In: B. Wallace, A. Ross, J. Davies, & T. Anderson (Eds.), *The mind, the body and the world* (pp. 327-345). Exeter: Imprint Academic.

Bishop, J. (2009). Increasing membership in online communities: The five principles of managing virtual club economies. *Proceedings of the 3rd International Conference on Internet Technologies and Applications - ITA09,* Glyndwr University, Wrexham.

Bishop, J. (2010). *Multiculturalism in intergenerational contexts: Implications for the design of virtual worlds.* Paper Presented to the Reconstructing Multiculturalism Conference, Cardiff, UK.

Bishop, J. (2011). Transforming lurkers into posters: The role of the participation continuum. *Proceedings of the Fourth International Conference on Internet Technologies and Applications (ITA11),* Glyndwr University.

Bishop, J. (2003). Factors shaping the form of and participation in online communities. *Digital Matrix, 85,* 22–24.

Bishop, J. (2007). Ecological cognition: A new dynamic for human-computer interaction. In Wallace, B., Ross, A., Davies, J., & Anderson, T. (Eds.), *The mind, the body and the world: Psychology after cognitivism* (pp. 327–345). Exeter, UK: Imprint Academic.

Bishop, J. (2007). Increasing participation in online communities: A framework for human–computer interaction. *Computers in Human Behavior, 23*(4), 1881–1893. doi:10.1016/j.chb.2005.11.004

Bishop, J. (2009). Enhancing the understanding of genres of web-based communities: The role of the ecological cognition framework. *International Journal of Web Based Communities, 5*(1), 4–17. doi:10.1504/IJWBC.2009.021558

Bishop, J. (2009). Increasing capital revenue in social networking communities: Building social and economic relationships through avatars and characters. In Dasgupta, S. (Ed.), *Social computing: Concepts, methodologies, tools, and applications* (pp. 1987–2004). Hershey, PA: IGI Global. doi:10.4018/978-1-60566-984-7.ch131

Bishop, J. (2009). Increasing the economic sustainability of online communities: An empirical investigation. In Hindsworth, M. F., & Lang, T. B. (Eds.), *Community participation: Empowerment, diversity and sustainability.* New York, NY: Nova Science Publishers.

Bishop, J. (2011). *The equatrics of intergenerational knowledge transformation in techno-cultures: Towards a model for enhancing information management in virtual worlds. Unpublished MScEcon.* Aberystwyth, UK: Aberystwyth University.

Bitney, J., & Title, B. (1997). *No bullying program: Preventing bullying/victim violence at school (programs directors manual).* Central City, MN: Hazelden and Johnson Institute.

Bjørnstad and Ellingsen. (2004). *Onliners: A report about youth and the Internet.* Norwegian Board of Film Classification.

Blanchard, A. L., & Markus, M. L. (2004). The experienced 'sense' of a virtual community: Characteristics and processes. *The Data Base for Advances in Information Systems, 35*(1), 65–79.

Blaze, M., Feigenbaum, J., & Lacy, J. (1996). Decentralized trust management. *Proceedings of IEEE Symposium on Security and Privacy,* May, 164–173.

Blignaut, A. S., & Trollip, S. R. (2003). Measuring faculty participation in asynchronous discussion forums. *Journal of Education for Business, 78*(6), 347–354. doi:10.1080/08832320309598625

Bloom, M. (2003, April 2). *E-learning in Canada, findings from 2003 e-survey.*

Bocij, P. (2004). *Corporate cyberstalking: An invitation to build theory.* http://www.firstmonday.dk/issues/issues7_11/bocij/

Bocij, P., & McFarlane, L. (2002, February). Online harassment: Towards a definition of cyberstalking. *Prison Service Journal.* HM Prison Service, London, 139, 31-38.

Bock, G. W., Zmud, R. W., Kim, Y. G., & Lee, J. N. (2005). Behavioral intention formation in knowledge sharing: Examining the roles of extrinsic motivators, social-psychological forces, and organizational climate. *Management Information Systems Quarterly, 29*(1), 87–111.

Boin, A. (2005). From crisis to disaster. In R.W. Perry & E.L. Quarantelli (Eds.), *What is a disaster? New answers to old questions* (pp. 153-172). Philadelphia: XLibris.

Bolton, G., & Ockenfels, A. (2000). ERC: A theory of equity, reciprocity, and cooperation. *The American Economic Review, 90*(1), 166–193.

Boon, S., & Holmes, J. (1991). The dynamics of interpersonal trust: Resolving uncertainty in the face of risk. In R. Hinde & J. Groebel (Eds.), *Cooperation and prosocial behavior,* pp. 190-211. Cambridge, UK: Cambridge University Press.

Boyd, D. (2004). Friendster and publicly articulated social networking. In *CHI '04: CHI '04 extended abstracts on human factors in computing systems* (pp. 1279-1282). New York: ACM Press.

Boyd, D. M., & Ellison, N. B. (2007). Social network sites: Definition, history, and scholarship. *Journal of Computer-Mediated Communication, 13*(1), article 11. Retrieved August 3, 2009 from http://jcmc.indiana.edu/vol13/issue1/boyd.ellison.html

Braidotti, R. (2003). Cyberteratologies: Female monsters negotiate the other's participation in humanity's far future. In: M. Barr (Ed.), *Envisioning the future: Science fiction and the next millennium.* Middletown, CT: Wesleyan University Press.

Bressler, S., & Grantham, C. (2000). *Communities of commerce.* New York, NY: McGraw-Hill.

Brin, S., & Page, L. (1998). The anatomy of a largescale hypertextual Web search engine. In *WWW7: Proceedings of the Seventh International Conference on World Wide Web 7.* Elsevier Science Publishers.

Brown, J. S. & Duguid, P. (1999). Organizing knowledge. *The society for organizational learning, 1*(2), 28-44.

Brown, J. S., & Gray, E. S. (1995). The people are the company. *FastCompany.* Retrieved from http://www.fastcompany.com/online/01/people.html

Brown, J. (2000). Employee turnover costs billions annually. *Computing Canada, 26,* 25.

Brown, J. S., & Duguid, P. (2000). Balancing act: How to capture knowledge without killing it. *Harvard Business Review,* 73–80.

Brown, K., Jackson, M., & Cassidy, W. (2006). Cyberbullying: Developing policy to direct responses that are equitable and effective in addressing this special form of bullying. *Canadian Journal of Educational Administration and Policy, 57*.

Brownstein, A. (2000). In the campus shadows, women are stalkers as well as the stalked. *The Chronicle of Higher Education, 47*(15), 4042.

Bruner, J. (1966). *Toward a theory of instruction.* New York: Norton.

Brynjolfsson, E., & Smith, M. (2000). Frictionless commerce? A comparison of Internet and conventional retailers. *Management Science, 46*(4), 563–585. doi:10.1287/mnsc.46.4.563.12061

Buchegger, S., & Le Boudec, J. Y. (2004). A robust reputation system for P2P and mobile ad-hoc networks. *Proceedings of the 2nd Workshop Economics of Peer-to-Peer Systems.*

Bucher, H. G. (2002). Crisis communication and the internet: Risk and trust in a global media. *First Monday, 7*(4). Retrieved May 15, 2009, from http://outreach.lib.uic.edu/www/issues/issue7_4/bucher/index.html

Buckle, P. (2005). Disaster: Mandated definitions, local knowledge and complexity. In R.W. Perry & E.L. Quarantelli (Eds.), *What is a disaster? New answers to old questions* (pp. 173-200). Philadelphia: XLibris.

Bunnell, D. (2000). *The E-bay phenomenon.* New York: John Wiley.

Butler, B. (1999) *The dynamics of electronic communities.* Unpublished PhD dissertation, Graduate School of Industrial Administration, Carnegie Mellon University.

Butler, B. S. (2001). Membership size, communication activity and sustainability: A resource-based model of online social structures. *Information Systems Research, 12*(4), 346–362. doi:10.1287/isre.12.4.346.9703

Byrne, M., & Whitmore, C. (2008). Crisis informatics. *IAEM Bulletin, February 2008*, 8.

Cabezas, H. (2007). Detección de Conductas Agresivas "Bullyings" en escolares de Sexto a Octavo Año en Una Muestra Costarricense. *Revista Educación, 31*(1), 123–133.

Cahill, V. (2003). Using trust for secure collaboration in uncertain environments. *IEEE Pervasive Computing / IEEE Computer Society [and] IEEE Communications Society, 2*(3), 52–61. doi:10.1109/MPRV.2003.1228527

Callon, M. (1986). Some elements of a sociology of translation: Domestication of the scallops and the fishermen of St. Brieuc Bay. In Law, J. (Ed.), *Power, action and belief* (pp. 196–233). London: Routledge & Kegan Paul.

Campadello, S., Coutand, O., Del Rosso, C., Holtmanns, S., Kanter, T., Räck, C., et al. (2005). Trust and privacy in context-aware support for communication in mobile groups. *Proceedings of Context Awareness for Proactive Systems (CAPS) 2005,* Helsinki, Finland.

Campbell, J., Fletcher, G., & Greenhil, A. (2002). Tribalism, conflict and shape-shifting identities in online communities. *Proceedings of the 13th Australasia Conference on Information Systems.* Melbourne, Australia.

Campbell, J., Sherman, R. C., Kraan, E., & Birchmeier, Z. (2005, June). *Internet privacy awareness and concerns among college students.* Paper presented at APS, Toronto, Ontario, Canada.

Campbell, M. (2005). Cyberbullying: An older problem in a new guise? *Australian Journal of Guidance & Counselling, 15*(1), 68–76. doi:10.1375/ajgc.15.1.68

Carlson, P. J., & Davis, G. B. (1998). An investigation of media selection among directors and managers: From "self" to "other" orientation. *Management Information Systems Quarterly, 22*(3). doi:10.2307/249669

Carson, T. L. (2002). Ethical Issues in Selling and Advertising. In Bowie, N. E. (Ed.), *The Blackwell Guide to Business Ethics* (pp. 186–205). Malden, MA: Blackwell Publishing.

Cassidy, W., Brown, K., & Jackson, M. (2010). *Redirecting students from cyber-bullying to cyber-kindness.* Paper presented at the 2010 Hawaii International Conference on Education, Honolulu, Hawaii, January 7 to 10, 2010.

Cassidy, W., & Bates, A. (2005). "Drop-outs" and "pushouts": finding hope at a school that actualizes the ethic of care. *American Journal of Education, 112*(1), 66–102. doi:10.1086/444524

Cassidy, W., Jackson, M., & Brown, K. (2009). Sticks and stones can break my bones, but how can pixels hurt me? Students' experiences with cyber-bullying. *School Psychology International, 30*(4), 383–402. doi:10.1177/0143034309106948

Cavoukian, A. (2005). *Identity theft revisited: Security is not enough.* Toronto, Ontario, Canada: Information and Privacy Commissioner of Ontario.

Chae, B., Koch, H., Paradice, D., & Huy, V. (2005). Exploring knowledge management using network theories: Questions, paradoxes and prospects. *Journal of Computer Information Systems, 45*(4), 62–74.

Chak, A. (2003). *Submit now: Designing persuasive Web sites.* London: New Riders Publishing.

Chan, T.-S. (1999). *Consumer behavior in Asia.* New York, NY: Haworth Press.

Chatterjee, S. (2005). A model of unethical usage of information technology. *Proceedings of the 11th Americas Conference on Information Systems,* Omaha, NE.

Chayko, M. (2002). *Connecting: How we form social bonds and communities in the Internet age.* Albany, NY: State University of New York Press.

Chayko, M. (2008). *Portable communities: The social dynamics of online and mobile connectedness.* Albany, NY: State University of New York Press.

Chen, M., & Singh, J. P. (2001). Computing and using reputations for internet ratings. *Proceedings of 3rd ACM Conference on Electronic Commerce.*

Chen, R., & Yeager, W. (2001). *Poblano: A distributed trust model for peer-to-peer networks.* Technical report, Sun Microsystems.

Chen, Y., Harper, F. M., Konstan, J., & Li, S. X. (2009). Group identity and social preferences. *The American Economic Review, 99*(1). doi:10.1257/aer.99.1.431

Cheuk, B. W. (2006). Using social networking analysis to facilitate knowledge sharing in the British Council. *International Journal of Knowledge Management, 2*(4), 67–76.

Chidambaram, L., & Jones, B. (1993). Impact of communication medium and computer support on group perceptions and performance: A comparison of face-to-face and dispersed meetings. *Management Information Systems Quarterly, 17*(4). doi:10.2307/249588

Chou, C., Yu, S., Chen, C., & Wu, H. (2009). Tool, Toy, Telephone, Territory, or Treasure of Information: Elementary school students' attitudes toward the Internet. *Computers & Education, 53*(2), 308–316. doi:10.1016/j.compedu.2009.02.003

Coates, T. (2003). *(Weblogs and) the mass amateurisation of (nearly) everything.* Retrieved December 28, 2005, from http://www.plasticbag.org/archives/2003/09/weblogs_and_the_mass_amateurisation_of_nearly_everything.shtml

Cohen, B. (2003). Incentives build robustness in BitTorrent. In *Proceedings of the Workshop on Economics of Peer-to-Peer Systems,* Berkeley, CA.

Cohen, J. (1977). *Statistical power analysis for the behavioral sciences.* New York: Academic Press.

Cohen, D. (2006). What's your return on knowledge? *Harvard Business Review, 84*(12), 28–28.

Collins, H. (2009). *Emergency managers and first responders use twitter and facebook to update communities.* Retrieved August, 7, 2009, from http://www.emergencymgmt.com/safety/Emergency-Managers-and-First.html

Compeau, D. R., & Higgins, C. A. (1995). Computer self-efficacy: Development of a measure and initial test. *MIS Quarterly, 19,* 189–211. doi:10.2307/249688

Compeau, D. R., Higgins, C. A., & Huff, S. (1999). Social cognitive theory and individual reactions to computing technology: A longitudinal study. *MIS Quarterly, 23*(2), 145–158. doi:10.2307/249749

Computers and Crime. (2000). IT law lecture notes (Rev. ed.). http://www.law.auckland.ac.nz/itlaw/itlawhome.htm

Constant, D., Keisler, S., & Sproull, L. (1994). What's mine is ours, or is it? A study of attitudes about information sharing. *Information Systems Research, 5*(4), 400–421. doi:10.1287/isre.5.4.400

Conway, J. (2004). Theory of games to improve artificial societies. *International Journal of Intelligent Games & Simulation, 2*(1).

Cook-Sather, A. (2002). Authorizing students' perspectives: Toward trust, dialogue, change in Education. *Educational Researcher, 24*(June-July), 12–17.

Cooper, A. (1999). The inmates are running the asylum—Why high tech products drive us crazy and how to restore the sanity. USA.

Cornelli, F., Damiani, E., De Capitani di Vimercati, S., Paraboschi, S., & Samarati, P. (2002). Implementing a reputation-aware Gnutella servent. In *Proceedings of the International Workshop on Peer-to-Peer Computing*, Berkeley, CA.

Corritore, C. L., Kracher, B., & Wiedenbeck, S. (2003). Online trust: Concepts, evolving themes, a model. *International Journal of Human-Computer Studies. Trust and Technology, 58*(6), 737–758.

Crick, N., & Grotpeter, J. (1995). Relational aggression, gender, and social-psychological adjustment. *Child Development, 66*, 710–722. doi:10.2307/1131945

Crick, N., Werner, N., Casas, J., O'Brien, K., Nelson, D., Grotpeter, J., & Markon, K. (1999). Childhood aggression and gender: A new look at an old problem. In Bernstein, D. (Ed.), *Nebraska symposium on motivation* (pp. 75–141). Lincoln, NE: University of Nebraska Press.

Cross, R., Laseter, T., Parker, A., & Velasquez, G. (2006). Using social network analysis to improve communities of practice. *California Management Review, 49*(1), 32–60. doi:10.2307/41166370

Croughton, P. (2007). *Arena: The original men's style magazine*. London: Arena International.

Csicsery-Ronay, I. (2003). Marxist theory and science fiction. In: E. James & F. Mendlesohn (Eds.), *The Cambridge companion to science fiction*. Cambridge: Cambridge University Press.

Csikszentmihalyi, M. (1990). *Flow: The psychology of optimal experience*. New York, NY: Harper & Row.

Cummings, J. N., Butler, B., & Kraut, R. (2002). The quality of online social relationships. *Communications of the ACM, 45*(7), 103–108. doi:10.1145/514236.514242

Currion, P. (2005). *An ill wind? The role of accessible ITC following hurricane Katrina*. Retrieved February, 10, 2009, from http://www.humanitarian.info/itc-and_katrina

Cyber Crime in India. (2004). *Cyber stalking—online harassment*. http://www.indianchild.com/cyberstalking.htm

CyberAngels. (1999). http://cyberangels.org

Cyberbulling (2010). *What is cyberbulling?* Obtained May 1, 2010, from: http://www.cyberbullying.info/whatis/whatis.php

Daconta, M., Orbst, L., & Smith, K. (2003). *The semantic web*. Indianapolis: Wiley Publishing.

Daft, R. L., & Lengel, R. H. (1986). Organizational information requirements, media richness and structural design. Management Science, 32(5, May), 554-571.

Damodaran, L., & Olphert, W. (2000). Barriers and facilitators to the use of knowledge management systems. *Behaviour & Information Technology, 19*(6), 405–413. doi:10.1080/014492900750052660

Dandoulaki, M., & Andritsos, F. (2007). Autonomous sensors for just in time information supporting search and rescue in the event of a building collapse. *International Journal of Emergency Management, 4*(4), 704–725. doi:10.1504/IJEM.2007.015737

Davenport, T. H., De Long, D. W., & Beers, M. C. (1997). *Building successful knowledge management projects*. Center for Business Innovation Working Paper, Ernst and Young.

Davenport, T. H., De Long, D. W., & Beers, M. C. (1998). Successful knowledge management projects. *MIT Sloan Management Review, 39*(2), 43–57.

Davenport, T. H., & Prusak, L. (1997). *Working knowledge: How organizations manage what they know*. Cambridge, MA: Harvard Business School Press.

Davis, S. (2008). With a little help from my online friends: The health benefits of internet community participation. *The Journal of Education, Community and Values, 8*(3).

Davis, D. L., & Davis, D. F. (1990). The effect of training techniques and personal characteristics on training end users of information systems. *Journal of Management Information Systems, 7*(2), 93–110.

Davis, F. D. (1989). Perceived usefulness, perceived ease of use, and user acceptance of information technology. *Management Information Systems Quarterly*, *13*(3), 319–341. doi:10.2307/249008

Davis, K. E., Coker, L., & Sanderson, M. (2002, August). Physical and mental health effects of being stalked for men and women. *Violence and Victims*, *17*(4), 429–443. doi:10.1891/vivi.17.4.429.33682

De Rosa, C., Cantrell, J., Havens, A., Hawk, J., & Jenkins, L. (2007). *Sharing, privacy and trust in our networked world: A report to the OCLC membership*. Dublin, OH: OCLC Online Computer Library Center.

de Tocqueville, A. (1840). *Democracy in America*. (G. Lawrence, Trans.). New York: Doubleday.

Dean, K. (2000). The epidemic of cyberstalking. *Wired News*http://www.wired.com/news/politics/0,1283,35728,00.html

Decker, P. J., & Nathan, B. R. (1985). *Behavior modeling training*. New York: Praeger.

Dellarocas, C. (2000). Immunizing online reputation reporting systems against unfair ratings and discriminatory behavior. *Proceedings of the 2nd ACM Conference on Electronic Commerce*.

Dellarocas, C. (2003). The digitization of word of mouth: Promise and challenges of online feedback mechanisms. *Management Science*, *49*(10), 1407–1424. doi:10.1287/mnsc.49.10.1407.17308

Dennen, V. P. (2009). *Because I know you: Elements of trust among pseudonymous bloggers*. Paper presented at IADIS International Conference on Web-based Communities 2009, Algarve, Portugal.

Dennen, V. P. (2006). Blogademe: How a group of academics formed and normed an online community of practice. In Mendez-Vilas, A., Martin, A. S., Mesa Gonzalez, J. A., & Mesa Gonzalez, J. (Eds.), *Current developments in technology-assisted education* (*Vol. 1*, pp. 306–310). Badajoz, Spain: Formatex.

Dennen, V. P. (2009). Constructing academic alter-egos: Identity issues in a blog-based community. *Identity in the Information Society*, *2*(1), 23–38. doi:10.1007/s12394-009-0020-8

Dennen, V. P., & Pashnyak, T. G. (2008). Finding community in the comments: The role of reader and blogger responses in a weblog community of practice. *International Journal of Web-based Communities*, *4*(3), 272–283. doi:10.1504/IJWBC.2008.019189

Denning, D. E. (1993). A new paradigm for trusted systems. *Proceedings of the IEEE New Paradigms Workshop*.

Dennis, A. R., & Kinney, S. T. (1998). Testing media richness theory in the new media: The effects of cues, feedback, and task equivocality. *Information Systems Research*, *9*(3). doi:10.1287/isre.9.3.256

DeSanctis, G., Fayard, A.-L., Roach, M., & Jiang, L. (2003). Learning in online forums. *European Management Journal*, *21*(5), 565–578. doi:10.1016/S0263-2373(03)00106-3

di Pellegrino, G., Fadiga, L., Fogassi, L., Gallese, V., & Rizzolatti, G. (1992). Understanding motor events: A neurophysiological study. *Experimental Brain Research*, *91*, 176–180.

Dibbell, J. (1998). *My tiny life: Crime and passion in a virtual world*. New York: Henry Holt and Company.

Diener, E. (1980). Deindividuation: The absence of self-awareness and self-regulation in group members. In Paulus, P. B. (Ed.), *Psychology of group influence*. Hillsdale, NJ: Lawrence Erlbaum.

Doctorow, C. (2003). *Down and out in the Magic Kingdom*. Tor Books.

Donath, J. (2007). Signals in social supernets. *Journal of Computer-Mediated Communication, 13*(1), article 12. Retrieved August 3, 2009, from http://jcmc.indiana.edu/vol13/issue1/donath.html

Donath, S. J. (1998). *Identity and deception in the virtual community*. Cambridge, MA: MIT Media Lab.

Donath, J., & Boyd, D. (2004). Public displays of connection. *BT Technology Journal*, *22*(4), 71–82. doi:10.1023/B:BTTJ.0000047585.06264.cc

Douceur, J. (2002). The Sybil attack. In *Proceedings of the 1st International Peer-To-Peer Systems Workshop (IPTPS). eBay help: Feedback extortion*. (n.d.). Retrieved December 28, 2005, from http://pages.ebay.co.uk/help/policies/feedback-extortion.html

Douglas, K. M., & McGarty, C. (2001). Identifiability and self-presentation: Computer-mediated communication and intergroup interaction. *The British Journal of Social Psychology*, *40*, 399–416. doi:10.1348/014466601164894

Dowell, E., Burgess, A., & Cavanaugh, D. (2009). Clustering of Internet Risk Behaviors in a Middle School Student Population. *The Journal of School Health*, *79*(11), 547–553. doi:10.1111/j.1746-1561.2009.00447.x

Dretske, F. (1991). *Explaining behavior: Reasons in a world of causes*. Cambridge, MA: MIT Press. doi:10.2307/2108238

DSM-IV. (1995). *Manual diagnóstico y estadístico de los transtornos mentales (DSM-IV)*. Madrid, Spain: Masson, S.A.

Eagly, A., Ashmore, R., Makhijani, M., & Longo, L. (1991). What is beautiful is good, but…: A meta-analytic review of research on the physical attractiveness stereotype. *Psychological Bulletin*, *110*(1), 109–128. doi:10.1037/0033-2909.110.1.109

E-Bay. (2006). E-bay form 10-K, Feb. 24, 2006. Retrieved June 14, 2008, from http://investor.E-Bay.com/secfiling.cfm?filingID=950134-06-3678

E-Bay. (2007). Detailed seller ratings. Retrieved June 14, 2008, from http://pages.E-Bay.com/help/feedback/detailed-seller-ratings.html

EDUCAUSE Learning Initiative. (2006). *7 things you should know about... Facebook*. Retrieved February 25, 2009, from http://educause.edu/LibraryDetailPage/666?ID=ELI7017

Efimova, L. (2009). Weblog as a personal thinking space. *Proceedings of the 20th ACM Conference on Hypertext and Hypermedia*, (pp. 289-298).

Ellison, L. (1999). Cyberspace 1999: Criminal, criminal justice and the internet. *Fourteenth BILETA Conference*, York, UK. http://www.bileta.ac.uk/99papers/ellison.html

Ellison, L., & Akdeniz, Y. (1998). Cyber-stalking: The regulation of harassment on the internet (Special Edition: Crime, Criminal Justice and the Internet). *Criminal Law Review*, 2948. http://www.cyber-rights.org/documents/stalking

Ellonen, H. K., Kosonen, M., & Henttonen, K. (2007). The development of a sense of virtual community. *International Journal of Web Based Communities*, *3*(1), 114–130. doi:10.1504/IJWBC.2007.013778

Epinions.com earnings FAQ. (n.d.). Retrieved December 28, 2005, from http://www.epinions.com/help/faq/show_faq_earnings

Epinions.com web of trust FAQ. (n.d.). Retrieved December 28, 2005, from http://www.epinions.com/help/faq/?show=faq_wot

Epple, D., & Argote, L. (1996). An empirical investigation of the microstructure of knowledge acquisition and transfer through learning by doing. *Operations Research*, *44*(1), 77–86. doi:10.1287/opre.44.1.77

Eslinger, P. J., Moll, J., & de Oliveira-Souza, R. (2002). Emotional and cognitive processing in empathy and social behaviour. *The Behavioral and Brain Sciences*, *25*, 34–35.

Esposito, J. J. (2010). Creating a consolidated online catalogue for the university press community. *Journal of Scholarly Publishing*, *41*(4), 385–427. doi:10.3138/jsp.41.4.385

Fante, C. (2005). *Fenómeno Bullying. Como prevenir a violência nas escolas e educar para a paz*. Florianopolis, Brasil: Verus.

Faraj, S., & Wasko, M. M. (Forthcoming). The web of knowledge: An investigation of knowledge exchange in networks of practice. *AMR*.

Farnham, F. R., James, D. V., & Cantrell, P. (2000). Association between violence, psychosis, and relationship to victim in stalkers. *Lancet*, *355*(9199), 199. doi:10.1016/S0140-6736(99)04706-6

Fauske, J., & Wade, S. E. (2003/2004). Research to practice online: Conditions that foster democracy, community, and critical thinking in computer-mediated discussions. *Journal of Research on Technology in Education*, *36*(2), 137–154.

Feller, J. (2005). *Perspectives on free and open source software*. Cambridge, MA: The MIT Press.

Feng, J., Lazar, J., & Preece, J. (2004). (in press). Empathic and predictable communication influences online interpersonal trust. *Behaviour & Information Technology*.

Festinger, L. (1957). A theory of cognitive dissonance. Stanford University Press.

Figallo, C. (1998). *Hosting Web communities: Building relationships, increasing customer loyalty and maintaining a competitive edge.* Chichester: John Wiley & Sons.

Fischer, K. W. (1980). A theory of cognitive development: The control and construction of hierarchies of skills. *Psychological Review, 87,* 477–531. doi:10.1037/0033-295X.87.6.477

Fischhof, B. (1995). Risk Perception and Communication Unplugged: Twenty Years of Process. *Risk Analysis, 15*(2), 137–145. doi:10.1111/j.1539-6924.1995.tb00308.x

Fischhoff, B. (2006). Bevaviorally realistic risk management. In Daniels, R. J., Kettl, D. F., & Kunreuther, H. (Eds.), *On risk and disaster: Lessons from hurricane Katrina* (pp. 78–88). Philadelphia: University Pennsylvania Press.

Fishbein, M., & Ajzen, I. (1975). *Beliefs, attitude, intention, and behavior: An introduction to theory and research.* Addison-Wesley, Reading, MA.

Forester, T., & Morrison, P. (1994). Computer ethics. London: MIT Press.

Fox, S. (2008). The engaged e-patient population. *Pew Internet & American Life Project.* Retrieved January 20, 2009, from http://www.pewinternet.org/Reports/2008/The-Engaged-Epatient-Population.aspx

Fox, N., & Roberts, C. (1999). GPs in cyberspace: The sociology of a 'virtual community.'. *The Sociological Review, 47*(4), 643–672. doi:10.1111/1467-954X.00190

Fremauw, W. J., Westrup, D., & Pennypacker, J. (1997). Stalking on campus: The prevalence and strategies for coping with stalking. *Journal of Forensic Sciences, 42*(4), 666–669.

Freud, S. (1933). *New introductory lectures on psychoanalysis.* New York, NY: W.W. Norton & Company, Inc.

Friedman, E. J., & Resnick, P. (2001). The social cost of cheap pseudonyms. *Journal of Economics & Management Strategy, 10*(2), 173–199. doi:10.1162/105864001300122476

Froese-Germain, B. (2008). Bullying in the Digital Age: Using Technology to Harass Students and Teachers. *Our Schools. Our Selves, 17*(4), 45.

Fudenberg, D., & Tirole, J. (1991). *Game theory.* Cambridge: MIT Press.

Fukuyama, F. (1995). *Trust: The social virtues and the creation of prosperity.* New York: Free Press Paperbacks.

Fulk, J. (1993). Social construction of communication technology. *Academy of Management Journal, 36,* 921–950. doi:10.2307/256641

Gambetta, D. (1990). *Can we trust Trust? Trust: Making and breaking cooperative relations.* Oxford: Basil Blackwell.

Ganeriwal, S., & Srivastava, M. B. (2004). Reputation-based framework for high integrity sensor networks. *Proceedings of the ACM Security for Ad-Hoc and Sensor Networks,* pp. 66–67.

Garton, L., Haythornthwaite, C., & Wellman, B. (1997). Studying online social networks. *Journal of Computer-Mediated Communication, 3*(1), 75–105.

Gauze, C. F. (2003). I see you're writing an article. INK19. Retrieved on http://www.ink19.com/issues/march2003/webReviews/iSeeYoureWritingAn.html

Gavriilidis, A. (2009). Greek riots 2008: A mobile Tiananmen. In Economides, S., & Monastiriotis, V. (Eds.), *The return of street politics? Essays on the December riots in Greece* (pp. 15–19). London: LSE Reprographics Department.

Gefen, D. (2000). E-commerce: The role of familiarity and trust. *Omega, 28*(6), 725–737. doi:10.1016/S0305-0483(00)00021-9

Genuis, S., & Genuis, S. (2005). Implications for Cyberspace Communication: A Role for Physicians. *Southern Medical Journal, 98*(4), 451–455. doi:10.1097/01.SMJ.0000152885.90154.89

Geroimenko, V., & Chen, C. (2003). *Visualizing the semantic web.* London: Springer.

Gibson, J. J. (1986). *The ecological approach to visual perception.* Lawrence Erlbaum Associates.

Gil, Y., & Ratnakar, V. (2002). Trusting information sources one citizen at a time. *Proceedings of the 1st International Semantic Web Conference,* Italy, June.

Gist, M. E., Schwoerer, C., & Rosen, B. (1989). Effects of alternative training methods on self-efficacy and performance in computer software training. *The Journal of Applied Psychology, 74*(6), 884–891. doi:10.1037/0021-9010.74.6.884

Givens, D. (1978). The non-verbal basis of attraction: Flirtation, courtship and seduction. *Journal for the Study of Interpersonal Processes, 41*(4), 346–359.

Goel, L. & Mousavidin, E. (2007). vCRM: Virtual customer relationship management. forthcoming in DATABASE Special Issue on Virtual Worlds, November 2007.

Goettke, R., & Joseph, C. (2007). *Privacy and online social networking websites.* Retrieved from http://www.eecs.harvard.edu/cs199r/fp/RichJoe.pdf

Goffman, E. (1959). *The presentation of self in everyday life.* Garden City, NY: Doubleday.

Golbeck, J. (2005). *Web-based social network survey.* Retrieved December 28, 2005, from http://trust.mindswap.org/cgibin/relationshipTable.cgi

Golbeck, J., Hendler, J., & Parsia, B. (2003). Trust networks on the semantic Web. In *Proceedings of Cooperative Intelligent Agents.*

Goldsborough, R. (1999). Web-based discussion groups. *Link-Up, 16*(1), 23.

Goldsborough, R. (1999). Take the 'flame' out of online chats. *Computer Dealer News, 15*(8), 17. Gordon, R.S. (2000). Online discussion forums. *Link-Up, 17*(1), 12.

Goldstein, M. (1964). Perceptual reactions to threat under varying conditions of measurement. *Journal of Abnormal and Social Psychology, 69*(5), 563–567. doi:10.1037/h0043955

Goldstein, M., Alsio, G., & Werdenhoff, J. (2002). The media equation does not always apply: People are not polite to small computers. *Personal and Ubiquitous Computing, 6,* 87–96. doi:10.1007/s007790200008

Goleman, D., McKee, A., & Boyatzis, R. E. (2002). *Primal leadership: Realizing the power of emotional intelligence.* Harvard, MA: Harvard Business School Press.

Gollmann, D. (2007). Why trust is bad for security. Retrieved May 2007, from http://www.sics.se/policy2005/Policy_Pres1/dg-policy-trust.ppt

Goode, M. (1995). Stalking: Crime of the nineties? *Criminal Law Journal, 19,* 21–31.

Goodhue, D. L., & Thompson, R. L. (1995). Task-technology fit and individual performance. *Management Information Systems Quarterly, 19*(2), 213–236. doi:10.2307/249689

Google. (2009). *Orkut help forum.* Retrieved May 1 2010 from http://www.google.com/support/forum/p/orkut/thread?tid=1792d1c9901741fe&hl=en

Gordon, R., & Grant, D. (2005). Knowledge management or management of knowledge? Why people interested in knowledge management need to consider foucault and the construct of power. *Journal of Critical Postmodern Organization Science, 3*(2), 27–38.

Gormley, M. (2004, November 8). E-Bay sellers admit to phony bids. *Associated Press.*

Govani, T., & Pashley, H. (2005). *Student awareness of the privacy implications when using Facebook.* Retrieved from http://lorrie.cranor.org/courses/fa05/tubzhlp.pdf

Gradinger, P., Strohmeier, D., & Spiel, C. (2009). Traditional Bullying and Cyberbullying: Identification of Risk Groups for Adjustment Problems. *The Journal of Psychology, 217*(4), 205–213.

Grandison, T., & Sloman, M. (2000). A survey of trust in Internet applications. *IEEE Communications and Survey, 4th Quarter, 3*(4), 2–16.

Granovetter, M. S. (1973). The strength of weak ties. *American Journal of Sociology, 78*(6), 1360–1380. doi:10.1086/225469

Grant, R. M. (1996). Toward a knowledge-based theory of the firm. *Strategic Management Journal 17(Winter Special Issue),* 109-122.

Green, H., & Hannon, C. (2007). *TheirSpace: Education for a digital generation*. Demos, London. Retrieved July 6, 2009 from: http://www.demos.co.uk/publications/theirspace

Greenfield, G., & Campbell, J. (2006). Communicative practices in online communication: A case of agreeing to disagree. *Journal of Organizational Computing and Electronic Commerce, 16*(3-4), 267–277. doi:10.1207/s15327744joce1603&4_6

Gregory, M. (2000). Care as a goal of democratic education. *Journal of Moral Education, 29*(4), 445–461. doi:10.1080/713679392

Gross, B., & Acquisti, A. (2003). *Balances of power on eBay: Peers or unequals? The Berkeley Workshop on Economics of Peer-to-Peer Systems*, Berkeley, CA.

Guha, R. (2003). *Open rating systems*. Technical report, Stanford University, CA.

Guha, R., & Kumar, R. (2004). Propagation of trust and distrust. *Proceedings of the 13th international conference on World Wide Web*, ACM Press, pp. 403–412.

Guha, R., Kumar, R., Raghavan, P., & Tomkins, A. (2004). Propagation of trust and distrust. In *WWW '04: Proceedings of the 13th International Conference on World Wide Web* (pp. 403-412). ACM Press.

Gundel, S. (2005). Towards a new typology of crises. *Journal of Contingencies and Crisis Management, 13*(3), 106–115. doi:10.1111/j.1468-5973.2005.00465.x

Guo, W., Xiong, Z., & Li, Z. (2005). Dynamic trust evaluation based routing model for ad hoc networks. Proceedings of International Conference on Wireless Communications, Networking and Mobile Computing, Vol. 2, September, pp. 727–730.

Gyongyi, Z., Molina, H. G., & Pedersen, J. (2004). Combating Web spam with TrustRank. In *Proceedings of the Thirtieth International Conference on Very Large Data Bases (VLDB)* (pp. 576-587). Toronto, Canada: Morgan Kaufmann.

Habermas, J. (1984). *The theory of communicative action*. Boston: Beacon Press.

Hammer, M., Leonard, D., & Davenport, T. H. (2004). Why don't we know more about knowledge? *MIT Sloan Management Review, 45*(4), 14–18.

Hanisch, J., & Churchman, D. (2008). Virtual communities of practice: The communication of knowledge across cultural boundaries. *International Journal of Web-based Communities, 4*(4), 418–433. doi:10.1504/IJWBC.2008.019548

Hansen, M. T. (1999). The search-transfer problem: The role of weak ties in sharing knowledge across organizational subunits. *ASQ, 44*, 82–111. doi:10.2307/2667032

Hardwig, J. (1991). The role of trust in knowledge. *The Journal of Philosophy, 88*(12), 693–708. doi:10.2307/2027007

Harrald, J. R. (2002). Web enabled disaster and crisis response: What have we learned from the September 11th. In *15th Bled eCommerce Conference Proceedings "eReality: Constructing the eEconomy* (17-19th June 2002). Retrieved October 4, 2008, from http://domino.fov.uni-mb.si/proceedings.nsf/Proceedings/D3A6817 C6CC6C4B5C1256E9F003BB2BD/$File/Harrald.pdf

Harrald, J. R. (2009). Achieving agility in disaster management. *International Journal of Information Systems for Crisis Response Management, 1*(1), 1–11.

Hart, D., & Warne, L. (2006). Comparing cultural and political perspectives of data, information, and knowledge sharing in organisations. *International Journal of Knowledge Management, 2*(2), 1–15. doi:10.4018/jkm.2006040101

Harvey, L., & Myers, M. (1995). Scholarship and practice: The contribution of ethnographic research methods to bridging the gap. *Information Technology & People, 8*(3), 13–27. doi:10.1108/09593849510098244

Haveliwala, T., Kamvar, S., & Jeh, G. (2003). An analytical comparison of approaches to personalizing PageRank. In *WWW '02: Proceedings of the 11th International Conference on World Wide Web*. ACM Press.

Heer, J., & Boyd, D. (2005). Vizster: Visualizing online social networks. In *IEEE Symposium on Information Visualization (InfoViz)*.

Heller, M. (2003). Six ways to boost morale. *CIO Magazine,* (November 15), 1.

Herlocker, J. L., Konstan, J. A., & Riedl, J. (2000). Explaining collaborative filtering recommendations. *Proceedings of the 2000 ACM conference on Computer supported cooperative work.*

Hernández, M., & Solano, I. (2007). CiberBullying, Un Problema de Acoso Escolar. *RIED*, 17-36.

Heron, S. (2009). Online privacy and browser security. *Network Security*, (6): 4–7. doi:10.1016/S1353-4858(09)70061-3

Herring, S. C., Kouper, I., Paolillo, J. C., Scheidt, L. A., Tyworth, M., Welsch, P., et al. (2005). Conversations in the blogosphere: An analysis " from the bottom up". *Proceedings of the 38th Hawaii International Conference on System Sciences.* Los Alamitos: IEEE Press.

Herring, S., Job-Sluder, K., Scheckler, R., & Barab, S. (2002). Searching for safety online: Managing "trolling" in a feminist forum. *The Information Society*, *18*(5), 371–385. doi:10.1080/01972240290108186

Heskell, P. *Flirt Coach: Communication tips for friendship, love and professional success.* London: Harper Collins Publishers Limited.

Hewitt, K. (1998). Excluded perspectives in the social construction of disaster. In Quarantelli, E. L. (Ed.), *What is a disaster? Perspectives on the question* (pp. 75–91). London: Routledge.

Hill, W., Stead, L., Rosenstein, M., & Furnas, G. (1995). Recommending and evaluating choices in a virtual community of use. *SIGCHI Conference on Human factors in Computing systems,* Denver, Colorado.

Hiltz, S. R., & Turoff, M. (1978). *The network nation: Human communication via computer.* Reading, MA: Addison Wesley.

Hippel, E. V. (1994). Sticky information and the locus of problem solving: Implications for innovation. *Management Science*, *40*(4), 429–440. doi:10.1287/mnsc.40.4.429

Hitchcock, J. A. (2000). Cyberstalking. *Link-Up, 17*(4). http://www.infotoday.com/lu/ju100/hitchcock.htm

Hjelm, J., & Holtmanns, S. (2006). Privacy and trust visualization. Computer Human Interaction - CHI 2006 Conference Proceedings and Extended Abstracts, ACM Press, Montréal, Canada.

Hobbes, T. (1996). *Leviathan* (Rev. student ed.). Cambridge, UK: Cambridge University Press.

Hoffman, D. L., Novak, T. P., & Peralta, M. (1999). Building consumer trust online. *Communications of the ACM*, *42*(4), 80–87. doi:10.1145/299157.299175

Hollis, M. (1982). Dirty Hands. *British Journal of Political Science, 12*, 385–398. doi:10.1017/S0007123400003033

Hollis, M. (2002). *The Philosophy of Social Science: An Introduction (Revised and Updated).* Cambridge, UK: Cambridge University Press.

Horton, W. (2000). *Designing Web-based training.* New York: John Wiley & Sons.

Houser, D., & Wooders, J. (2001). Reputation in Internet auctions: Theory and evidence from E-Bay, Working Paper, University of Arizona. Retrieved June 14, 2008, from http://bpa.arizona.edu/~jwooders/E-Bay.pdf

Howard, T. W. (2010). *Design to thrive: Creating social networks and online communities that last.* Morgan Kaufmann.

Hsu, M., & Kuo, F. (2003). The effect of organization-based self-esteem and deindividuation in protecting personal information privacy. *Journal of Business Ethics*, *42*(4). doi:10.1023/A:1022500626298

Huber, G. (2001). Transfer of knowledge in knowledge management systems: unexplored issues and suggested studies. *European Journal of Information Systems (EJIS)*, *10*, 72–79. doi:10.1057/palgrave.ejis.3000399

Hughes, A. L., Palen, L., Sutton, J., & Vieweg, S. (2008). "Site-Seeing" in Disaster: An examination of on-line social convergence. In F. Fiedrich & B. Van de Walle, (Eds.), *Proceedings of the 5th International ISCRAM Conference.* Washington, DC, USA, May 2008.

Hurley, P. J. (2000). *A concise introduction to logic.* Belmont, CA: Wadsworth.

Husted, K., & Michailova, S. (2002). Diagnosing and fighting knowledge-sharing hostility. *Organizational Dynamics*, *31*(1), 60–73. doi:10.1016/S0090-2616(02)00072-4

Ingham, J., & Feldman, L. (1994). *African-American business leaders: A biographical dictionary.* Westport, CT: Greenwood Press.

Inkpen, A. C. (2005). Learning through alliances: General motors and NUMMI. *California Management Review*, *47*(4), 114–136. doi:10.2307/41166319

International Data Corporation. (2002, September 30). *While corporate training markets will not live up to earlier forecasts, IDC suggests reasons for optimism, particularly e-learning.* Author.

International Risk Governance Council. (2005). *Risk governance: Towards and integrative approach.* White paper no.1. Retrieved August 5, 2009, from http://www.irgc.org/IMG/pdf/IRGC_WP_No_1_Risk_Governance__reprinted_version_.pdf

I-SAFE. (2010). *Cyber Bullying: Statistics and Tips.* Retrieved May 1 2010, from http://www.isafe.org/channels/sub.php? ch=op&sub_id=media_cyber_bullying

ISE. (n.d.). *The internet no1 close protection resource.* http://www.intel-sec.demon.co.uk

Ives, B., Olson, M., & Baroudi, S. (1983). The measurement of user information satisfaction. *Communications of the ACM*, *26*, 785–793. doi:10.1145/358413.358430

Jackson, H. K. (2008). *The seven deadliest social network attacks.* Retrieved August 26, 2008, from http://www.darkreading.com/security/app-security/showArticle.jhtml?articleID=211201065

Jackson, M., Cassidy, W., & Brown, K. (2009). *"You were born ugly and youl die ugly too"*: Cyber-bullying as relational aggression. In Education (Special Issue on Technology and Social Media (Part 1), 15(2), December 2009.http://www.ineducation.ca/article/you-were-born-ugly-and-youl-die-ugly-too-cyber bullying-relational-aggression

Jackson, M., Cassidy, W., & Brown, K. (2009). Out of the mouths of babes: Students "voice" their opinions on cyber-bullying. *Long Island Education Review*, *8*(2), 24–30.

Jackson, P. W., Hansen, D., & Boomstrom, R. (1993). *The moral life of schools.* San Francisco: Jossey-Bass Publishers.

James, S. (2005). *Outfoxed: Trusted metadata distribution using social networks.* Retrieved December 28, 2005, from http://getoutfoxed.com/about

James, A. M., & Rashed, T. (2006). In their own words: Utilizing weblogs in quick response research. In Guibert, G. (Ed.), *Learning from Catastrophe: Quick Response Research in the Wake of Hurricane Katrina* (pp. 57–84). Boulder: Natural Hazards Center Press.

Jansen, E. (2002). *Netlingo: The Internet dictionary.* Ojai, CA: Independent Publishers Group.

Jarvenpaa, S. L., & Staples, D. S. (2000). The Use of Collaborative Electronic Media for Information Sharing: An Exploratory Study of Determinants. *The Journal of Strategic Information Systems*, *9*(2/3), 129–154. doi:10.1016/S0963-8687(00)00042-1

Jarvenpaa, S., Tractinsky, N., & Vitale, M. (2000). Consumer trust in an Internet store. *Information Technology and Management*, *1*(1/2), 45–71. doi:10.1023/A:1019104520776

Jaschik, S. (2005, January 5). Historians reject proposed boycott. *Inside Higher Ed.* Retrieved from http://www.insidehighered.com/news/2009/01/05/boycott

Jenkins, S. (2006). Concerning interruptions. *Computer*, (November): 114–116.

Jennex, M., & Olfman, L. (2000). Development recommendations for knowledge management/ organizational memory systems. In *Proceedings of the Information Systems Development Conference.*

Jennex, M., & Olfman, L. (2004). Assessing Knowledge Management success/Effectiveness Models. *Proceedings of the 37th Hawaii International Conference on System Sciences.*

Jennex, M. E. (2005). What is knowledge management? *International Journal of Knowledge Management*, *1*(4), i–iv.

Jennex, M. E., & Olfman, L. (2006). A model of knowledge management success. *International Journal of Knowledge Management*, *2*(3), 51–68. doi:10.4018/jkm.2006070104

Jennex, M., Smolnik, S., & Croasdell, D. (2007). Knowledge management success. *International Journal of Knowledge Management*, *3*(2), i–vi.

Jennings, S., & Gersie, A. (1987). *Drama therapy with disturbed adolescents* (pp. 162–182).

Jenson, B. (1996). *Cyberstalking: Crime, enforcement and personal responsibility of the on-line world.* S.G.R. MacMillan. http://www.sgrm.com/art-8.htm

Johnson, D. G. (1997). Ethics online. *Communications of the ACM, 40*(1), 60–65. doi:10.1145/242857.242875

Johnson, D. G. (2009). *Computer Ethics* (4th ed.). Upper Saddle River, NJ: Prentice Hall.

Joinson, A. (2003). *Understanding the psychology of Internet behavior: Virtual worlds, real lives.* New York: Palgrave MacMillan.

Jones, H., & Hiram, S. J. (2005). *Facebook: Threats to privacy.* Cambridge, MA: MIT.

Jones, Q. (1997). Virtual-communities, virtual settlements and cyber archeology: A theoretical outline. *Journal of Computer-Mediated Communication, 3*(3).

Jones, Q., & Rafaeli, S. (2000). Time to split, virtually: 'Discourse architecture' and 'community building' create vibrant virtual publics. *Electronic Markets, 10*(4), 214–223. doi:10.1080/101967800750050326

Jones, Q., Ravid, G., & Rafaeli, S. (2004). Information overload and the message dynamics of online interaction spaces: A theoretical model and empirical exploration. *Information Systems Research, 15*(2), 194–210. doi:10.1287/isre.1040.0023

Jones, S., Wilikens, M., Morris, P., & Masera, M. (2000). Trust requirements in e-business - A conceptual framework for understanding the needs and concerns of different stakeholders. *Communications of the ACM, 43*(12), 80–87. doi:10.1145/355112.355128

Jordan, T. (1999). *Cyberpower: An introduction to the politics of cyberspace.* London: Routledge.

Jøsang, A. (1999). An algebra for assessing trust in certification chains. *Proceedings of the Networking Distributed System Security Symposium.*

Jøsang, A., & Ismail, R. (2002). The beta reputation system. *Proceedings of the 15th Bled Electronic Commerce Conference*, June.

Jøsang, A., & Tran, N. (2003). Simulating the effect of reputation systems on e-markets. *Proceedings of the First International Conference on Trust Management.*

Jøsang, A., Ismail, R., & Boyd, C. (2005). A survey of trust and reputation systems for online service provision. In *Decision support systems.*

Jøsang, A. (2001). A logic for uncertain probabilities. *International Journal of Uncertainty. Fuzziness and Knowledge-Based Systems, 9*(3), 279–311.

Josefsson, U. (2005). Coping with illness online: The case of patients' online communities. *The Information Society, 21*, 141–153. doi:10.1080/01972240590925357

Joyce, M. (2008). *Campaign: Digital tools and the Greek riots.* Retrieved August 1, 2009, from http://www.digiactive.org/2008/12/22/digital-tools-and-the-greek_riots

Juvonen, J., & Gross, E. F. (2008). Extending the school grounds? Bullying experiences in cyberspace. *The Journal of School Health, 78*(9), 496–505. doi:10.1111/j.1746-1561.2008.00335.x

Kamphuis, J. H., & Emmelkamp, P. M. G. (2000). Stalking—A contemporary challenge for forensic and clinical psychiatry. *The British Journal of Psychiatry, 176*, 206–209. doi:10.1192/bjp.176.3.206

Kamvar, S. D., Schlosser, M. T., & Garcia-Molina, H. (2003). The Eigentrust algorithm for reputation management in P2P Networks. In *WWW'03 Conference.*

Kankanhalli, A., Tan, B. C. Y., & Kwok-Kei, W. (2005). Contributing knowledge to electronic knowledge repositories: An empirical investigation. *Management Information Systems Quarterly, 29*(1), 113–143.

Kant, I. (1981). *Grounding for the metaphysics for morals.* Indianapolis, IN: Hackett. (Original work published 1804)

Karau, S. J., & Williams, K. D. (1993). Social loafing: A meta-analytic review and theoretical integration. *Journal of Personality and Social Psychology, 65*, 681–706. doi:10.1037/0022-3514.65.4.681

Karlsson, L. (2007). Desperately seeking sameness. *Feminist Media Studies, 7*(2), 137–153. doi:10.1080/14680770701287019

Kerlinger, F.N., & Lee. H.B. (2000). *Foundations of behavioral research.* New York: Harcourt Brace.

Khare, R., & Rifkin, A. (1998). Trust management on the World Wide Web. *First Monday, 3*(6).

Kiesler, S., Siegel, J., & McGuire, T. W. (1984). Social psychological aspects of computer-mediated communication. *The American Psychologist*, *39*(10), 1123–1134. doi:10.1037/0003-066X.39.10.1123

Kiesler, S., & Sproull, L. (1992). Group decision making and communication technology. *Organizational Behavior and Human Decision Processes*, *52*(1), 96–123. doi:10.1016/0749-5978(92)90047-B

Kil, S. H. (2010). Telling stories: The use of personal narratives in the social sciences and history. *Journal of Ethnic and Migration Studies*, *36*(3), 539–540. doi:10.1080/13691831003651754

Kim, A. (2000). *Community building on the Web: Secret strategies for successful online communities*. Berkeley, CA: Peachpit Press.

King, W. R. (2006). The critical role of information processing in creating an effective knowledge organization. *Journal of Database Management*, *17*(1), 1–15. doi:10.4018/jdm.2006010101

Kinney, S. T., & Dennis, A. R. (1994). Re-evaluating media richness: Cues, feedback and task. *Proceedings of the 27th Annual Hawaii International Conference on System Sciences.*

Kirpatrick, D. L. (Ed.). (1967). *Evaluation of training. Training and development handbook*. New York: McGraw-Hill.

Kittur, A., Suh, B., & Chi, E. H. (2008). Can you ever trust a wiki?: impacting perceived trustworthiness in wikipedia. In *Proceedings of the ACM 2008 Conference on Computer Supported Cooperative Work (San Diego, CA, USA, November 08 - 12, 2008). CSCW '08.* ACM, New York, NY, (pp.477-480). Retrieved August 7, 2009, from http://doi.acm.org/10.1145/1460563.1460639

Kling, R., & Courtright, C. (2003). Group behavior and learning in electronic forums: A sociotechnical approach. *The Information Society*, *19*, 221–235. doi:10.1080/01972240309465

Kling, R., McKim, G., & King, A. (2003). A bit more to it: Scholarly communication forums as socio-technical interaction networks. *Journal of the American Society for Information Science and Technology*, *54*(1), 47–67. doi:10.1002/asi.10154

Koeglreiter, G., Smith, R., & Torlina, L. (2006). The role of informal groups in organisational knowledge work: Understanding an emerging community of practice. *International Journal of Knowledge Management*, *2*(1), 6–23. doi:10.4018/jkm.2006010102

Kohlhase, M., & Anghelache, R. (2004). Towards collaborative content management and version control for structured mathematical knowledge. *In Proceedings Mathematical Knowledge Management: 2nd International Conference, MKM 2003*, Bertinoro, Italy

Koman, R. (2005). *Stewart Butterfield on Flickr*. Retrieved December 28, 2005, from www.oreillynet.com/pub/a/network/2005/02/04/sb_flckr.html

Kosko, B. (1986). Fuzzy cognitive maps. *International Journal of Man-Machine Studies*, *24*, 65–75. doi:10.1016/S0020-7373(86)80040-2

Kosonen, M. (2009). Knowledge sharing in virtual communities: A review of the empirical research. *International Journal of Web Based Communities*, *5*(2), 144–163. doi:10.1504/IJWBC.2009.023962

Kowalski, R. M., Limber, S. E., & Agatston, P. W. (2008). *Cyber bullying: Bullying in the digital age*. Malden, MA: Blackwell Publishers. doi:10.1002/9780470694176

Kowalski, R., & Limber, S. (2007). Electronic Bullying Among Middle School Students. *The Journal of Adolescent Health*, *41*(6), S22–S30. doi:10.1016/j.jadohealth.2007.08.017

Kress, N. (2004). *Dynamic characters: How to create personalities that keep readers captivated*. Cincinnati, OH: Writer's Digest Books.

Krimsky, S. (2007). Risk communication in the internet age: The rise of disorganized skepticism. *Environmental Hazards*, *7*, 157–164. doi:10.1016/j.envhaz.2007.05.006

Kuchinskas, S. (2005). *Amazon gets patents on consumer reviews*. Retrieved December 28, 2005, from http://www.internetnews.com/bus-news/article.php/3563396

Kumar, N., & Benbasat, I. (2002). Para-social presence and communication capabilities of a Website: A theoretical perspective. *E-Service Journal, 1*(3).

Kumar, R., Novak, J., Raghavan, P., & Tomkins, A. (2004). Structure and evolution of blogspace. *Communications of the ACM, 47*(12), 35–39. doi:10.1145/1035134.1035162

Kurzweil, R. (1999). The age of spiritual machines. Toronto: Penguin Books.

Kyttä, M. (2003). *Children in outdoor contexts: Affordances and independent mobility in the assessment of environmental child friendliness.* Doctoral dissertation presented at Helsinki University of Technology, Espoo, Finland.

Laertius, D. (1985). *The Lives and Opinions of Eminent Philosophers.* In C. D. Yonge (Eds.), Retrieved from http://classicpersuasion.org/pw/diogenes/

Lagadec, P. (2005). *Crisis management in the 21st century: "Unthinkable" events in "inconceivable" contexts.* Ecole Polytechnique - Centre National de la Recherche Scientifique, Cahier No 2005-003. Retrieved August 5, 2009, from http://hal.archives-ouvertes.fr/docs/00/24/29/62/PDF/2005-03-14-219.pdf

Lampe, C., & Resnick, P. (2004). Slash (dot) and burn: Distributed moderation in a large online conversation space. *Proceedings of the SIGCHI Conference on Human Factors in Computing Systems,* (pp. 543-550).

Lancaster, J. (1998, June). Cyber-stalkers: The scariest growth crime of the 90's is now rife on the net. *The Weekend Australian,* 20-21.

Landon, S., & Smith, C. E. (1997). The use of quality and reputation indicators by consumers: The case of Bordeaux wine. *Journal of Consumer Policy, 20,* 289–323. doi:10.1023/A:1006830218392

Lapre, M. A., & Van Wassenhove, L. N. (2003). Managing learning curves in factories by creating and transferring knowledge. *California Management Review, 46*(1), 53–71. doi:10.2307/41166231

Larkin, E. (2005). PC World, December, 28.

Laughren, J. (2000). *Cyberstalking awareness and education.* http://www.acs.ucalgary.ca/~dabrent/380/webproj/jessica.html

Lave, J., & Wenger, E. (1991). *Situated learning. Legitimate peripheral participation.* Cambridge: Cambridge University Press. doi:10.1017/CBO9780511815355

Lee, S., Sherwood, R., & Bhattacharjee, B. (2003). Cooperative peer groups in NICE. In *IEEE Infocom.*

Lee, D. H., Im, S., & Taylor, C. R. (2008). Voluntary self-disclosure of information on the internet: A multimethod study of the motivations and consequences of disclosing information on blogs. *Psychology and Marketing, 25*(7), 692–710. doi:10.1002/mar.20232

Lee, E., & Nass, C. (2002). Experimental tests of normative group influence and representation effects in computer-mediated communication: When interacting via computers differs from interacting with computers. *Human Communication Research, 28,* 349–381. doi:10.1093/hcr/28.3.349

Lee, S. (2009). Online Communication and Adolescent Social Ties: Who benefits more from Internet use? *Journal of Computer-Mediated Communication, 14*(3), 509–531. doi:10.1111/j.1083-6101.2009.01451.x

Leidner, D. E., & Jarvenpaa, S. L. (1995). The use of information technology to enhance management school education: A theoretical view. *MIS Quarterly, 19,* 265–291. doi:10.2307/249596

Leimeister, J. M., Ebner, W., & Krcmar, H. (2005). Design, implementation, and evaluation of trust-supporting components in virtual communities for patients. *Journal of Management Information Systems, 21*(4), 101–135.

Lem, A. (2008). *Letter from Athens: Greek riots and the news media in the age of twitter.* Retrieved August 3, 2009, from http://www.alternet.org/media/113389/letter_from_athens:_greek_riots_and_the_news_media_in_the_age_of_twitter

Lenhart, A., & Fox, S. (2006). Bloggers: A portrait of the internet's new storytellers. *Pew Internet & American Life Projects.* Retrieved October 20, 2009, from http://www.pewinternet.org/Reports/2006/Bloggers.aspx

Lenhart, A., & Madden, M. (2007). Teens, Privacy & Online Social Networks: How teens manage their online identities and personal information in the age of MySpace. *Pew Internet & American Life Project.* Retrieved from: http://www.pewinternet.org

Lenhart, A., Madden, M., & Macgill, A. (2007). Teens and Social Media: The use of social media gains a greater foothold in teen life as they embrace the conversational nature of interactive online media. *Pew Internet & American Life Project*. Retrieved from: http://www.pewinternet.org

Lenhart, A., Rainie, L., & Lewis, O. (2001). *Teenage life online: The rise of the instant-message generation and the Internet's impact on friendships and family relationships*. Retrieved December 28, 2005, from http://www.pewinternet.org/report_display.asp?r=36

Lessig, L. (1999). Code and other laws of cyberspace. New York: Basic Books.

Leung, C. H. (2010). Critical factors of implementing knowledge management in school environment: A qualitative study in Hong Kong. *Research Journal of Information Technology*, *2*(2), 66–80. doi:10.3923/rjit.2010.66.80

Levenson, R. W., & Ruef, A. M. (1992). Empathy: A physiological substrate. *Journal of Personality and Social Psychology*, *63*(2), 234–246. doi:10.1037/0022-3514.63.2.234

Levien, R. (n.d.). *Attack resistant trust metrics*. Retrieved December 28, 2005, from http://www.advogato.org/trust-metric.html

Lewis, S. F., Fremouw, W. J., Ben, K. D., & Farr, C. (2001). An investigation of the psychological characteristics of stalkers: Empathy, problem-solving, attachment and borderline personality features. *Journal of Forensic Sciences*, *46*(1), 8084.

Li, X., Valacich, J. S., & Hess, T. J. (2004). Predicting user trust in information systems: a comparison of competing trust models. Proceedings of the 37th Annual Hawaii International Conference on System Sciences

Liang, Z., & Shi, W. (2005). PET: a PErsonalized Trust model with reputation and risk evaluation for P2P resource sharing. Proceedings of the 38thAnnual Hawaii International Conference on System Sciences, January, pp. 201b–201b.

Liang, T. P., & Huang, J. S. (1998). An empirical study on consumer acceptance of products in electronic markets: A transaction cost model. *Decision Support Systems*, *24*, 29–43. doi:10.1016/S0167-9236(98)00061-X

Liau, A., Khoo, A., & Ang, P. (2008). Parental awareness and monitoring of adolescent Internet use. *Current Psychology (New Brunswick, N.J.)*, *27*(4), 217–233. doi:10.1007/s12144-008-9038-6

Licklider, J., & Taylor, R. (1968). The computer as a communication device. *Sciences et Techniques (Paris)*.

Li, D., Li, J., & Lin, Z. (2007). Online consumer-to-consumer market in China—a comparative study of Taobao and eBay. *Electronic Commerce Research and Applications*. doi:.doi:10.1016/j.elerap.2007.02.010

Li, H., & Singhal, M. (2007). Trust management in distributed systems. *Computer*, *40*(2), 45–53. doi:10.1109/MC.2007.76

Lim, K., Sia, C., Lee, M., & Benbasat, I. (2006). Do I trust you online, and if so, will I buy? An empirical study of two trust-building strategies. *Journal of Management Information Systems*, *23*(2), 233–266. doi:10.2753/MIS0742-1222230210

Lin, C., Varadharajan, V., Wang, Y., & Pruthi, V. (2004). Enhancing GRID security with trust management. Proceedings of the IEEE International Conference on Services Computing (SCC 2004), pp. 303–310.

Lin, T. C., Hsu, M. H., Kuo, F. Y., & Sun, P. C. (1999). An intention model based study of software piracy. *Proceedings of the 32nd Annual Hawaii International Conference on System Sciences*.

Lindahl, C., & Blount, E. (2003). Weblogs: Simplifying Web publishing. *IEEE Computer*, *36*(11), 114–116.

Linder, R. (2006). *Wikis, webs, and networks: Creating connections for conflict-prone settings*. Washington: Council for Strategic and International Studies. Retrieved August 5, 2009, from http://www.csis.org/component/option,com_csis_pubs/task,view/id,3542/type,1/

Lippman, S. A., & Rumelt, R. P. (1982). Uncertain imitability: An analysis of interfirm differences in efficiency under competition. *The Bell Journal of Economics*, *13*, 418–438. doi:10.2307/3003464

Lipps, T. (1903). Einfühlung, Innere Nachahmung und Organempfindung. *Archiv für die Gesamte Psychologie*, *1*, 465–519.

Li, Q. (2007). New bottle but old wine: A research of cyberbullying in schools. *Computers in Human Behavior*, *23*(4), 1777–1791. doi:10.1016/j.chb.2005.10.005

Liu, S. B., Palen, L., Sutton, J., Hughes, A. L., & Vieweg, S. (2008). In search of the bigger picture: The emergent role of on-line photo sharing in times of disaster. In F. Fiedrich & B. Van de Walle, (Eds.), *Proceedings of the 5th International ISCRAM Conference – Washington, DC, USA, May 2008*.

Liu, Z., Joy, A. W., & Thompson, R. A. (2004). A dynamic trust model for mobile ad hoc networks. Proceedings of the 10th IEEE International Workshop on Future Trends of Distributed Computing Systems (FTDCS 2004), May, pp. 80–85.

Livingstone, S., & Bober, M. (2006). *Children go online, 2003-2005*. Colchester, Essex: UK Data Archive.

Lucking-Reiley, D., Bryan, D., Prasad, N., & Reeves, D. (2005). *Pennies from E-Bay: The determinants of price in online auctions*. Working Paper, Vanderbilt University. Retrieved June 14, 2008, from http://www.u.arizona.edu/~dreiley/papers/PenniesFromE-Bay.pdf

Macy, M., & Willer, R. (2002). From factors to actors. England. *Annual Review of Sociology*, *28*.

Malhotra, A., Gosain, S., & Hars, A. (1997). Evolution of a virtual community: Understanding design issues through a longitudinal study. *International Conference on Information Systems (ICIS), AIS*.

Malik, S. (2002). *Representing black Britain: A history of black and Asian images on British television*. London: Sage Publications.

Manchala, D. W. (2000). E-commerce trust metrics and models. *IEEE Internet Computing*, *4*(2), 36–44. doi:10.1109/4236.832944

Mann, C., & Stewart, F. (2000). *Internet communication and qualitative research: A handbook for Research Online*. London: Sage Publications.

Mantovani, F. (2001). Networked seduction: A test-bed for the study of strategic communication on the Internet. *Cyberpsychology & Behavior*, *4*(1), 147–154. doi:10.1089/10949310151088532

Markus, L. M. (1987). Towards a critical mass theory of interactive media: Universal access, interdependence and diffusion. *Communication Research*, *14*, 491–511. doi:10.1177/009365087014005003

Markus, L. M. (2001). Towards a theory of knowledge reuse: Types of knowledge reuse situations and factors in reuse success. *Journal of Management Information Systems*, *18*(1), 57–94.

Markus, L. M., Majchrzak, A., & Gasser, L. (2002). A design theory for systems that support emergent knowledge processes. *Management Information Systems Quarterly*, *26*(3), 179–212.

Marsh, S. (1994). *Formalising trust as a computational concept*. (Doctoral dissertation, University of Stirling, 1994).

Maslow, A. H. (1943). A theory of motivation. *Psychological Review*, *50*(4), 370–396. doi:10.1037/h0054346

Massa, P. (2005). *Identity burro: Making social sites more social*. Retrieved December 28, 2005, from http://moloko.itc.it/paoloblog/archives/2005/07/17/identity_burro_greasemonkey_extension_for_social_sites.html

Massa, P., & Avesani, P. (2004). Trust-aware collaborative filtering for recommender systems. In *Proceedings of Federated International Conference on the Move to Meaningful Internet: CoopIS, DOA, ODBASE*.

Massa, P., & Avesani, P. (2005). Controversial users demand local trust metrics: An experimental study on Epinions.com community. In *Proceedings of 25th AAAI Conference*.

Massa, P., & Hayes, C. (2005). Page-rerank: Using trusted links to re-rank authority. In *Proceedings of Web Intelligence Conference*.

Maurer, U. (1996). Modeling a public-key infrastructure. *Proceedings of the European Symposium of Research on Computer Security, LNCS*, *1146*, 325–350.

Maxwell, A. (2001). *Cyberstalking*. Masters' thesis, http://www.netsafe.org.nz/ie/downloads/cyberstalking.pdf

Maxwell, J. A., & Miller, B. A. (2008). Categorizing and connecting strategies in qualitative data analysis. In Leavy, P., & Hesse-Biber, S. (Eds.), *Handbook of emergent methods* (pp. 461–477).

Mayer, R. C., Davis, J. H., & Schoorman, F. D. (1995). An integrative model of organizational trust. *Academy of Management Review*, *20*(3), 709–734. doi:10.2307/258792

Mayer, R. E., Johnson, W. L., Shaw, E., & Sandhu, S. (2006). Constructing computer-based tutors that are socially sensitive: Politeness in educational software. *International Journal of Human-Computer Studies*, *64*(1), 36–42. doi:10.1016/j.ijhcs.2005.07.001

McAfee, R. P., & McMillan, J. (1987). Auctions and bidding. *Journal of Economic Literature*, *25*(2), 699–738.

McCandlish, S. (2005). *EFF's top 12 ways to protect your online privacy*. Electronic Frontier Foundation. Retrieved from http://www.eff.org/Privacy/ eff_privacy_top_12.html

McCann, J. T. (2000). A descriptive study of child and adolescent obsessional followers. *Journal of Forensic Sciences*, *45*(1), 195–199.

McDermott, R. (1999). How to get the most out of human networks: Nurturing three-dimensional communities of practice. *Knowledge Management Review*, *2*(5), 26–29.

McDermott, R. (1999). Why information inspired but cannot deliver knowledge management. *California Management Review*, *41*(4), 103–117. doi:10.2307/41166012

McDougall, W. (1908). *An introduction to social psychology*. London: Methuen. doi:10.1037/12261-000

McEvoy, G. M., & Cascio, W. F. (1987). Do good or poor performers leave? A meta-analysis of the relationship between performance and turnover. *Academy of Management Journal*, *30*(4), 744–762. doi:10.2307/256158

McGehee, W., & Tullar, W. (1978). A note on evaluating behavior modification and behavior modeling as industrial training techniques. *Personnel Psychology*, *31*, 477–484. doi:10.1111/j.1744-6570.1978.tb00457.x

McKnight, D. H., & Chervany, N. L. (2000). What is Trust? a conceptual analysis and an interdisciplinary model. *Proceedings of the 2000 Americas Conference on Information Systems (AMCI2000). AIS*, August, Long Beach, CA.

McKnight, D. H., & Chervany, N. L. (2003). The meanings of trust. UMN university report. Retrieved December 2006, from http://misrc.umn.edu/wpaper/WorkingPapers/9604.pdf

McKnight, D. H., Choudhury, V., & Kacmar, C. (2002). Developing and validating trust measures for e-commerce: An integrative typology. *Information Systems Research*, *13*(3), 334–359. doi:10.1287/isre.13.3.334.81

Meloy, J. R. (1996). Stalking (obsessional following): A review of some preliminary studies. *Aggression and Violent Behavior*, *1*(2), 147–162. doi:10.1016/1359-1789(95)00013-5

Meloy, J. R., & Gothard, S. (1995). Demographic and clinical comparison of obsessional followers and offenders with mental disorders. *The American Journal of Psychiatry*, *152*(2), 25826.

Memory Alpha. (2009). *Memory Alpha wiki (Star Trek World)*. Retrieved May 1 2010, from: http://memory-alpha.org/ wiki/Portal:Main

Mena, J. (1999). *Data mining your Web site*. Oxford: Digital Press.

Meyrowitz, J. (1985). No sense of place: The impact of electronic media on social behavior. New York: Oxford University Press.

Michaelides, G. (2009). *The project page of the CWIW*. Budapest University of Technology and Economics, Department of Electron Devices. Retrieved February 18, 2009, from http://turul.eet.bme.hu/~adam30

Mill, J. S. (1859). *On liberty. History of economic thought books*. McMaster University Archive for the History of Economic Thought.

Millen, D. R., Fontaine, M. A., & Muller, M. J. (2002). Understanding the benefit and costs of communities of practice. *Communications of the ACM*, *45*(4), 69–74. doi:10.1145/505248.505276

Miller, G. (1999). Gore to release cyberstalking report, call for tougher laws. *Latimes.com*. http://www.latimes.com/news/ploitics/elect2000/pres/gore

Miller, N. H., Resnick, P., & Zeckhauser, R. J. (2002). *Eliciting honest feedback in electronic markets*. KSG Working Paper Series RWP02-039.

Miller, C. A. (2004). Human-computer etiquette: Managing expectations with intentional agents. *Communications of the ACM*, *47*(4), 31–34. doi:10.1145/975817.975840

Mill, J. S. (1979). *Utilitarianism*. Indianapolis, IN: Hackett. (Original work published 1861)

Ministerio (2009). Ministério público pede que Google explique crimes no Orkut (in Portuguese). *Folha Online*. March 10, 2006.

Mitchell, W. (2005). *What do pictures want?: The lives and loves of images*. Chicago, IL: University of Chicago Press.

Montano, B. R., Porter, D. C., Malaga, R. A., & Ord, J. K. (2005). Enhanced reputation scoring for online auctions. In *Twenty-sixth International Conference on Information Systems, Proceedings*.

Mook, D. G. (1987). *Motivation: The organization of action*. London, UK: W.W. Norton & Company Ltd.

Moran, J. (2007). Generating more heat than light? Debates on civil liberties in the UK. *Policing*, *1*(1), 80. doi:10.1093/police/pam009

Moss, M. L., & Townsend, A. M. (2006). Disaster forensics: Leveraging crisis information systems for social science. In Van de Walle, B. and Turoff, M. (Eds.) *Proceedings of the 3rd International ISCRAM Conference*, Newark, NJ (USA), May 2006.

Mui, L. (2003). *Computational models of trust and reputation: agents, evolutionary games, and social networks*. Doctoral dissertation, Massachusetts Institute of Technology.

Mullen, P. E., & Pathe, M. (1997). The impact of stalkers on their victims. *The British Journal of Psychiatry*, *170*, 12–17. doi:10.1192/bjp.170.1.12

Mullen, P. E., Pathe, M., Purcell, R., & Stuart, G. W. (1999). Study of stalkers. *The American Journal of Psychiatry*, *156*(8), 1244–1249.

Murray, S., & Mobasser, A. (2007). Is the Internet killing everything? In: P. Croughton (Ed.), *Arena: The original men's style magazine*. London: Arena International.

Narayanan, A., & Shmatikov, V. (2009). De-anonymizing social networks. *In IEEE Symposium on Security and Privacy*.

Nass, C. (2004). Etiquette equality: Exhibitions and expectations of computer politeness. *Communications of the ACM*, *47*(4), 35–37. doi:10.1145/975817.975841

Nidumolu, S. R., Subramani, M., & Aldrich, A. (2001). Situated learning and the situated knowledge web: Exploring the ground beneath knowledge management. *Journal of Management Information Systems (JMIS)*, *18*(1), 115–150.

Nielsen, J. (1998, February 8). The reputation manager. Retrieved June 14, 2008, from http://www.useit.com/alertbox/980208.html

Ninad. (2007). *Feedback 2.0 – a conversation with Brian Burke*. Retrieved June 14, 2008, from http://www.E-Baychatter.com/the_chatter/2007/02/feedback_20_a_c.html

No Law to Tackle Cyberstalking. (2004). *The Economic Times*. http://ecoonomictimes.indiatimes.com/articleshow/43871804.cms

Noddings, N. (2002). *Educating moral people: A caring alternative to character education*. New York: Teachers College Press.

Noddings, N. (2005). *The challenge to care in schools: An alternative approach to education*. New York: Teachers College Press.

Nonaka, I. (1994). A dynamic theory of organizational knowledge creation. *Organization Science*, *5*(1), 14–37. doi:10.1287/orsc.5.1.14

Norman, D. A. (1991). Cognitive artifacts. In Carroll, J. M. (Ed.), *Designing interaction: Psychology at the human-computer interface* (pp. 17–38). New York, NY: Cambridge University Press.

Ochoa, A., et al. (2007). Baharastar – simulador de algoritmos culturales para la minería de datos social. *In Proceedings of COMCEV'2007*.

Ochoa, A., et al. (2009) Six degrees of separation in a graph to a social networking. *In Proceedings of ASNA'2009*, Zürich, Switzerland.

Ochoa, A. (2008). Social data mining to improve bioinspired intelligent systems. In Giannopoulou, E. G. (Ed.), *Data Mining in Medical and Biological Research*. Vienna, Austria: In-Tech.

Ogilvie, E. (2000). *Cyberstalking, trends and issues in crime and criminal justice.* 166. http://www.aic.gov.au

Olweus, D. (1998). *Bullying at school: What we know and what we can do* [Spanish edition: Conductas de acoso y amenaza entre escolares]. Madrid, Spain: Ediciones Morata.

Omidyar, P. (1996). *eBay founders letter to eBay community.* Retrieved December 28, 2005, from http://pages.ebay.com/services/forum/feedback-foundersnote.html

Oosterhof, N. N., & Todorov, A. (2008). The Functional Basis of Face Evaluation. *Proceedings of the National Academy of Sciences of the United States of America, 105,* 11087–11092. doi:10.1073/pnas.0805664105

Oram, A. (Ed.). (2001). *Peer-to-peer: Harnessing the power of disruptive technologies.* O'Reilly and Associates.

Orczy, E. (1905). *The scarlet pimpernel.* Binding Unknown.

Orengo Castellá, V., Zornoza Abad, A., Prieto Alonso, F., & Peiró Silla, J. (2000). The influence of familiarity among group members, group atmosphere and assertiveness on uninhibited behavior through three different communication media. *Computers in Human Behavior, 16,* 141–159. doi:10.1016/S0747-5632(00)00012-1

Orlikowski, W. J. (2002). Knowing in practice: Enacting a collective capability in distributed organizing. *Organization Science, 13*(3), 249–273. doi:10.1287/orsc.13.3.249.2776

Orr, J. E. (1989). Sharing knowledge, celebrating identity: War stories and community memory among service technicians. In Middleton, D. S., & Edwards, D. (Eds.), *Collective remembering: memory in society.* Newbury Park, CA: Sage Publications.

Owen, L., & Ennis, C. (2005). The ethic of care in teaching: An overview of supportive literature. *Quest, 57*(4), 392–425.

Page, S. E. (1999). Computational models from A to Z. *Complexity, 5*(1), 35–41. doi:10.1002/(SICI)1099-0526(199909/10)5:1<35::AID-CPLX5>3.0.CO;2-B

Palen, L., & Liu, S. B. (2007). Citizen communications in crisis: Anticipating the future of ITC-supported public participation. In *CHI 2007 Proceedings* (pp. 727-735), San Jose, CA, USA.

Palen, L., Vieweg, S., Sutton, J., Liu, S. B., & Hughes, A. (2007). *Crisis informatics: Studying crisis in a networked world.* Paper presented at the Third International Conference on e-Social Science (e-SS). Retrieved August 5, 2009, from http://www.cs.colorado.edu/~palen/Papers/iscram08/CollectiveIntelligenceISCRAM08.pdf

Palen, L. (2008). On line social media in crisis events. *Educase Quarterly, 3,* 76–78.

Palen, L., Hiltz, S. R., & Liu, S. B. (2007). On line forums supporting grassroots participation in emergency preparedness and response. *Communications of the ACM, 50*(3), 54–58. doi:10.1145/1226736.1226766

Pallatto, J. (2007, January 22). Monthly Microsoft patch hides tricky IE 7 download. Retrieved on http://www.eweek.com/article2/0,1895,2086423,00.asp

Palmer, J. (2008, May 3). Emergency 2.0 is coming to a website near you. *New Scientist,* 24–25. doi:10.1016/S0262-4079(08)61097-0

Palmer, P. J. (1998). *The courage to teach: Exploring the inner landscape of a teacher's Life.* New York: Jossey-Bass.

Palys, T. (2003). *Research Decisions: Quantitative and Qualitative Perspectives* (3rd ed.). New York: Thomson Nelson.

Pan, S. L., & Leidner, D. E. (2003). Bridging communities of practice with information technology in the pursuit of global knowledge sharing. *The Journal of Strategic Information Systems, 12,* 71–88. doi:10.1016/S0963-8687(02)00023-9

Papacharissi, Z. (2006). Audiences as media producers: Content analysis of 260 blogs. In Tremayne, P. (Ed.), *Blogging, citizenship, and the future of media* (pp. 21–38). New York: Routledge.

Patchin, J. W., & Hinduja, S. (2006). Bullies move beyond the school yard: A preliminary look at cyber bullying. *Youth Violence and Juvenile Justice, 4*(2), 148–169. doi:10.1177/1541204006286288

Pavlou, P., & Gefen, D. (2004). Building effective online marketplaces with institution-based trust. *Information Systems Research, 15*(1), 37–59. doi:10.1287/isre.1040.0015

PCMagazine. (2001). 20th anniversary of the PC survey results. Retrieved on http://www.pcmag.com/article2/0,1759,57454,00.asp.

Peace and War in Taking IT Global. (2004). Retrieved November 5, 2004, from www.takingitglobal.org/

Pennington, R., Wilcox, H. D., & Grover, V. (2004). The role of system trust in business-to-consumer transactions. *Journal of Management Information Systems, 20*(3), 197–226.

Pérez, C., & Santín, D. (2006). *Data mining. Soluciones con Enterprise Manager*. Alfa Omega Grupo Editor, S.A. de C.V., Mexico

Perlman, R. (1999). An overview of PKI trust models. *IEEE Network, 13*(6), 38–43. doi:10.1109/65.806987

Perry, R. W., & Quarantelli, E. L. (Eds.). (2005). *What is a disaster? New answers to old questions*. Philadelphia: XLibris.

Piccoli, G., Ahmad, R., & Ives, B. (2001). Web-based virtual learning environments: A research framework and a preliminary assessment of effectiveness in basic IT skills training. *MIS Quarterly, 25*(4). doi:10.2307/3250989

Pino-Silva, J., & Mayora, C. A. (2010). English teachers' moderating and participating in OCPs. *System, 38*(2). doi:10.1016/j.system.2010.01.002

Polanyi, M. (1958). *Personal knowledge, towards a post-critical philosophy*. Chicago, IL: University of Chicago Press.

Ponce, J., Hernández, A., Ochoa, A., Padilla, F., Padilla, A., Alvarez, F., & Ponce de León, E. (2009). Data mining in web applications. In Ponce, J., & Karahoka, A. (Eds.), *Data Mining and Discovery Knowledge in Real Life Application*. Vienna, Austria: In-Tech.

Poor, N. (2005). Mechanisms of an online public sphere: The website slashdot. *Journal of Computer-Mediated Communication, 10*(2).

Porra, J., & Goel, L. (2006, November 18-21). Importance of Power in the Implementation Process of a Successful KMS: A Case Study. *In 37th Annual Meeting of the Decision Sciences Institute,* San Antonio, TX.

Porra, J., & Parks, M.S. (2006). Sustaining virtual communities: Suggestions from the colonial model. *Information Systems and e-Business Management, 4*(4), 309-341.

Postmes, T., Spears, R., & Lea, M. (2000). The formation of group norms in computer-mediated communication. *Human Communication Research, 26*(3), 341–371. doi:10.1111/j.1468-2958.2000.tb00761.x

Powazek, D. M. (2002). *Design for community: The art of connecting real people in virtual places*. New Riders.

Power, R. (2000). Tangled Web: Tales of digital crime from the shadows of cyberspace. Indianapolis: QUE Corporation.

Prandelli, E., Verona, G., & Raccagni, D. (2006). Diffusion of web-based product innovation. *California Management Review, 48*(4), 109–135. doi:10.2307/41166363

Pratley, C. (2004). Chris_Pratley's OneNote WebLog. Retrieved http://weblogs.asp.net/chris_pratley/archive/2004/05/05/126888.aspx

Preece, J. (2001). *Online communities: Designing usability, supporting sociability*. Chichester, UK: John Wiley & Sons.

Preece, J. (2001). Sociability and usability in online communities: Determining and measuring success. *Behaviour & Information Technology, 20*(5), 347–356. doi:10.1080/01449290110084683

Preece, J., & Ghozati, K. (2001). Observations and explorations of empathy online. In Rice, R. R., & Katz, J. E. (Eds.), *The Internet and health communication: Experience and expectations* (pp. 237–260). Thousand Oaks, CA: Sage Publications. doi:10.4135/9781452233277.n11

Preece, J., & Maloney-Krichmar, D. (2002). Online communities: Focusing on sociability and usability. In Jacko, J., & Sears, A. (Eds.), *The human-computer interaction handbook* (pp. 596–620). Mahwah: Lawrence Earlbaum Associates.

Preece, J., Nonnecke, B., & Andrews, D. (2003). (in press). The top five reasons for lurking: improving community experiences for everyone. *Computers in Human Behavior*.

Preece, J., Nonnecke, B., & Andrews, D. (2004). The top 5 reasons for lurking: Improving community experiences for everyone. *Computers in Human Behavior, 2*(1), 42.

Prensky, M. (2001). Digital natives, digital immigrants. *Horizon, 9*(5), 1–6. doi:10.1108/10748120110424816

Preston, S. D., & de Waal, B. M. (2002). Empathy: Its ultimate and proximate bases. *The Behavioral and Brain Sciences, 25*, 1–72.

Propp, V. (1969). *Morphology of the folk tale.* Austin, TX: University of Texas Press.

Protégeles. (2009). *Protegeles línea de denuncia.* Retrieved May 1, 2010, from http://www.protegeles.com/

Prusak, L. (2001). Where did knowledge management come from? *IBM Systems Journal, 40*(4), 1002–1007. doi:10.1147/sj.404.01002

Puccinelli, N. (2006). Putting your best face forward: The impact of customer mood on salesperson evaluation. *Journal of Consumer Psychology, 16*(2), 156–162. doi:10.1207/s15327663jcp1602_6

Pujol, J. M., Sanguesa, R., & Delgado, J. (2002). Extracting reputation in multi-agent systems by means of social network topology. *Proceedings of the First International Joint Conf. Autonomous Agents and Multiagent Systems.*

Qian, H., & Scott, C. R. (2007). Anonymity and self-disclosure on Weblogs. *Journal of Computer-Mediated Communication, 12*(4). Retrieved from http://jcmc.indiana.edu/vol12/issue4/qian.html. doi:10.1111/j.1083-6101.2007.00380.x

Quarantelli, E. L. (1998). *What is a disaster? Perspectives on the question.* New York: Routledge.

Quarantelli, E. L. (2007). Problematical aspects of the information/ communication revolution for disaster planning and research: ten non-technical issues and questions. *Disaster Prevention and Management, 6*(2), 94–106. doi:10.1108/09653569710164053

Racism in Brazilian. (2009). *Racism in Brazilian Orkut.* Retrieved May 1, 2010, from http://zonaeuropa.com/20050326_2.htm

Racismo na internet (2006). *Racismo na internet chega à Justiça* (in Portuguese). Estadão. Retrieved July 10, 2007, from http://www.estadao.com.br/tecnologia/ internet/ noticias/2006/fev/01/97.htm.

Rao, A. S., & Georgeff, M. P. (1998). Decision procedures for BDI logics. *Journal of Logic and Computation, 8*(3), 293. doi:10.1093/logcom/8.3.293

Rashid, A. M., Ling, K., Tassone, R. D., Resnick, P., Kraut, R., & Riedl, J. (2006). Motivating participation by displaying the value of contribution. *Proceedings of the SIGCHI Conference on Human Factors in Computing Systems,* (p. 958).

Raskauskas, J., & Stoltz, A. D. (2007). Involvement in traditional and electronic bullying among adolescents. *Developmental Psychology, 43*(3), 564–575. doi:10.1037/0012-1649.43.3.564

Raskin, J. (2000). The humane interface. Boston: Addison-Wesley.

Rauner, D. (2000). *They still pick me up when I fall: The role of caring in youth development and community life.* New York: Columbia University Press.

Rawls, J. (2001). Justice as fairness. Cambridge, MA: Harvard University Press.

Reeves, B., & Nass, C. (1996). The media equation: How people treat computers, television, and new media like real people and places. New York: Cambridge University Press/ICSLI.

Reigeluth, C. M., & Frick, T. W. (1999). Formative research: A methodology for creating and improving design theories. In Reigeluth, C. M. (Ed.), *Instructional design theories and models: A new paradigm of instructional theory* (pp. 633–652). Mahwah, NJ: Lawrence Erlbaum.

Reiss, S. (2004). Multifaceted nature of intrinsic motivation: The theory of 16 basic desires. *Review of General Psychology, 8*(3), 179–193. doi:10.1037/1089-2680.8.3.179

Reiter, M. K., & Stubblebine, S. G. (1998). Resilient authentication using path independence. *IEEE Transactions on Computers, 47*(12), 1351–1362. doi:10.1109/12.737682

Renn, O., & Walker, K. (Eds.). (2008). *Global Risk Governance: Concept and Practice Using the IRGC Framework*. Dordrecht, The Netherlands: Springer. doi:10.1007/978-1-4020-6799-0

Report on Cyberstalking. (1999, August). *Cyberstalking: A new challenge for law enforcement and industry*. A Report from the Attorney General to The Vice President. http://www.usdoj.gov/criminal/cybercrime/cyberstalking.htm

Resnick, P., & Zeckhauser, R. (2002). Trust among strangers in Internet transactions: Empirical analysis of eBay's reputation system. In M. Baye, (Ed.), *Advances in applied microeconomics: The economics of the Internet and e-commerce, 11*, 127–157.

Resnick, P., Zeckhauser, R., Swanson, J., & Lockwood, K. (2003). *The value of reputation on eBay: A controlled experiment*.

Resnick, P., Zeckhauser, R., Friedman, E., & Kuwabara, K. (2000). Reputation systems: Facilitating trust in Internet interactions. *Communications of the ACM, 43*(12), 45–48. doi:10.1145/355112.355122

Rettew, J. (2009). *Crisis communication and social media*. Paper presented at the AIM2009 Conference. Retrieved August 4, 2009, from http://www.slideshare.net/AIM_Conference/crisis-communications-and-social-media-jim-rettew-the-red-cross-2009-aim-conference

Reynolds, R. (1998). An *introduction to cultural algorithms. Cultural algorithms repository*. Retrieved May 1, 2010, from http://www.cs.wayne.edu/~jcc/car.html

Rheingold, H. (2000). *The virtual community: Homesteading on the electronic frontier* (2nd ed.). London, UK: MIT Press.

Rice, R. E. (1992). Task analyzability, use of new media, and effectiveness: A multi-site exploration of media richness. *Organization Science, 3*(4). doi:10.1287/orsc.3.4.475

Ridings, C. M., & Gefen, D. (2004). Virtual community attraction: Why people hang out online. *Journal of Computer-Mediated Communication, 10*(1).

Rigby, K. (2005). What children tell us about bullying in schools. *Children Australia Journal of Guidance and Counselling, 15*(2), 195–208. doi:10.1375/ajgc.15.2.195

Rizzolatti, G., & Arbib, M. A. (1998). Language within our grasp. *Trends in Neurosciences, 21*, 188–194. doi:10.1016/S0166-2236(98)01260-0

Rodríguez, R., Seoane, A., & Pereira, J. (2006). Niños contra niños: El bullying como transtorno emergente. *Anal Pediátrico, 64*(2), 162–166. doi:10.1157/13084177

Rogers, E. M., & Agarwala-Rogers, R. (1975). Organizational communication. G. L. Hanneman, & W. J. McEwen, (Eds.), Communication behaviour (pp. 218–236). Reading, MA: Addision Wesley.

Romm, C. T., & Setzekom, K. (2008). *Social network communities and E-dating services: Concepts and implications*. London, UK: Information Science Reference. doi:10.4018/978-1-60566-104-9

Rose, G., Khoo, H., & Straub, D. (1999). Current technological impediments to business-to-consumer electronic commerce. Communications of the AIS, I(5).

Rosen, L., Cheever, N., & Carrier, L. (2008). The association of parenting style and child age with parental limit setting and adolescent MySpace behavior. *Journal of Applied Developmental Psychology, 29*(6), 459–471. doi:10.1016/j.appdev.2008.07.005

Rousseau, D. M., Sitkin, S. B., Burt, R. S., & Camerer, C. (1998). Not so different after all: A cross-discipline view of trust. *Academy of Management Review, 23*(3), 393–404.

Rutter, J. (2001). From the sociology of trust towards a sociology of "e-trust". *International Journal of New Product Development & Innovation Management, 2*(4), 371–385.

Ryle, G. (1949/1984). *The concept of mind*. Chicago, IL: University of Chicago Press.

Sabater, J., & Sierra, C. (2002). Reputation and social network analysis in multi-agent systems. *Proceedings of the First International Joint Conference on Autonomous Agents and Multiagent Systems*.

Salam, A. F., Iyer, L., Palvia, P., & Singh, R. (2005). Trust in e-commerce. *Communications of the ACM, 48*(2), 73–77. doi:10.1145/1042091.1042093

Salton, G., & McGill, M. J. (1986). *Introduction to modern information retrieval*. New York: McGraw-Hill, Inc.

Sandoval, G., & Wolverton, T. (1999). *eBay files suit against auction site bidder's edge.* Retrieved December 28, 2005, from http://news.com.com/2100-1017-234462.html

Sarkio, K., & Holtmanns, S. (2007). Tailored trustworthiness estimations in peer to peer networks. To appear in International Journal of Internet Technology and Secured Transactions IJITST, Inderscience.

Schmugar, C. (2008). *The future of social networking sites.* Retrieved December 13, 2008 from http://www.mcafee.com

Scholing, A., & Emmelkamp, P. M. G. (1993). Exposure with and without cognitive therapy for generalized social phobia: Effects of individual and group treatment. *Behaviour Research and Therapy, 31*(7), 667–681. doi:10.1016/0005-7967(93)90120-J

Schultze, U., & Leidner, D. (2002). Studying knowledge management in information systems research: Discourses and theoretical assumptions. *MISQ, 26*(3), 213–242. doi:10.2307/4132331

Schultze, U., & Stabell, C. (2004). Knowing what you don't know? Discourses and contradictions in knowledge management research. *Journal of Management Studies, 41*(4), 549–573. doi:10.1111/j.1467-6486.2004.00444.x

Scott-Joynt, J. (2005). *What Myspace means to Murdoch.* London: BBC Online. Retrieved from http://news.bbc.co.uk/1/hi/business/4697671.stm.

Sen, S., & Sajja, N. (2002). Robustness of reputation-based trust: Boolean case. *Proceedings of the First International Joint Conference on Autonomous Agents and Multiagent Systems.*

Shankar, K. (2008). Wind, Water, and Wi-Fi: New Trends in Community Informatics and Disaster Management. *The Information Society, 24*(2), 116–120. doi:10.1080/01972240701883963

Shariff, S., & Gouin, R. (2005). *Cyber-Dilemmas: Gendered Hierarchies, Free Expression and Cyber-Safety in Schools.*paper presented at Oxford Internet Institute, Oxford University, U.K. International Conference on Cyber-Safety. Paper available at: www.oii.ox.ac.uk/cybersafety

Shariff, S. (2006). Cyber-Dilemmas: Balancing Free Expression and Learning in a Virtual School Environment. *International Journal of Learning, 12*(4), 269–278.

Shariff, S. (2008). *Cyber-bullying: Issues and solutions for the school, the classroom and the home.* New York: Routledge.

Shariff, S., & Johnny, L. (2007). Cyber-Libel and Cyber-bullying: Can Schools Protect Student Reputations and Free Expression in Virtual Environments? *Education Law Journal, 16*(3), 307.

Sharples, M., Graber, R., Harrison, C., & Logan, K. (2009). E-safety and Web 2.0 for children aged 11-16. *Journal of Computer Assisted Learning, 25*(1), 70–84. doi:10.1111/j.1365-2729.2008.00304.x

Shaw, M. (1996). *Civil society and media in global crisis: Representing distant violence.* London: Pinter.

Shectman, N., & Horowitz, L. M. (2003). Media inequality in conversation: How people behave differently when interacting with computers and people. Paper presented at the CHI (Computer Human Interaction) 2003, Ft Lauderdale, Florida.

Shen, X., Radakrishnan, T., & Georganas, N. (2002). vCOM: Electronic commerce in a collaborative virtual worlds. *Electronic Commerce Research and Applications, 1*, 281–300. doi:10.1016/S1567-4223(02)00021-2

Sheridan, L., Davies, G. M., & Boon, J. C. W. (2001). Stalking: Perceptions and prevalence. *Journal of Interpersonal Violence, 16*(2), 151–167. doi:10.1177/088626001016002004

Sherif, K., & Xing, B. (2006). Adaptive processes for knowledge creation in complex systems: the case of a global IT consulting firm. *Information & Management, 43*(4), 530–540. doi:10.1016/j.im.2005.12.003

Sherwin, A. (2006, April 3). A family of Welsh sheep - The new stars of Al-Jazeera. *Times (London, England)*, (n.d), 7.

Shin, M. (2004). A framework for evaluating economics of knowledge management systems. *Information & Management, 42*, 179–196. doi:10.1016/j.im.2003.06.006

Shirky, C. (2000). *What is P2P... and what isn't?* Retrieved December 28, 2005 from http://www.openp2p.com/pub/a/p2p/2000/11/24/shirky1-whatisp2p.html

Shirky, C. (2004). *Multiply and social spam: Time for a boycott*. Retrieved December 28, 2005, from http://many.corante.com/archives/2004/08/20/multiply_and_social_spam_time_for_a_boycott.php

Silva, J. A. (2001). Do hipertexto ao algo mais: usos e abusos do conceito de hipermídia pelo jornalismo online. In: A. Lemos, & M. Palacios (orgs.) *Janelas do Ciberespaço - Comunicação e Cibercultura*. Sao Paulo, Brasil: Editora Sulina, pp. 128-139.

Silva, L., Mousavidin, E., & Goel, L. (2006). Weblogging: Implementing Communities of Practice. *In Social Inclusion: Societal and Organizational Implications for Information Systems: IFIP TC8 WG 8.2*, Limirick, Ireland.

Simmons, L. L., & Clayton, R. W. (2010). The impact of small business B2B virtual community commitment on brand loyalty. *International Journal of Business and Systems Research*, *4*(4), 451–468. doi:10.1504/IJBSR.2010.033423

Simon, S. J., Grover, V., Teng, J. T. C., & Whitcomb, K. (1996). The relationship of information system training methods and cognitive ability to end-user satisfaction, comprehension, and skill transfer: A longitudinal field study. *Information Systems Research*, *7*(4), 466–490. doi:10.1287/isre.7.4.466

Singh, A., & Liu, L. (2003). TrustMe: Anonymous management of trust relationships in decentralized P2P systems. *Proceedings of the IEEE International Conference on Peer-to-Peer Computing*, pp. 142–149.

Sinwelski, S., & Vinton, L. (2001). Stalking: The constant threat of violence. *Affilia*, *16*, 46–65. doi:10.1177/08861090122094136

Skinner, B. F. (1938). *The behavior of organisms: An experimental analysis*. B. F. Skinner Foundation.

Slonje, R., & Smith, P. (2008). Cyberbullying: Another main type of bullying? *Scandinavian Journal of Psychology*, *49*(2), 147–154. doi:10.1111/j.1467-9450.2007.00611.x

Smith, A., Schlozman, K. L., Verba, S., & Brady, H. (2009). The Internet and civic engagement. *Pew Internet & American Life Project*. Retrieved November 17, 2009, from http://www.pewinternet.org/Reports/2009/15--The-Internet-and-Civic-Engagement.aspx

Smith, C. (2000). Content analysis and narrative analysis. In: H. Reis & C. Judd (Eds.), *Handbook of research methods in social and personal psychology*. Cambridge: Cambridge University Press.

Smith, G. J. H. (1996). Building the lawyer-proof web site. Paper presented at the *Aslib Proceedings, 48,* (pp. 161-168).

Smith, P., Mahdavi, J., Carvalho, M., Fisher, S., Russell, S., & Tippett, N. (2008). Cyberbullying: its nature and impact on secondary school pupils. *Journal of Child Psychology and Psychiatry, and Allied Disciplines*, *49*(4), 376–385. doi:10.1111/j.1469-7610.2007.01846.x

Song, S., Hwang, K., Zhou, R., & Kwok, Y. K. (2005). Trusted P2P transactions with fuzzy reputation aggregation. *IEEE Internet Computing*, *9*(6), 24–34. doi:10.1109/MIC.2005.136

Sorensen, J. H., Sorensen, B. V., Smith, A., & Williams, Z. (2009). *Results of An Investigation of the Effectiveness of Using Reverse Telephone Emergency Warning Systems in the October 2007 San Diego Wildfires*. Report prepared for the U.S. Department of Homeland Security. Retrieved August 4, 2009, from http://galainsolutions.com/resources/San$2520DiegoWildfires$2520Report.pdf

Spertus, E. (1996). *Social and technical means for fighting on-line harassment*. http://ai.mit.edu./people/ellens/Gender/glc

Sproull, L., & Kiesler, S. (1991). *Connections: New ways of working in the networked organization*. Cambridge, MA: MIT Press.

Stahl, C., & Fritz, N. (2002). Internet safety: adolescents' self-report. *The Journal of Adolescent Health*, *31*(1), 7–10. doi:10.1016/S1054-139X(02)00369-5

Stake, R. E. (1978). The case study method in social inquiry. *Educational Researcher*, *7*(2), 5–8.

Stake, R. E. (1995). *The art of case study research*. Thousand Oaks, CA: Sage.

Starr, C. (1969). Social benefits versus technological risks. *Science*, *165*(3899), 1232–1238. doi:10.1126/science.165.3899.1232

State Goverment of Victoria, Department of Justice. (2008). *Privacy and security tips for using social network sites*. Retrieved in 2009 from http://www.justice.vic.gov.au

Stein, E. W. (2006). A qualitative study of the characteristics of a community of practice for knowledge management and its success factors. *International Journal of Knowledge Management, 1*(4), 1–24.

Steiner, D. (2004). *Auctionbytes survey results: Your feedback on eBay's feedback system*. Retrieved December 28, 2005, from http://www.auctionbytes.com/cab/abu/y204/m11/abu0131/s02

Stephenson, W. D., & Bonabeau, E. (2007). Expecting the unexpected: The need for a networked terrorism and disaster response strategy. *Homeland Security Affairs, III*(1). Retrieved August 5, 2009, from http://www.hsaj.org/?article=3.1.3

Sternberg, R. (1986). A triangular theory of love. *Psychological Review, 93*(2), 119–135. doi:10.1037/0033-295X.93.2.119

Stevens, V. (2004). Webhead communities: Writing tasks interleaved with synchronous online communication and Web page development. In: J. Willis & B. Leaver (Eds.), *Task-based instruction in foreign language education: Practices and programs.* Georgetown, VA: Georgetown University Press.

Stoehr, T. (2002). *Managing e-business projects: 99 key success factors.* London: Springer.

Stutzman, F. (2005). *An evaluation of identity-sharing behavior in social network communities*. Retrieved from http://www.ibiblio.org/fred/pubs/stutzman_pub4.pdf

Sun, P., Unger, J., Palmer, P., Gallaher, P., Chou, P., & Baezconde-Garbanati, L. (2005). Internet accessibility and usage among urban adolescents in southern California: Implications for web-based health research. *Cyberpsychology & Behavior, 8*(5), 441–453. doi:10.1089/cpb.2005.8.441

Sunstein, C. (1999). *Republic.com*. Princeton, NJ: Princeton University Press.

Sun, Y., Yu, W., Han, Z., & Liu, K. J. R. (2006). Information theoretic framework of trust modeling and evaluation for ad hoc networks. *IEEE Journal on Selected Areas in Communications, 24*(2), 305–317. doi:10.1109/JSAC.2005.861389

Sutton, J., Palen, L., & Shklovski, I. (2008). Backchannels on the front lines: Emergent uses of social media in the 2007 Southern California Wildfires. In F. Fiedrich & B. Van de Walle (Eds.), *Proceedings of the 5th International ISCRAM Conference*, Washington, DC, USA, May 2008.

Taba, H. (1963). Learning by discovery: Psychological and educational rationale. *Elementary School,* 308-316.

Tajeddine, A., Kayssi, A., Chehab, A., & Artail, H. (2005, September). A comprehensive reputation-based trust model for distributed systems. Proceedings of the Workshop of the 1st International Conference on Security and Privacy for Emerging Areas in Communication Networks, pp. 118–127.

Tajfel, H., & Turner, J. C. (1986). The *social* identity theory of intergroup behavior. In Worchel, S., & Austin, W. G. (Eds.), *Psychology of intergroup relations* (pp. 7–24). Chicago: Nelson-Hall.

Takhteyev, Y., & Hall, J. (2005). *Blogging together: Digital expression in a real-life community*. Paper presented at the Social Software in the Academy Workshop, Los Angeles, CA.

Tan, Y., & Thoen, W. (1998). Toward a generic model of trust for electronic commerce. *International Journal of Electronic Commerce, 5*(2), 61–74.

Tapscott, D., & Williams, A. D. (2010). *Macrowikinomics: Rebooting business and the world*. Canada: Penguin Group.

TCG. Trusted Computing Group, Trusted Platform Module - TPM Specification v1.2, 2003. Retrieved May, 2006, from https://www.trustedcomputinggroup.org/specs/TPM/

Technology threats to privacy. (2002, February 24). New York Times, Section 4, p. 12.

Technorati.com. (n.d.). *VoteLinks*. Retrieved December 28, 2005, from http://microformats.org/wiki/votelinks

The Risk Communicator. (2008). Social media and your emergency communication efforts. The *Risk Communicator, 1*, 3-6. Retrieved June 15, 2009, from http://emergency.cdc.gov

Thelwall, M. (2008). Social networks, gender, and friending: An analysis of MySpace member profiles. *Journal of the American Society for Information Science and Technology, 59*(8), 1321–1330. doi:10.1002/asi.20835

Thelwall, M. (2009). Social network sites: Users and uses. In Zelkowitz, M. (Ed.), *Advances in computers: Social networking and the web* (p. 19). London, UK: Academic Press. doi:10.1016/S0065-2458(09)01002-X

Theodorakopoulos, G., & Baras, J. S. (2006). On trust models and trust evaluation metrics for ad hoc networks. *IEEE Journal on Selected Areas in Communications, 24*(2), 318–328. doi:10.1109/JSAC.2005.861390

Tinker, T., & Fouse, D. (Eds.). (2009). *Expert round table on social media and risk communication during times of crisis: Strategic challenges and opportunities.* Special report. Retrieved August 6, 2009, from http://www.boozallen.com/media/file/Risk_Communications_Times_of_Crisis.pdf

Tjaden, P., & Thoennes, N. (1997). Stalking in America: Findings from the National Violence. Retrieved May 25, 2007 from http://www.ncjrs.gov/txtfiles/169592.txt

Transparency-International. (2001). Corruption perceptions. Retrieved on www.transparency.org

Tullock, G. (1997). Adam Smith and the prisoners' dilemma. In D.B. Klein (Ed.), *Reputation: Studies in the voluntary elicitation of good conduct* (pp. 21-28). Ann Arbor, MI: University of Michigan Press.

Turkle. (1997). *Life on the screen: Identity in the age of the Internet.* New York, NY: Touchstone.

Tynes, B. (2007). Internet safety gone wild? Sacrificing the educational and psychosocial benefits of online social environments. *Journal of Adolescent Research, 22*(6), 575–584. doi:10.1177/0743558407303979

Tzeng, J. (2004). Toward a more civilized design: studying the effects of computers that apologize. *International Journal of Human-Computer Studies, 61*(3), 319–345. doi:10.1016/j.ijhcs.2004.01.002

Tziros, T. (2008). *The riots in new media.* Article in Newspaper "Makedonia" on December 15, 2008. [In Greek]. Retrieved August 5, 2009, from http://www.makthes.gr/index.php?name=News&file=article&sid=30225

Usoro, A., & Kuofie, M. H. S. (2006). Conceptualisation of cultural dimensions as a major influence on knowledge sharing. *International Journal of Knowledge Management, 2*(2), 16–25. doi:10.4018/jkm.2006040102

Ustaoglu, Y. (2009). Simulating the behavior of a minority in Turkish society. *In Proceedings of ASNA'2009.*

Valcke, M., Schellens, T., Van Keer, H., & Gerarts, M. (2007). Primary school children's safe and unsafe use of the Internet at home and at school: An exploratory study. *Computers in Human Behavior, 23*(6), 2838–2850. doi:10.1016/j.chb.2006.05.008

Valdez, S. I., Hernández, A., & Botello, S. (2009). Approximating the search distribution to the selection distribution. *In Proceedings of GECCO 2009,* 461-468.

Vidal, J. (2007). *Fundamentals of multiagent systems with NetLogo examples.* Retrieved May 1, 2010, from: http://jmvidal.cse.sc.edu/papers/mas.pdf

Vieweg, S. (2008). Social networking sites: Reinterpretation in crisis situations. *Workshop on Social Networking in Organizations CSCW08,* San Diego USA, November 9. Retrieved August 5, 2009, from http://research.ihost.com/cscw08-socialnetworkinginorgs/papers/vieweg_cscw08_workshop.pdf

Vieweg, S., Palen, L., Liu, S. B., Hughes, A. L., & Sutton, J. (2008). Collective intelligence in disaster: Examination of the phenomenon in the aftermath of the 2007 Virginia Tech shooting. In F. Fiedrich & B. Van de Walle (Eds.), *Proceedings of the 5th International ISCRAM Conference,* Washington, DC, USA, May 2008.

Vogel, D. (1999). *Financial investigations: A financial approach to detecting and resolving crimes.* London: DIANE Publishing.

von Lubitz, D. K. J. E., Beakley, J. E., & Patricelli, F. (2008). Disaster management: The structure, function, and significance of network-centric operations. *Journal of Homeland Security and Emergency Management, 5*(1). Retrieved August 5, 2009, from http://www.bepress.com/jhsem/vol5/iss1/42/

von Lubitz, D. K. J. E., Beakley, J. E., & Patricelli, F. (2008). All hazards approach' to disaster management: the role of information and knowledge management, Boyd's OODA Loop, and network-centricity. *Disasters*, *32*(4), 561–585.

Vygotsky, L. S. (1930). *Mind in society*. Cambridge, MA: Waiton, S. (2009). Policing after the crisis: Crime, safety and the vulnerable public. *Punishment and Society*, *11*(3), 359.

Vygotsky, L. S. (1978). *Mind in society: The development of higher psychological processes*. Cambridge, MA: Harvard University Press.

Wallace, B. (2000, July 10). Stalkers find a new tool—The Internet e-mail is increasingly used to threaten and harass, authorities say. *SF Gate News*. http://sfgate.com/cgi-bin/article_cgi?file=/chronicle/archive/2000/07/10/MN39633.DTL

Wallace, P. (2001). *The psychology of the Internet*. Cambridge: Cambridge University Press.

Wang, Y., & Varadharajan, V. (2005). Trust2: developing trust in peer-to-peer environments. *Proceedings of the IEEE International Conference on Services Computing*, *1*(July), 24–31.

Wasko, M. M., & Faraj, S. (2000). 'It is what one does:' Why people participate and help others in electronic communities of practice. *JSIS*, *9*(2/3), 155–173.

Wasko, M. M., & Faraj, S. (2005). Why should I share? Examining knowledge contribution in networks of practice. *Management Information Systems Quarterly*, *29*(1), 35–57.

Wasko, M. M., Faraj, S., & Teigland, R. (2004). Collective action and knowledge contribution in electronic networks of practice. *JAIS*, *5*(11-12), 493–513.

Weiler, J. H. H. (2002). A constitution for Europe? Some hard choices. *Journal of Common Market Studies*, *40*(4), 563–580. doi:10.1111/1468-5965.00388

Wellman, B. (1996). *For a Social Network Analysis of Computer Networks: A Sociological Perspective on Collaborative Work and Virtual Community*. Denver, Colorado: SIGCPR/SIGMIS.

Wellman, B., & Gulia, M. (1997). *Net surfers don't ride alone: Virtual communities as communities. Communities and cyberspace*. New York: Routledge.

Wellman, B., Salaff, J., Dimitrova, D., Garton, L., Gulia, M., & Haythornthwaite, C. (1996). Computer networks as social networks: Collaborative work, telework, and virtual community. *Annual Review of Sociology*, *22*, 213–238. doi:10.1146/annurev.soc.22.1.213

Wenger, E., White, N., & Smith, J. D. (2009). *Digital habitats: Stewarding technology for communities*. Portland, OR: CPSquare.

Wenger, E. (1998). *Communities of practice: learning, meaning and identity*. Cambridge: Cambridge University Press.

Werry, C. (2001). Imagined electronic community: Representations of online community in business texts. In: C. Werry & M. Mowbray (Eds.), *Online communities: Commerce, community action and the virtual university*. Upper Saddle River, NJ: Prentice Hall.

Wertsch, J. V. (1985). *Vygotsky and the social formation of mind*. Cambridge, MA: Harvard University Press.

Westrup, D., Fremouw, W. J., Thompson, R. N., & Lewis, S. F. (1999). The psychological impact of stalking on female undergraduates. *Journal of Forensic Sciences*, *44*, 554–557.

Wexley, K. N., & Baldwin, T. T. (1986). Post-training strategies for facilitating positive transfer: An empirical exploration. *Academy of Management Journal*, *29*, 503–520. doi:10.2307/256221

Whitworth, B. (2005). Polite computing. Behaviour & Information Technology, September 5, 353–363. Retrieved on http://brianwhitworth.com/polite05.pdf

Whitworth, B. (2006). Spam as a symptom of electronic communication technologies that ignore social requirements. In C. Ghaoui (Ed.), Encyclopaedia of human computer interaction (pp. 559-566). London: Idea Group Reference.

Whitworth, B. Aldo de Moor, & Liu, T. (2006, Nov 2-3). Towards a theory of online social rights. Paper presented at the International Workshop on Community Informatics (COMINF'06), Montpellier, France.

Whitworth, B., & deMoor, A. (2003). Legitimate by design: Towards trusted virtual community environments. *Behaviour & Information Technology*, *22*(1), 31–51. doi:10.1080/01449290301783

Whitworth, B., Gallupe, B., & McQueen, R. (2001). Generating agreement in computer-mediated groups. *Small Group Research*, *32*(5), 621–661. doi:10.1177/104649640103200506

Whitworth, B., & Whitworth, E. (2004). Reducing spam by closing the social-technical gap. Retrieved on http://brianwhitworth.com/papers.html. *Computer*, (October): 38–45. doi:10.1109/MC.2004.177

Wilkinson, G. (2006). Commercial breaks: An overview of corporate opportunities for commercializing education in US and English schools. *London Review of Education*, *4*(3), 253–269. doi:10.1080/14748460601043932

Willard, N. (2006). Flame Retardant. *School Library Journal*, *52*(4), 55–56.

Willard, N. (2007). The authority and responsibility of school officials in responding to cyberbullying. *The Journal of Adolescent Health*, *41*(6), 64–65. doi:10.1016/j.jadohealth.2007.08.013

Williams, K. R., & Guerra, N. G. (2007). Prevalence and predictors of internet bullying. *The Journal of Adolescent Health*, *41*, S14–S21. doi:10.1016/j.jadohealth.2007.08.018

Wilson, R. (1985). Reputations in games and markets. In A. Roth (Ed.), *Game theoretic models of bargaining* (pp. 27-62). Cambridge: Cambridge University Press.

Wilson, J. M., Straus, & McEvily, B. (2006). All In due time: the development of trust in computer-mediated and face-to-face teams. *Organizational Behavior and Human Decision Processes*, *99*, 16–33. doi:10.1016/j.obhdp.2005.08.001

Wishart, J., Oades, C., & Morris, M. (2007). Using online role play to teach internet safety awareness. *Computers & Education*, *48*(3), 460–473. doi:10.1016/j.compedu.2005.03.003

Wolverton, T. (2000). *EBay, authorities probe fraud allegations*. Retrieved December 28, 2005, from http://news.com.com/2100-1017_3-238489.html

Woodfill, W. (2009, October 1). The transporters: Discover the world of emotions. *School Library Journal Reviews*, 59.

Working to Halt Online Abuse. (WHOA). (2003). *Online harrassment statistics*. Retrieved May 25, 2007 from http://www.haltabuse.org/resources/stats/index.shtml

Worthen, M. (2007). Education policy implications from the expert panel on electronic media and youth violence. *The Journal of Adolescent Health*, *41*(6), 61–62. doi:10.1016/j.jadohealth.2007.09.009

Wright, E. R., & Lawson, A. H. (2004). *Computer-mediated Communication and Student Learning in Large Introductory Sociology Courses*. Paper presented at the Annual Meeting of the American Sociological Association, Hilton San Francisco & Renaissance Parc 55 Hotel, San Francisco, CA. Retrieved from: http://www.allacademic.com/meta/p108968_index.html

Wright, R. (2001). Nonzero: The logic of human destiny. New York: Vintage Books.

XHTML Friends Network. (n.d.). Retrieved December 28, 2005, from http://gmpg.org/xfn/

Xiong, L., & Liu, L. (2004). PeerTrust: Supporting reputation-based trust for peer-to-peer electronic communities. *IEEE Transactions on Knowledge and Data Engineering*, *16*(7), 843–857. doi:10.1109/TKDE.2004.1318566

Yan, Z. (2006). A conceptual architecture of a trusted mobile environment. *Proceedings of IEEE SecPerU06*, France, pp. 75–81.

Yan, Z., & MacLaverty, R. (2006). Autonomic trust management in a component based software system. *Proceedings of 3rd International Conference on Autonomic and Trusted Computing (ATC06), LNCS*, Vol. 4158, September, pp. 279–292.

Yan, Z., & Prehofer, C. (2007). An adaptive trust control model for a trustworthy software component platform. *Proceedings of the 4th International Conference on Autonomic and Trusted Computing (ATC2007), LNCS*, Vol. 4610, pp. 226-238.

Yan, Z., Zhang, P., & Virtanen, T. (2003). Trust evaluation based security solution in ad hoc networks. *Proceedings of the Seventh Nordic Workshop on Secure IT Systems (NordSec03)*, Norway.

Yang, G. (2003). The Internet and the rise of a transnational Chinese cultural sphere. *Media Culture & Society*, *25*, 469–490. doi:10.1177/01634437030254003

Ybarra, M., Diener-West, M., & Leaf, P. (2007). Examining the overlap in Internet harassment and school bullying: Implications for school intervention. *The Journal of Adolescent Health*, *41*(6), 42–50. doi:10.1016/j.jadohealth.2007.09.004

Ybarra, M., & Mitchell, K. (2004). Youth engaging in online harassment: associations with caregiver-child relationships, Internet use and personal characteristics. *Journal of Adolescence*, *27*(3), 319–336. doi:10.1016/j.adolescence.2004.03.007

Ybarra, M., & Mitchell, K. (2008). How Risky are Social Networking Sites? A comparison of places online where youth sexual solicitation and harassment occurs. *Pediatrics*, *121*(2), 350–357. doi:10.1542/peds.2007-0693

Yi, Y. (Ed.). (1990). *A critical review of consumer satisfaction. Review of marketing.* Chicago: American Marketing Association.

Yi, M. Y., & Davis, F. D. (2001). Improving computer training effectiveness for decision technologies: Behavior modeling and retention enhancement. *Decision Sciences*, *32*(3), 521–544. doi:10.1111/j.1540-5915.2001.tb00970.x

Yu, B., & Singh, M. P. (2000). A social mechanism of reputation management in electronic communities. *Proceedings of the Seventh International Conference on Cooperative Information Agents.*

Yue Wah, C., Menkhoff, T., Loh, B., & Evers, H. D. (2007). Social capital and knowledge sharing in knowledge-based organizations: An empirical study. *International Journal of Knowledge Management*, *3*(1), 29–38. doi:10.4018/jkm.2007010103

Yukawa, J. (2005). Story-lines: A case study of online learning using narrative analysis. *Proceedings of the 2005 Conference on Computer Support for Collaborative Learning: Learning 2005: The Next 10 Years!* (p. 736).

Zacharia, G., & Maes, P. (2000). Trust management through reputation mechanisms. *Applied Artificial Intelligence*, *14*(8), 881–908.

Zajonc, R. (1962). Response suppression in perceptual defense. *Journal of Experimental Psychology*, *64*, 206–214. doi:10.1037/h0047568

Zhang, Z., Wang, X., & Wang, Y. (2005). A P2P global trust model based on recommendation. *Proceedings of the 2005 International Conference on Machine Learning and Cybernetics,* Vol. 7, August, pp. 3975–3980.

Zhang, P., Ma, X., Pan, Z., Li, X., & Xie, K. (2010). Multi-agent cooperative reinforcement learning in 3D virtual world. In *Advances in swarm intelligence* (pp. 731–739). London, UK: Springer. doi:10.1007/978-3-642-13495-1_90

Zhou, M., Mei, H., & Zhang, L. (2005). A multi-property trust model for reconfiguring component software. *Proceedings of the Fifth International Conference on Quality Software QAIC2005*, September, pp. 142–149.

Ziegler, C., & Lausen, G. (2004). Spreading activation models for trust propagation. In *Proceedings of the IEEE International Conference on E-Technology, E-Commerce, and E-Service (EEE'04).*

Zimbardo, P. G. (1969). The human choice: Individuation, reason, and order vs. deindividuation, impulse and chaos. In Arnold, W. J., & Levine, D. (Eds.), *Nebraska symposium on motivation* (Vol. 17, pp. 237–307). Lincoln, NE: University of Nebraska Press.

Zona, M. A., Sharma, K. K., & Lone, J. (1993). A comparative study of erotomanic and obsessional subjects in a forensic sample. *Journal of Forensic Sciences*, *38*, 894–903.

About the Contributors

Jonathan Bishop has been writing about Internet trolling since 2008. Known to those who like him as having a wicked and mischievous sense of humour, it is easy to see why the topic is so appealing to him. He has accolades on trolling from sources that appeal to all reading levels and intelligences. He has featured in articles in newspapers aimed all abilities such as The Sun, The Star as well as in more highbrow newspapers like The Western Mail, Wales on Sunday and Daily Mail, winning Letter of the Week for his debut letter to the editor on Internet trolling in the latter, which was also a featured letter. His academic credentials on trolling are also without question. He has dozens of research papers on Internet trolling published and other data misuse published in journals as learned as *The Statute Law Review, the International Review of Law,* and *Computers and Technology.* In 2011, he founded The Trolling Academy, as part of the Centre for Research into Online Communities and E-Learning Systems, a global research initiative with offices in over 5 countries.

* * *

Karen Brown, Ph.D. (Cand.), is an instructor in the School of Criminology at Simon Fraser University and the Projector Coordinator for the Centre for Education, Law and Society in the Faculty of Education at Simon Fraser University. Her primary research focuses on violence and threats against lawyers, law and legal citizenship in schools, cyber-bullying and cyber-crimes in schools, and studying the counterpoint to cyber-bullying – cyber-kindness, and reviewing the positive aspects of online exchange and interaction.

Wanda Cassidy, Ph.D., is Associate Professor of Education and Director of the Centre for Education, Law and Society at Simon Fraser University in British Columbia, Canada, a centre which seeks to improve the legal literacy of youth through a program of research, teaching, program development and community-based initiatives. Her work focuses on the relationship between law/citizenship education and social justice, and the values that underpin a civil society, including the ethics of care, inclusion and social responsibility. Recently she completed a four-year SSHRC study examining the ethics of care in schools and is currently working on a project that examines students' and educators' understanding of human rights, identity, citizenship and environmental sustainability. She also researches cyber-bullying and cyber-kindness.

Charlie Chen is an assistant professor in the Department of Computer Information Systems at Appalachian State University. He received his PhD in management information systems from Claremont Graduate University. Dr. Chen likes to view the field of management information systems from infrastructural, managerial, and operational perspectives. He is working on improving information system solutions in each of these three areas. His current main research areas are online learning, social learning technology, mobile commerce, and security awareness.

Miranda Dandoulaki is an expert in risk management with considerable field experience. She has studied civil engineering (NTUA 1981) and owns a MSc in regional development and a PhD in spatial planning. She has worked for the Greek Earthquake Planning and Protection Organization (1994-2002) and the Joint Research Centre of the European Commission (2004-2008) and has served as Vice Director of the European Centre for the Prevention and Forecasting of Earthquakes. She is presently appointed by the Greek National Centre for Public Administration and Local Government as a scientific and research officer in the Institute of Training. Dr.Dandoulaki has significant research experience and has published in books and scientific journals. Her current research interests lay in multidisciplinarity of disaster management and in spatial aspects of risk management.

Vanessa Paz Dennen is an Associate Professor of Instructional Systems in the Department of Educational Psychology and Learning Systems at Florida State University. Prior to joining the faculty at Florida State University she was an Assistant Professor of Educational Technology at San Diego State University and earned a PhD in Instructional Systems Technology from Indiana University. Her research, broadly described, investigates the nexus of cognitive, motivational, and social elements in computer-mediated communication. Within this larger area, her work has concentrated on two major problems: the study of learner participation in computer-supported collaborative learning activities; and, more recently, the study of interactions, norm development, and support networks within online communities of practice.

Lakshmi Goel is a PhD candidate in the Decision and Information Sciences Department at the C.T. Bauer College of Business, University of Houston. She holds an MS in computer science from the University of Houston. Her research interests include knowledge sharing, learning, and collaboration through information technologies such as blogs, wikis, knowledge management systems, and virtual worlds.

Călin Gurău is Associate Professor of Marketing at GSCM - Montpellier Business School, France, since September 2004. His present research interests are focused on marketing strategies for high-technology firms, internet marketing and sustainable development. He has published more than 50 papers in internationally refereed journals, such as *International Marketing Review, Journal of Consumer Marketing, Journal of Marketing Communications*, et cetera.

Matina Halkia is an architect engineer and human computer interaction expert. Her core research interests lie in augmented reality/ICT applications in physical/architectural space. She is currently working on emergency response map validation and urban analysis for remote sensing applications in human security contexts. She has studied History of Art and Architecture at Tufts University under a Fulbright Scholarship and Media Arts and Sciences at the Massachusetts Institute of Technology. She has worked as an interaction designer in Boston (1997-1998), and as principal investigator at Starlab Laboratories, a technology think-tank, in Brussels (1998-2001). Since 2001 she works for the Joint Research Centre of the European Commission, currently at the Global Security and Crisis Management Unit, as a scientific officer.

José Alberto Hernández Aguilar received PhD Grade in 2008 at Universidad Autónoma del Estado de Morelos (UAEM). He received the MBA degree in 2003 at Universidad de las Américas (UDLA), A.C. Since 2002, he is part time professor at UDLA, México, D.F., for the IT Career, and since 2007 year, he is part time professor at the Sciences Faculty at UAEM. Areas of interest: Databases, Artificial Intelligence, Online Assessment Systems and Marketing Research.

Silke Holtmanns received a PhD in mathematics from the University of Paderborn (Germany), Department of Computer Science and Mathematics. She is a principal member of the research staff at Nokia Research Center since 2004. Before she was working in Ericsson Research Lab Aachen (Germany) as a master research engineer and at the University of Paderborn as a scientific assistant. She has more than 30 publications and is rapporteur of six 3GPP security specifications and reports and involved in various standardization activities. Her research activities involve trust, privacy, identity management and mobile application security.

Gábor Hosszú received the MSc degree from Technical University of Budapest in electrical engineering and the Academic degree of Technical Sciences (PhD) in 1992. After graduation he received a three-year grant of the Hungarian Academy of Sciences. Currently he is a full-time associate professor at the Budapest University of Technology and Economics. He published several technical papers, chapters and books. In 2001 he received the three-year Bolyai János Research Grant of the Hungarian Academy of Sciences. He led number of research projects. His main interests are Internet-based media-communication, multicasting, P2P communication, network intrusion detection, character encoding, and script identification, and the VHDL based system design.

Margaret Jackson, Ph.D. is Professor in the School of Criminology at Simon Fraser University, Co-Director of the SFU Institute for Studies in Criminal Justice Policy, and Director of FREDA, which undertakes research on violence against women and children. She was principal investigator for a Ministry of Justice study on child abuse. Other research areas of interest include bullying/cyber-bullying, problem-solving courts and sociocultural factors impacting marginalized girls and youth. These involvements resulted in her serving as a co-investigator/principal investigator for SSHRC, Status of Women, Heritage Canada, Metropolis-RIIM and NCPC grants, as well as currently for two CIHR grants.

Niki Lambropoulos is an experienced e-learning expert, researcher, consultant, HCI designer, and online communities manager. Her interests fall in the fields of E-Learning, and Idea Management for Distributed Leadership and User Innovation Networks. She was born in Ancient Olympia, Greece. She holds two BAs and a Diploma in Education from the University of Athens, Greece and an MA in ICT in Education from the Institute of Education, University of London. She finished her PhD at London South Bank University, UK. She started working as a Greek language teacher in Greece in 1989. She has worked as a Greek language, ICT teacher and ICT coordinator, and from 2002-2006 as a Project Manager mostly over the Net. She now works as a researcher in EU projects. Outside her office she likes Yoga, reading, arts, swimming, and cloud watching. She enjoys working collaboratively over the Net.

Tong Liu (t.liu@massey.ac.nz) has degrees in engineering design from Sichuan University, China, an Honour's degree in information science, and a master's degree in computer science, both from Massey University, Albany, New Zealand. Her current research interest is the design of online systems that better support human and social interactions, with e-mail being just one example. Others include online auctions, online learning systems, wikipedias and blogs, and online voting systems. Her goal is to help develop more socially aware software.

Ross A. Malaga received his PhD in 1998 from George Mason University. He is currently an associate professor of Management and Information Systems in the School of Business at Montclair State University. His research focuses on search engine optimization, online auctions, and electronic commerce. He has published articles in *Communications of the ACM, Decision Support Systems, Journal of Organizational Computing and Electronic Commerce, Electronic Commerce Research*, and the *Journal of Electronic Commerce in Organizations*. Dr. Malaga serves on the editorial review boards of the *Information Resources Management Journal* and the *Journal of Electronic Commerce in Organizations*.

Paolo Massa is a researcher at the Institute for Scientific and Technological Research (IRST) in Trento, Italy. He received his PhD from ICT International Graduate School of University of Trento in March 2006 defending a thesis titled "Trust-aware Decentralized Recommender Systems." He has a master's in computer science from Eastern Piedmont University, Italy. His research interests are trust and reputation, recommender systems and social software. Contact him at massa@itc.it.

Georgios Michaelides received the BSc and MSc degree in Software Engineering, specialized on System Developments, from the Budapest University of Technology and Economics in 2007 and 2009. His main interests are working with web development and web services, Internet-based media-communication, mobile devices and communication network protocols. He speaks fluent Greek and English.

Elham Mousavidin is a PhD candidate in the Decision and Information Sciences Department at the C.T. Bauer College of Business, University of Houston. She received her MBA from C. T. Bauer College of Business. Her research interest lies in the area of information systems development (ISD). Currently she is studying technological frames of reference in the process of ISD. She has also been involved in research on web-enabled communities (e.g., Weblogs) and virtual worlds, and their potential applications in organizations.

Carlos Alberto Ochoa Ortiz-Zezzatti (Bs'94-Eng. Master'00-PhD'04-Postdoctoral Academia Research' 06 & Industrial Postdoctoral Research'09). He has 1 book, and 7 chapters in books related with AI. He has supervised 12 thesis of PhD, 13 thesis of master and 27 thesis of Bachelor. He participated in the organization of HAIS'07, HAIS'08, HAIS'09, HAIS'10, ENC'06, ENC'07, ENC'08, MICAI'08, MICAI'09 & MICAI'10. He is a permanent reviewer at Computers in Human Behavior Journal of Elsevier.

Lorne Olfman is dean of the School of Information Systems and Technology and Fletcher Jones chair in Technology Management at Claremont Graduate University. He came to Claremont in 1987 after graduating with a PhD in business (management information systems) from Indiana University. Dr. Lorne's research interests include how software can be learned and used in organizations, the impact of computer-based systems on knowledge management, and the design and adoption of systems used for group work.

Felipe Padilla Díaz received the B.S. degree in computer science engineering from the Instituto Tecnológico de Queretaro in 1985, received the M.S. degree in enterprise management from the Universidad Autónoma de Aguascalientes in 1999, and the PhD. in engineering (Artificial Intelligence) from the UNAM. His research interests include Evolutionary Computation, Data Mining and learning classifier systems.

Julio Cesar Ponce Gallegos received the B.S. degree in computer science engineering from the Universidad Autónoma de Aguascalientes in 2003, and the M.S. degree in computer sciences from the Universidad Autónoma de Aguascalientes in 2007. He is currently working towards the PhD. Degree in Ant Colony Optimization from the UAA. He is currently an professor in the Universidad Autónoma de Aguascalientes. His research interests include Evolutionary Computation, and Data Mining.

Eric M. Rovie is currently a Visiting Instructor in the Philosophy Department at Agnes Scott College in Atlanta. He holds graduate degrees in Philosophy from Georgia State University and Washington University in St Louis, and is co-editor (with Larry May and Steve Viner) of *The Morality of War: Classical and Contemporary Readings* (Prentice-Hall). He works in normative and applied ethics, with particular interests in the problem of dirty hands and virtue ethics. Recent publications include "Reevaluating the History of Double Effect: Anscombe, Aquinas, and the Principle of Side Effects" (*Studies in the History of Ethics*) and "Tortured Knowledge: Epistemological Problems with Ticking Bomb Cases" (forthcoming in the *International Journal of Applied Philosophy*).

Terry Ryan is an associate professor in the School of Information Systems and Technology at Claremont Graduate University. His teaching and research interests are in the design, development, and evaluation of information systems to support teaching and learning, online discussions and dialogues, and preparing for and responding to emergencies.

Brian Whitworth completed his psychology master's thesis on the implications of split-brain research. After being the New Zealand Army senior psychologist, he became a defense computer analyst, then worked in operational simulations. After "retiring" his doctorate on how online groups create agreement sparked an interest in the design and evaluation of "social-technical systems." He has published in journals like *Small Group Research*, *Group Decision & Negotiation*, *THE DATABASE for Advances in IS*, *Communications of the AIS*, *IEEE Computer*, *Behavior and Information Technology (BIT)*, and *Communications of the ACM*. He is now at Massey University (Albany), New Zealand. For more information, visit http://brianwhitworth.com.

Zheng Yan holds two master's degrees from Xi'an Jiaotong University, China and National University of Singapore, respectively. She received her licentiate degree at the Helsinki University of Technology. She will defend her doctoral dissertation about trust management for mobile computing platforms in 2007. Zheng is a senior member of research staff at the Nokia Research Center, Helsinki. Before joining in the NRC in 2000, she worked as a research scholar at I2R and software engineer at IBM SingaLab, Singapore. She first-authored more than 20 paper publications. Her research interests are trust modeling and management; trusted computing; trustworthy software, applications and services; usable security/ trust and DRM.

Index

U

unethical IT behavior 1, 6-7
unique problem 127, 130-132, 134, 141
usability 17, 23, 84, 86-87, 90, 102, 111-112, 114, 122, 297-298

V

version control system 234
victim 35-36
Virginia Tech Shootings 149
virtual community 10
virtual world 19, 44, 46, 54, 56-57, 77, 123, 144, 190, 263
Vygotsky, Lev 117-118, 122, 140, 145

W

Web 2.0 82, 146-148, 156-160, 166
WebCT 98, 100
wiki 45, 78, 92, 150, 162, 165, 190-191, 196, 269, 275
Windows 34, 89, 91, 93, 95-96, 98, 101, 174, 186, 268
wizard 48, 53-54, 57
Women Halting Online Abuse (WHOA) 37
World Trade Center (WTC) 149

Y

Yahoo! 45, 175-176, 190

Z

zone of proximal development 118